Monte Carlo Applications in Systems Engineering

Monte Carlo Applications in Systems Engineering

A. Dubi
Ben Gurion University of the Negev,
Beer-Sheva, Israel

John Wiley & Sons, Ltd
Chichester • New York • Weinheim • Brisbane • Singapore • Toronto

Baffins Lane, Chichester,
West Sussex, PO19 IUD, England

National 01243 779777
International (+44) 1243 779777

e-mail (for orders and customer service enquiries): cs-books@wiley.co.uk

Visit our Home Page on http://www.wiley.co.uk
or
http://www.wiley.com

Other Wiley Editorial Offices

John Wiley & Sons, Inc., 605 Third Avenue,
New York, NY 10158-0012, USA

Weinheim • Brisbane • Singapore • Toronto

Library of Congress Cataloging-in-Publication Data

Dubi, A.
Monte Carlo Applications in Systems Engineering/A. Dubi.
p. cm.
Includes bibliographical references and index.
ISBN 0-471-98172-9 (alk. paper)
1. Monte Carlo method. 2. Systems engineering. I. Title.
TA168.D83 1999 99-35831
620′.001′1—dc21 CIP

British Library Cataloguing in Publication Data

A catalogue record for this book is available from the British Library

ISBN 0 471 98172 9

Typeset in 10/12pt Times by Keyword Typesetting Services Ltd, Wallington, Surrey.
Printed and bound in Great Britain by Antony Rowe Ltd, Chippenham.
This book is printed on acid-free paper responsibly manufactured from sustainable forestry, in which at least two trees are planted for each one used for paper production.

"To all those whose souls have been bent by the burden of the search for the truth and to she who has spent a lifetime straightening mine, my wife, Nehama."

" – What will you give the music?
she then asked.
– My life. Yes, I will give my life!
– What will the music give you in return?
– My life!"

Quoted and translated from Natan Altermann: the greatest Israeli poet of the millenium.

"Electrons in a cloud, a beautiful equation,
a tiny glimpse of truth – and so much music"

A. Dubi

Contents

Preface and General Discussion

This book focuses on the industry and use of mathematical methods, in particular the Monte Carlo method as a tool that can support and improve the engineering of systems. It is in the nature of the subject and the related problems that a fair amount of mathematical discussion and theory is involved to allow a deeper understanding of the underlying structure. Many of the examples and methods presented in the book are supplemented by computer programs. These programs can be downloaded from the ftp site `ftp://ftp.wiley.co.uk/pub/books/dubi/`.

A number of years ago the author took part in a meeting in which a design improvement problem was discussed. The problem involved two data transfer computers, (A) and (B), operating in parallel. (A), a better and more expensive machine, was used as the main system. (B) served as a back up. Upon failure of (A), (B) was activated with a delay. Both (A) and (B) were shipped for repairs to an external location. This process could take up to a month. Both (A) and (B) are not new and aging caused the time between their failures to gradually decrease. The performance of the system was below specification and the question was, what should be done to improve the performance?

A lively debate developed. A number of suggestions were made as to the optimal way through which performance might be improved. These included:

- Set unit (B) active all the time to eliminate the delay.
- Replace (B) with a new (A) since (A) is considerably more reliable.
- Add a spare unit of type (A) on location.
- Add a new (B) unit as spare on location.
- Replace the system with an identical new system.
- Add a third active unit (B) in parallel.

Each participant brought convincing arguments in favor of his suggestion. Experience, intuition and engineering insight were used in varying measures to justify each anticipated improvement. There was, however, something wrong in the debate. The very fact that there was a debate! The problem, should not have ever been a subject for discussion. It is a subject for calculation and numerical evaluation. In fact, the result of the debate was that the first suggestion was adopted. This proved to be the worst possible solution yielding an even lower performance than before. A short set of calculations could have easily identified the optimal choice.

In 1990 a large utility in the southern hemisphere constructed a dedicated electrical substation to supply power to a big chemical industrial complex. A substation is basically a complex structure of bus bars, circuit breakers and transformers intended for transferring power from high voltage transmission lines to a lower voltage line at the production site. It involves a large number of redundancies to ensure an undisturbed energy supply. At the same time, multiple redundancies increase the danger of a severe propagating failure that may cause damage either to the customer or to the power generator. Upon its completion it was verified that the substation performance was far below expectations. The client of the substation sued the utility for $150M in damages. A lengthy debate then started as to how to improve the system. A trial and error approach was considered. This meant, first a debate about which was apparently the best way to improve the system (presumably no one really knew how to do it, or the problem would not have arisen in the first place). Changes were to be introduced, at a cost of millions of dollars and many months delay to be followed by a process of "wait, watch and hope". This process was unacceptable to the client. A Monte Carlo calculation of the performance eventually identified the optimal design and since then, the system has been operating, in a satisfactory manner.

In a large waste burning facility in Europe a serious problem was identified as a result of corrosion caused by aggressive waste materials in the container in which burning took place. Repeated failures in the plant forced lengthy shut-downs and expensive repairs. The possibility of preventive maintenance to diminish the probability of severe failures was considered. Since maintenance itself causes shut-down and costs money it was not at all clear what the optimal maintenance approach would be or even if optimal maintenance would improve the overall situation. A procedure of debates, studies and trial and error would be costly and long. Again, the ability to calculate the effect of maintenance on the system performance proved to be the least costly and most effective approach.

The above are examples from hundreds of cases encountered by the author in industry over the past 15 years. They point at the definition of Systems Engineering. Systems Engineering is about *predicting the future reality of a system*. When one designs a new system, or considers changes in an existing system, it is desirable to be able to predict the behavior of the system upon its becoming a reality. Such a prediction should, in principle, include answers to any question that may arise. For example:

- What would the availability of the system be as a function of time?
- How many and what type of failures might occur?
- How many spare parts should be procured?
- What type of preventive maintenance should be applied and when?
- What will the performance of the system (production, mission success and other criteria) be?
- What is the expected cost of operation of the system?
- How many repair teams should be assigned?
- What kind of logistics should be used, how will failures be detected and how will they be handled?

This book is an attempt to present the Monte Carlo method and its potential to support this process.

Can one predict the future and realistic behavior of an industrial system?

Every science deals with the rules and methods that enable prediction. One may consider, in this respect, two extremes. Physics on the one hand and social or political sciences on the other. It could be argued that these represent two edges of the predictability spectrum. It is commonly agreed that predictions calculated in physics are superbly accurate. They are so accurate that the Heisenberg Uncertainty Principle is needed to put a lower limit of the order of 10^{-34} to that accuracy. At the same time it is quite clear that predicting the future of even a single human being, let alone communities and organizations, is beyond the ability of any science. Psycho-History, the science that the late great I. Asimov invented for predicting and manipulating human history, is still science fiction.

The fascinating capacity of physics to predict accurately the energy levels of the hydrogen atom or the existence and properties of the neutrino particle or the thermal motion of atoms and its relation to temperature, is not a result of the wisdom of physicists (although this may be of some value). Mainly, it is because the questions asked in physics are *extremely simple*. Very few things can be more simple than a single proton and a single electron holding each other in an electrostatic grip in the hydrogen atom. Yet, much of quantum physics is devoted to that subject. A living being contains about 10^{23} atoms per cubic centimeter, huge molecules, DNA strings, cells, organs, a brain, instincts and desires, free will and possibly, a soul, most of which we can hardly define or understand. Industrial, man-made, systems lie somewhere in the middle of this spectrum. Systems usually operate according to coherent laws. These are partially physical laws and partially man-made rules. Systems do not operate in the controlled environment of a laboratory. They are affected by internal and external random events and interactions, environmental factors, human mistakes, aging of equipment and many other elements that are cloaked in a veil of uncertainty. Their behavior is inherently stochastic in the sense that two identical systems will have different 'life histories' and different events. The use of mathematical methods, and in particular the Monte Carlo method, in the prediction process is an attempt to push system engineering significantly toward the left end of the predictability spectrum as is shown in Figure 1.

Interestingly, although it is evident that higher predictability improves every aspect of systems engineering, still, experience has shown that it is not always a welcomed process. Clearly, no one would dare to have an opinion on the law of gravity. It is hardly imaginable to hear someone say "My opinion is that the refraction coefficient of light is such and such". One can not have an opinion about things that can be accurately calculated, predicted and measured. Yet, everyone has an opinion about social and political matters. This leads to the interesting observation that: "the more complex the problem the less one needs to learn in order to have an opinion". Predictability deprives many from the right to have an opinion.

One of the major items of concern in systems engineering is resources. Resources are all the elements, apart from the system itself, that are needed to support the operation of the system. This includes, diagnostic equipment, skilled repair teams, repair facilities, spare parts, inspections, maintenance operations etc. It is clear, by now, that the cost of

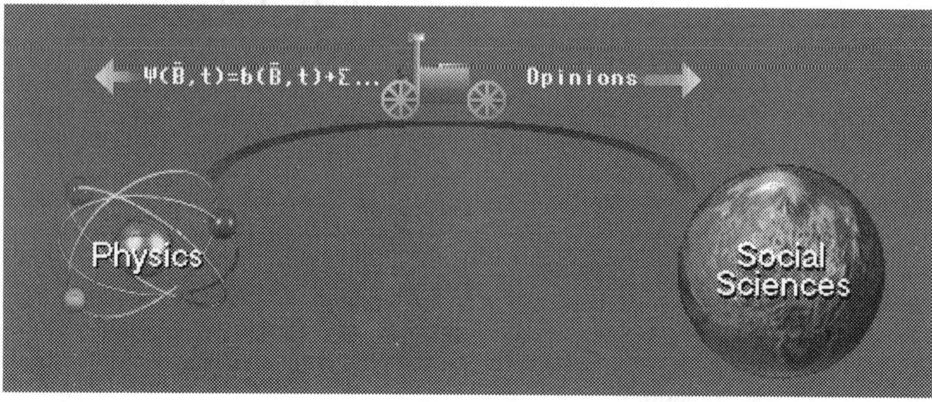

Figure 1 The predictability spectrum

resources is far from negligible. It often equals and sometimes is considerably higher than that of the system itself.

Consider the problem of spare parts. One of the important and most frequently asked questions is: How many spare parts should be procured?

Here are some simple thoughts and observations about spare parts. The answer to the "spare parts" question requires a numerical answer, a number. Usually a number with a policy such as what to buy, how many and when. But, it must be a number. Any question for which the answer is a number is a mathematical question. A mathematical problem can be addressed only if it is well defined. Spare parts, as well as any other resource, are bought and used for one single purpose: to support the operation of the system. If the spare parts question is not related explicitly and coherently to the performance of the system it is ill-defined.

Thus, it is essential to first define the target performance of the system and then to derive the number of spare parts (and other resources) that will support the performance target (see Figure 2). The definition of the performance requirements may not be a simple question in itself and it must be addressed before anything else is considered. Otherwise, it is not at all clear for what purpose the system was built. For example, in a production line one may set the performance target: the line should operate at full production for at least 85% of the year. For a squadron of aircraft, at least 90% of the aircraft should be available at all times with a probability not below 0.98. Some systems may have multiple performance definitions or complex ones such as: the average semi-annual power production of a gas turbine should not be below 170 MGW and the probability for producing an average of less than 150 MGW at any month should be less then 0.5%. It is critical to understand that a system is bought, designed and operated only for its performance.

This point is stressed because the industrial reality, to date, is that the problem is almost never addressed in this fashion. Instead, the number of required spare parts is linked to the number of expected failures. This seems reasonable but could be misleading because the link between failures, spare parts and system performance is not at all simple.

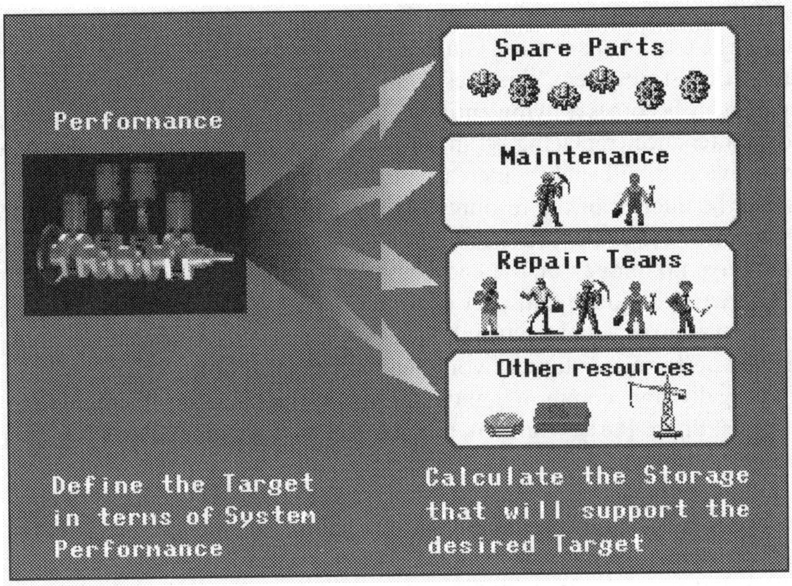

Figure 2 Resources are defined so as to support the single most important element: system performance

On one occasion, a manager of the customer service department of a large medical equipment company stated in a policy paper that: "The target of the service organization is to reduce the number of failures in our systems". – This statement sounds plausible. Yet upon closer examination one observes that there are two extremely cheap solutions that will immediately achieve the stated goal. One solution is never to repair any failed system. Obviously each system will fail only once. An even smaller number of failures will be achieved if the systems are never used or sold. This too, will result in zero failures. One can not ask for fewer failures. Indeed, both solutions suffer from a deficiency, there will be no system. But, the system or its performance are not mentioned in the policy statement. The policy is ill-defined because it does not include the system's performance.

Furthermore, if one could significantly reduce the number of failures and in the process reduce slightly the availability of the systems (say due to prolonged diagnostics and maintenance), would it be acceptable? – The system performance target must be part of the policy statement.

The result of missing the link between resources and performance is, often, overstocking of spare parts. A rough, yet conservative, estimate of the amounts of spare parts in all air forces in Europe would amount to 10–20 billion dollars in spare parts that will never be used. This excessive amount serves the purpose of the industries producing the spare parts and the interest of the organization in charge of these spare parts. It should be noted that in most organizations the spare parts department is separated from the system and its performance. The only interaction occurs when a spare part is demanded. Naturally, as long as spares are available no complaints or questions will be raised. It is, thus, also in the organization's interest to have excess spare parts. The cost of spare parts is not just that of purchasing them; there is a whole organization that stores them, classifies them and maintains them. Very soon the spare parts department becomes a living entity of its own. An overstocking of spare parts or excessive maintenance (replacing oil in a car every 1000 km – would cause no damage except excessive cost and slightly increased down time) or too many inspections almost always serve the interest of the resource organization. Indeed, someone pays for those excess spare parts. It is, often, the anonymous customer that pays the electric bill or the citizen that pays tax money or the end customer that buys the product. It is unlikely that paying party will notice that part of his money is wasted on too many spare parts.

Adopting the simple, almost natural, idea that the resources needed are derived from a well defined performance target, would eliminate the problem.

The existing division, common in industry, of separate and hardly interacting departments, such as: Spare Parts, Maintenance, Logistics, Operations Research, Inspection, Production etc. also negates the simple concept that the system performance is the only concern and everything else is derived from its demands. These departments operate almost independently, rarely exchanging information. The only drawback in the simple, almost self evident – yet rarely realized idea of a link between performance and resources, is that its realization involves complex mathematical problems, sometimes prohibitively difficult. The Monte Carlo method is suited to solving such problems.

It will be shown that the spare parts problem and the modeling system's behavior in general, can be compared to the behavior of neutral particles in matter. The basic equations governing the behavior of neutrons in a medium are quite similar to those that govern the behavior of systems in time. Yet, the question: "What is the flux of neutrons in a nuclear reactor shield?" seems to be a serious problem deserving the attention and dedication of scientists, nuclear engineers, large computers and complex mathematical methods. Indeed this is the case. Hundreds of millions of dollars have been invested in the development of methods and computer codes that address this problem. Relative to the neutron flux problem the "spare parts" problem seems rather mundane and undeserving of extensive scientific attention and effort. Why?

From a pure mathematical point of view the spare parts problem is considerably more complicated. Although the equations are of the same general type there are three reasons why the spare parts situation is more complicated. Firstly, the neutron problem is a six dimensional problem (location in space, direction and energy); whereas the spare parts problem is multi-dimensional and its dimensions depend on the number of components in the system. Secondly, the laws that govern the events of a neutron are universal and limited in number. A neutron can undergo elastic or inelastic scattering, can be absorbed and produce other particles etc. These interactions are universal and apply everywhere in the universe. They are not man made. It is possible to measure the probabilities of each reaction and store the data for all to use. The situation with systems is dramatically different. If one considers an event to be a change in the state of a system, then, the rules upon which events take place are, at least partially, man made. For example, the process of failure in a lamp can be regarded as being controlled by physical laws. But, the events that take place upon such a failure, depend on design and human decisions. Should the lamp be replaced? How does one identify that it has failed? Does one have a backup lamp that can be operated, etc.? These rules are part of the event. There is no closed, limited or even finite set of rules that describes all the possible events and contingencies in systems. The third point is that components in systems suffer from aging. The properties of neutrons do not depend on time, whereas those of components, in particular the tendency to fail, increases with time. Incorporating aging in the process, or even understanding how it should be incorporated in a mathematical model is not a trivial matter. It adds a great deal to the complexity of the spare parts problem.

From an economic viewpoint the importance of the spare parts problem, and resources in general, is not less and probably far greater than that of the neutron problem. Why is it then that the spare parts problem is not treated as a serious scientific problem?

It has already been mentioned that when the complexity of a problem exceeds a certain level anyone may have an opinion about it. Moreover, neutrons are something rare and exotic. They are not part of daily life. They exist in the realm of scientists. Spare parts, on the other hand, are a rather mundane matter. Every mechanic deals with spare parts. When one buys eggs one deals, in fact, with spare parts. It is not an exclusive domain of scientists. No one is expected to solve a multi-dimensional integral equation in order to decide whether to buy 3 or 20 eggs. Yet, one manages, in most cases, to buy the amount needed. This is, mainly, because the cost of a mistake is low. Too many eggs or not enough eggs will have only a small impact on life and can be corrected easily. The use of eggs is so frequent and regular that in a short while, one learns what the amount consumed is, even if this amount varies with days of the week and periods in the year. Trial and error as an approach to resolving the egg inventory problem, is sufficient. If, however, lack of an egg would be extremely dangerous, say, put the life of a family member in danger, then the problem is not that simple. Trial and error may then become very unsuitable. In such a case one would, most probably, tend to buy a surplus of eggs. But, if then there is a need to buy more than just eggs, such as butter and bread etc., and if the total amount of money does not suffice to buy a surplus of all items, then there is a problem.

The properties of spare parts resemble those of eggs, i.e. the cost of a mistake, on either side, is low and can be corrected easily and the consumption is regular so that the necessary quantities can be readily learned and do not require special mathematical attention. Trial and error as well as life experience will do. But, if the spare parts are not of that type, if a mistake is costly, and if no simple regular consumption can be observed, then the problem deserves serious attention that will yield savings on a national scale.

Separation of departments was mentioned above. This separation is also a result of the mathematical complexity of the "true" problem, namely: realistic prediction of performance as a function of resources. Separation of variables is a commonly used technique in three-dimensional mathematical problems. Instead of looking at a single three-dimensional function $f(x, y, z)$ as a solution of a differential equation, one assumes that the function can be written as a product of three, one-dimensional functions, namely: $f(x, y, z) = f_x(x) f_y(y) f_z(z)$. Substituting this assumption into the differential equation breaks the single three-dimensional equation into three one-dimensional equations. This is a far simpler problem to solve. The mathematical procedure is justified by the use of uniqueness theorems that guarantee that if a solution is found, then it is the correct solution. A similar approach is applied in industry without a similar justification. The overall performance is considered to be a "product" of a number of functions: design, spare parts, maintenance, inspection, repair facilities and test equipment, operation research, etc. Accordingly, respective departments are established to cover each subject. This separation has no justification except for the fact that it is extremely difficult to do things otherwise. It is very easy to see, for example, that separating preventive maintenance from spare parts is incorrect. Maintenance is intended to reduce the number of failures, thus to reduce the number of spare parts required. Obviously maintenance and spare parts are closely linked. Also design is linked with spare parts. A design that includes redundancies may increase the tolerance of the system to failures i.e. the system can go on operating while a number of units are failed. This could increase the time one could wait until a spare part arrives. All the aspects of systems are linked together. Separating them simplifies the solution but yields a non optimal solution. Separation is not a good way of doing things, it is a reflection of our ignorance.

The following is an example (one of many) of the results of separation. A large, defense related system has to operate on demand. Schematically it contains a main control unit and two branches. Each branch contains an additional execution/control unit connected to a set of four units. The general structure is shown in Figure 3. In response to a demand, the system uses one or more units which are consumed and may be replaced by new ones. The probability of successful response to a demand depends on the number of consumable units used. The higher the number the larger is the probability of success. Yet, if too many units are used per demand, the number of available units may decrease so that not enough will be left for future demands, if demands arise faster than the rate of re-deployment of new units. This is a classical optimization problem that will naturally be directed to the operation research department. The operation research department considered the problem taking into account the number of available units, the expected rate of arrival of demands and the rate of renewal of consumed units and suggests an optimal number of units, say 2, to be used upon each demand.

The reliability of the system refers to its ability to operate when a demand appears. This problem is referred to the reliability department. They look at the probability that the main control and at least one of the LC units will be operational. Each department solved its problem as if the other does not exist. In the testing phase of the system two versions are tested. One with advanced, low failure LC units and one with lower grade LC units which have about twice as many failures. The amazing result is that the performance of the low grade system was considerably better. This, incidentally, agrees with a prediction obtained with a "true" integrated model calculation carried out using the Monte Carlo method. These results do not seem reasonable. Something must have gone wrong. They negate the deeply-rooted (and wrong) belief that "If one improves every, or any, element in a system, the system performance is improved". To understand what has happened, consider a scenario with the high grade LC units assuming that during a set of demands there will be a single failure in the LC units. Upon the arrival of the first demand, every thing is working and consumable units 1 & 2 are used. Upon the arrival of the second demand, units 3 & 4 are used. No renewal of the consumable units takes place. Just before the arrival of the third demand, LC unit number 2 fails. The third demand appears before the LC unit is repaired. The system is unable to respond to the demands because units 1, 2, 3 & 4 are missing and LC 2 is not operational. Consider now the response of the low grade system to the same scenario with twice as many LC failures. Upon the arrival of the first demand, units

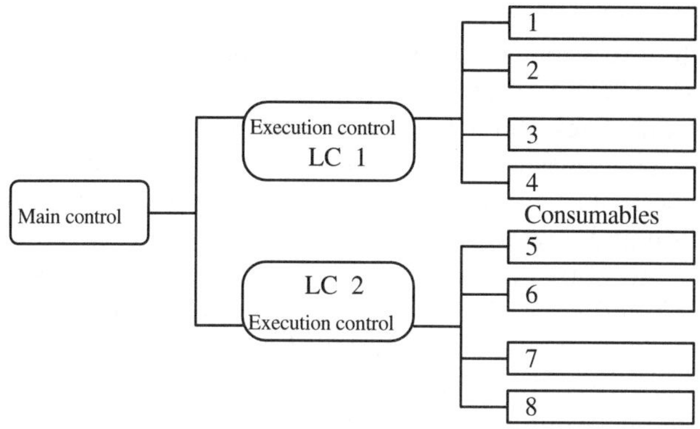

Figure 3 Schematic description of a demand driven system

1 & 2 are used. Just before the arrival of the second demand, unit LC 1 fails. Upon arrival of the second demand it can not be used but unit LC 2 is available. Thus units 5 & 6 are used. Unit LC 1 is repaired and then, just before the arrival of the third demand, unit LC 2 fails. The system can respond to the third demand with LC 1 using units 3 & 4. Thus, the low grade system responds well to three demands whereas the high grade system does not. Choosing the high-grade system would mean paying more for less performance. This is a direct result of the separation of the departments.

The assertion that systems have a "monotone" behavior i.e. that improving any unit in the system improves the system, is unfounded in reality. It becomes true only if one can assume that the system performance is a simple product or a linear composition of the various elements. The reality is that system performance is a very complex function of its constituents and there is no reason to assume that it is a monatomic increasing function.

A general view of system engineering is summarized in Figure 4. When one designs a new system or attempts to improve an existing operating system it is basically an attempt to weigh future realities of such a system with varying options. In order to make any type of prediction one looks at a model of reality rather than reality itself. This is simply because scientific prediction is not prophecy. Science, in general, does not look at the absolute reality, but rather, on a model of reality. We have never "seen" an electron. Physics does not claim to have a knowledge of what a "true" electron is. A free quantum mechanical electron is actually described as: $e^{i(\vec{k}\vec{r}-wt)}$ which is a plain wave with momentum $(h/2\pi)|\vec{k}|$. Is this really the form of an electron? All one can really say is, that with this description one obtains results that are consistent with measurements. This is the profound meaning of "truth" in the scientific context. After all, measurements (in a broad sense) are the only way in which we view and interact with reality. The blocks shown in Figure 3 are not a "control unit". They are a model of a control unit. It will be a satisfactory model if the predictions one makes with it closely describe the measured future reality. Thus, in the prediction process, one relies on a model of reality. Since this is not prophecy, there will always be a distance between the model and reality. The model will never fit reality in all its infinite details. It will always be at some "distance" from it. That distance is called Distance To Reality (DTR). The model is based on the information one has about the structure of the system, the interactions between the elements of the system, the interaction of the system with its environment, data related to the properties of the system components and how they move among their possible states. Once the model is established, a set of questions are generated (one seems to ask the questions about the real system; but, in fact, they should be asked about the model). The questions are those already mentioned relating to the future performance of the system, its failures, spare parts, repair teams, inspections, maintenance, production and anything else that is of interest.

The next step is to obtain answers to the questions. It is at this point that mathematical methods should be used. This concept is shown graphically in Figure 4.

It is essential to note that the mathematical methods available are the key to the entire process. They basically define the kind of models used and the DTR of such models. After all, one would not construct a model for which no answers can be obtained.

Once the answers are obtained, one reviews them. If they are satisfactory the design is considered to be satisfactory. However, it should be noted that the answers are related to the model and not to the reality of the system. Although the DTR is difficult to measure it is clear that a large DTR will result a situation in which the relevancy of the answers to the future reality of the system is cast in doubt. Some common approximations made are: in almost all reliability and availability calculations it is assumed that components do not age; in almost all maintenance and renewal calculations, immediate repair is assumed; in almost all operation research analysis full reliability

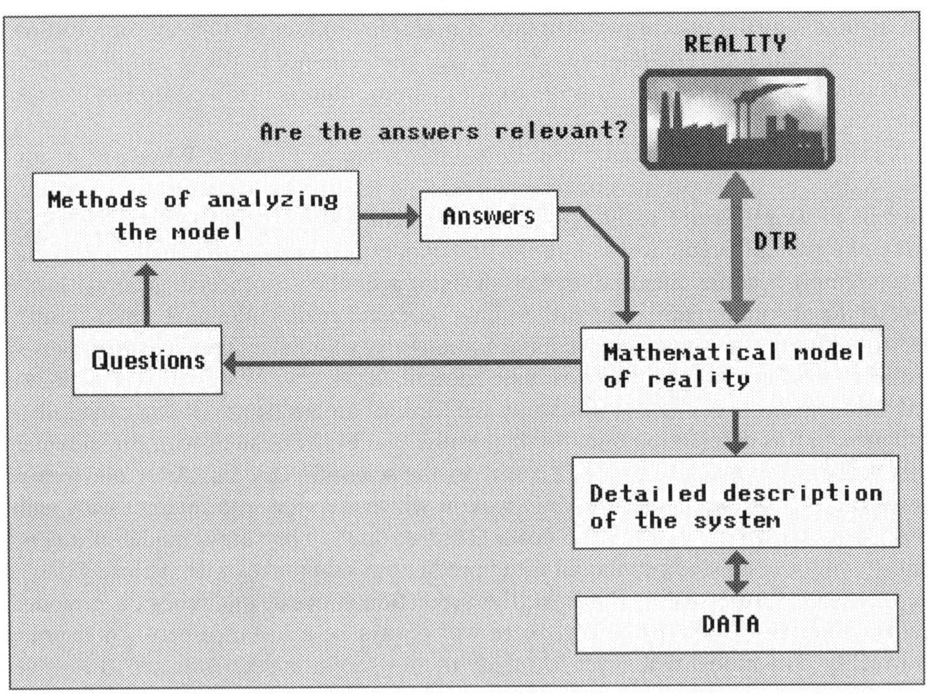

Figure 4 The relation between a Model and Reality

is assumed. Different aspects of the system are treated as if other aspects are non existent, as already discussed. Complex structures of components are grouped together and viewed as a single component with simplified properties. All these approximations can be applied to the model. They should be done simply, because otherwise the cycle "Model – Questions – Methods – Answers" can not be closed. Instead of fitting the applied methods to the needs of the realistic system, the system is modeled to fit the available methods. The only penalty is in the increase of the DTR which in turn causes the answers to be of little relevance to reality.

The Monte Carlo method is a powerful method that can be a valuable aid in reducing the DTR. Monte Carlo methods were invented and utilized in the early forties in the context of nuclear technology and the analysis of neutron behavior in matter. They provided the only option to solve a six-dimensional integral equation in order to design shielding for nuclear reactors and other nuclear devices. It was probably the first case in human history in which trial and error was clearly too risky to attempt. The Monte Carlo method, as will be shown, is the only method that could yield solutions to complex multi-dimensional problems. For about three decades the Monte Carlo was used almost exclusively yet extensively in nuclear technology. Presumably, the main reason for this was lack of suitable computing power. The method is computer time and memory intensive, and one should recall that the problems in systems engineering are more complex than those of nuclear technology. Yet, with the increasing availability of fast computers the method becomes more and more feasible. Monte Carlo methods are extremely powerful. They enable solutions to models that are very close to reality and yield useful and relevant results. Thus, they have the potential of amending many of the problems raised in the above discussion. Resources can be evaluated to meet performance targets and the separation of categories can be avoided. Detailed integrated models can be analyzed. Experience over the past fifteen years indicates that this technique permits profound improvements both in terms of reducing the life cycle cost of systems, providing better designs and even going as far and deep as the very structure of the industry.

Although the main purpose of this book is to present the Monte Carlo method and its applications, a considerable part is dedicated to the main analytical methods. This is because analytical methods still form the basis for systems engineering and provide a conceptual basis to the Monte Carlo theory and applications. Furthermore, analytical methods do not "compete" with the Monte Carlo method. In any situation in which analytical methods can provide satisfactory answers they are preferred because they are cheap in terms of computer time and provide accurate and more general solutions. The methods are complementary. The Monte Carlo method can and should be used in those cases in which a complex model is required and can not be addressed by analytical methods.

1

Introduction – Probability and Statistics

In this chapter a statistical introduction is presented. It is assumed throughout this book that the reader has prior knowledge of statistical and probability concepts. Even so, for the purpose of completeness and convenience, the essential concepts, both for system engineering and Monte Carlo methods, are discussed.

1.1 Experiments, Events and Probability

1.1.1 Elementary events and the sample space

Any type of measurement is considered to be an experiment. Experiments can be broadly divided into two categories. Experiments for which the same result is obtained each time the same experiment is repeated are called *deterministic* experiments. For example, if a small mass is dropped from a given height and the time of fall and the point of impact are measured, then the same results will be obtained each time the experiment is repeated. An experiment for which different results are obtained each time the experiment is repeated is called a *statistical* experiment. If the number of customers entering a bank is counted, different results may be expected each time the experiment is repeated. Statistical experiments are at the core of every aspect of reality because most events in reality are of a statistical nature. A measurement of the number of failures in a batch of identical lamps is of a statistical nature, and so is almost everything else in reality. Consider the throw of a die. This experiment is used in almost every text on statistics and probability to demonstrate a statistical experiment. If considered carefully, it reveals the source of one of the interesting debates that took place at the early years of quantum mechanics. It may be argued that, when a throw of a die is repeated it is not exactly the same experiment. The motion of the throwing hand differs every time, the direction changes, the position of the die in the hand, the velocity of the throw, etc. These changes give rise to different results at each throw. If a mechanism is to be designed such that every parameter of the throw is kept exactly the same at each throw, the position of the die, the direction, height, velocity of the throw, etc., then the same result will be repeated and the experiment will become a deterministic experiment. Thus, the statistical nature of the experiments results from the existence of many parameters affecting the final results which change in a random way each time the experiment is repeated. Such parameters may be called the 'hidden parameters' of the experiment.

Are there any 'true' statistical experiments? Indeed there are. If a neutron with a given energy collides with a nuclei, the result may be elastic scattering. If the same experiment is repeat under *exactly the same conditions*, the result may be the absorption of the neutron. In fact, all subatomic events are of an inherent statistical nature. Two *identical* excited nuclei will decay (by emitting a photon) at two different times. Subatomic particles are not like the throw of a die. There are no 'hidden parameters' that, by their variation, cause each experiment to yield a different result. The inherent statistical nature of physics is one of the essential aspects of quantum mechanics. Interestingly enough, it was Einstein that insisted on the concept of 'hidden parameters' with his famous statement "God does not play dice". It is told that Nills Bohr replied: "Do not tell God what he may or may not do".

From the point of view of probability theory, the source of the statistical behavior is irrelevant. Once the same experiment yields different results, the experiment is considered to be a statistical experiment. Every possible outcome of the experiment is called an 'event'. For example, if in the experiment of dropping a mass the height and location are changed in a random manner, the outcome will vary in each measurement. It will become a statistical experiment.

The distinct possible outcomes of a statistical experiment are called *elementary events*. They will be denoted by Greek letters such that Λ_i is a single distinct possible outcome of an experiment. The event that includes all the possible outcomes of the experiment is called the *sample space*, and is denoted by the Greece letter Ω. Ω is sometimes called also the 'universal event' or the 'universe'. It is the event that will happen every time the experiment is conducted. It is convenient to write $\Lambda \subset \Omega$, meaning that event Λ belongs to, or is included in, Ω. Two events A and B are considered equal, i.e. $A = B$, if $A \subset B$ and $B \subset A$. Likewise, $A \subseteq B$ means that event A is either included in or is equal to event B.

Events may be combined with other events using the operations of set theory. Such combined events are denoted by capital letters. The event $A = \Gamma \cup \Lambda$ is the union of the events Γ and Λ, and is the event that occurs if either Γ or Λ occur. The event $B = \Gamma \cap \Lambda$ is called the *intersection* of Γ and Λ, and is the event that occurs if both Γ *and* Λ occur. The event $\overline{A}$ is called the *complementary* of A. It is the event that occurs when A does not occur. A union of n events $A_1 \cup A_2 \cup \cdots \cup A_n$ will be denoted as $\bigcup_{i=1}^{n} A_i$ with a similar notation for the intersection of n events. The Null event is the event that cannot occur in an experiment. It is also called the *zero event* or the *empty event*, and is denoted by the letter ϕ. If two events are such that $A \cap B = \phi$, they are called *disjoint* or *mutually exclusive*. This means that the events A *and* B cannot occur simultaneously in an experiments.

The definition of the elementary events is not unique. In a throw of a die the 'natural' elementary events are the six numbers on the face of the die. However, the events 'An even outcome' or 'an odd outcome' can also be defined as the elementary events. In other cases, the definition of the elementary events is not that evident. Consider a set of n cells and a group of k balls. Assume that the balls are entered in a random manner, one after the other, into the cells. An elementary event can then be a set of k numbers such as $(2,2,3,1,4,5,2,6,\ldots,5)$, where the ith number in the list indicates the cell number into which the ith ball was entered. This is called an *assignment vector*. A different type of elementary event is a set of n numbers such as $(1,1,2,3,\ldots,2)$, where the ith entry indicates the number of balls in the ith cell. This is called an *occupancy vector*. The two sets of events are not identical. There are n^k different assignment vector events, because for each ball there are n possible choices of cells. There are $\binom{n+k-1}{k}$ different occupancy vectors. This can be seen by regarding the occupancy vector as a set of $n+1$ walls separating the cells, with k balls scattered between the walls. To change the occupancy, one may change the location of any of the $n-1$ walls (the two external walls cannot be moved) or any of the k balls. This can be done in $(n-1+k)$ ways. However, the $(n-1)!$ permutations involving changes of the walls amongst themselves and the $k!$ permutations in which balls are exchanged produce no new occupancy. Thus, the total number is

$$\frac{(n-1+k)!}{(n-1)!k!} = \binom{n+k-1}{k}.$$

1.1.2 The probability function

Definition 1.1.1: Probability function A probability function $P(\mathrm{A})$, is an assignment of a number to each event $A \subseteq \Omega$ such that the following properties are fulfilled:

$$P(\mathrm{A}) \geq 0 \tag{1.1.1}$$
$$P(\Omega) = 1 \tag{1.1.2}$$

Countable additivity: if $\mathrm{A}_1, \mathrm{A}_2, \mathrm{A}_3, \ldots$ is a set of mutually exclusive events (i.e. $\mathrm{A}_i \cap \mathrm{A}_j = \phi \ (i \neq j)$) then

$$P\left(\bigcup_{i=1}^{\infty} \mathrm{A}_i\right) = \sum_{i=1}^{\infty} P(\mathrm{A}_i). \tag{1.1.3}$$

Every function defined on the events space with the above properties is a legitimate probability function.

A number of important properties of the probability function can be verified:

$$P(\phi) = 0. \tag{1.1.4}$$

To prove this relation, define $\mathrm{A}_n = \phi \ (n = 1,2,\ldots)$. The union of these events is also the null event and by the countable additivity (1.1.3),

$$P(\phi) = P\left(\bigcup_{i=1}^{\infty} \mathrm{A}_i\right) = \sum_{i=1}^{\infty} P(\mathrm{A}_i) = \sum_{i=1}^{\infty} P(\phi) \quad \text{resulting in } P(\phi) = 0.$$

Finite additivity: Property 1.1.3 also holds for a finite set of mutually exclusive events, namely,

$$P\left(\bigcup_{i=1}^{n} \mathrm{A}_i\right) = \sum_{i=1}^{n} P(\mathrm{A}_i). \tag{1.1.5}$$

This can be shown by simply choosing the null event in 1.1.3 for all the events from $(n+1)$ and on.

If $\mathrm{A} \subseteq \mathrm{B}$ then $P(\mathrm{A}) \leq P(\mathrm{B})$. This can be shown by expressing B as the union of A and the intersection of B with the complement of A, namely, $\mathrm{B} = \mathrm{A} \cup (\mathrm{B} \cap \overline{\mathrm{A}})$, since A and $\mathrm{B} \cap \overline{\mathrm{A}}$ are mutually exclusive by Property 1.1.5 one obtains:

$$P(\mathrm{B}) = P(\mathrm{A}) + P(\mathrm{B} \cap \overline{\mathrm{A}}), \tag{1.1.6}$$

which proves the property.

$P(\mathrm{A}) + P(\overline{\mathrm{A}}) = 1$. The union of any event with its complementary is the universal event, and since they are disjoint,

$$P(\Omega) = P(\mathrm{A} \cup \overline{\mathrm{A}}) = P(\mathrm{A}) + P(\overline{\mathrm{A}}) = 1. \tag{1.1.7}$$

Property 1.1.5 indicates that if the events are disjoint, then the probability of their union is the sum of their individual probabilities. If events A and B are not mutually exclusive, then the following considerations apply: let $\mathrm{C} = \mathrm{A} \cap \mathrm{B}$, then $\overline{\mathrm{C}}$ is the event outside the intersection of A and B. Also, $\mathrm{A} \cap \overline{\mathrm{C}}$ includes the part of A which is not included in B, therefore $\mathrm{A} \cap \overline{\mathrm{C}}$ and B are disjoint and $\mathrm{A} \cup \mathrm{B} = (\mathrm{A} \cap \overline{\mathrm{C}}) \cup \mathrm{B}$, so that $P(\mathrm{A} \cup \mathrm{B}) = P(\mathrm{A} \cap \overline{\mathrm{C}}) + P(\mathrm{B})$.

Also, $\mathrm{A} = (\mathrm{A} \cap \overline{\mathrm{C}}) \cup \mathrm{C}$ and $\mathrm{A} \cap \overline{\mathrm{C}}$ and C are disjoint, so that $P(\mathrm{A}) = P(\mathrm{A} \cap \overline{\mathrm{C}}) + P(\mathrm{C})$. Substituting $P(\mathrm{A} \cap \overline{\mathrm{C}})$ into $P(\mathrm{A} \cup \mathrm{B})$ yields:

$$P(\mathrm{A} \cup \mathrm{B}) = P(\mathrm{A}) + P(\mathrm{B}) - P(\mathrm{A} \cap \mathrm{B}). \tag{1.1.8}$$

Property 1.1.8 can be understood as follows: when the probabilities of A and B are summed, the probability of the intersection between them (the common part) is summed twice. Therefore, it needs to be reduced from the sum. If the events are disjoint the intersection is null and Property 1.1.5 holds. For three events it is straightforward to show that the relation is:

$$\begin{aligned} P(\mathrm{A} \cup \mathrm{B} \cup \mathrm{C}) = {} & P(\mathrm{A}) + P(\mathrm{B}) + P(\mathrm{C}) - P(\mathrm{A} \cap \mathrm{B}) - P(\mathrm{A} \cap \mathrm{C}) - P(\mathrm{B} \cap \mathrm{C}) \\ & + P(\mathrm{A} \cap \mathrm{B} \cap \mathrm{C}). \end{aligned} \tag{1.1.9}$$

This expression can be extended using induction to any finite number of events, yielding a general expression for the probability of a union of n events:

$$P\left(\bigcup_{i=1}^{n} \mathrm{A}_i\right) = S_1 - S_2 + S_3 - \cdots + (-1)^{n+1} S_n = \sum_{i=1}^{n} (-1)^{i+1} S_i \tag{1.1.10}$$

where:

$$S_1 = \sum_{j=1}^{n} P(A_j)$$

$$S_2 = \sum_{1 \le j_1 < j_2}^{n} P(A_{j_1} \cap A_{j_2})$$

$$S_3 = \sum_{1 \le j_1 < j_2 < j_3}^{n} P(A_{j_1} \cap A_{j_2} \cap A_{j_3})$$

$$\vdots$$

$$S_k = \sum_{1 \le j_1 < j_2 \cdots < j_k}^{n} P(A_{j_1} \cap A_{j_2} \cap \cdots \cap A_{j_k})$$

and

$$S_n = P(A_1 \cap A_2 \cap \cdots \cap A_n).$$

The probability function is an abstract concept defined on a space of events. From the mathematical point of view, there is no such thing as a 'correct' probability function. Intuitively, one is inclined to see the probability function as a measure of the frequency of appearance of a specific result. Thus, if the number of experiments is N, and $n(\Lambda)$ is the number of appearances of event Λ, one is intuitively looking for the relation:

$$\lim_{N \to \infty} \frac{n(\Lambda)}{N} = P(\Lambda). \tag{1.1.11}$$

Expression (1.1.11) is sometimes mistakenly considered as the definition of the probability function. It is not, if only for the reason that one would never conduct an infinite number of experiments in order to obtain $P(\Lambda)$. The only definition of the probability function is that of expressions (1.1.1–1.1.3). It will later be shown that some specific probability functions fulfil equation (1.1.11). Such probability functions may be considered 'correct' or 'useful' in relation to a certain model, but the definition of a probability function, for any given sample space, has nothing to do with being 'correct' or 'useful'. It is simply a definition.

In the experiment of throwing a die, one 'feels' intuitively that the probability of obtaining any face up is $1/6$. This is not because a die experiment was ever conducted an infinite number of times. This result can be obtained if we look for a probability function with some additional physical properties which are assumed to be correct. Let Λ_i be the elementary event {the die has the number i on its face}. Obviously there are six elementary events. Now, add the condition that the elementary events are symmetric, i.e. they are all the same in every respect. This implies that $P(\Lambda_i) = P(\Lambda_j)$ $(i \neq j)$. Substituting this condition in 1.1.3 yields $\sum_{i=1}^{6} P(\Lambda_i) = 6 \times P(\Lambda_j) = 1 \,\forall j$ (the symbol $\forall$ means 'for every'), so that $P(\Lambda_j) = \frac{1}{6} \forall j$. Thus, the probability of any face to appear is $1/6$, not because the frequency of appearance is approaching it, but because we added the assumption that all events are symmetric.

Many problems in probability are addressed with the assumption that elementary events are equally probable. In such cases care must be exercised, since different definitions of the elementary events lead to different results. Thus, in the example of putting k balls into n cells, if the assignment vectors are considered to be the symmetric elementary events, the probability of every such event is $1/n^k$, whereas if the occupancy vectors are the symmetric elementary events the probability of each event will be $1/\binom{n+k-1}{k}$.

In fact, for the throw of a die experiment, any six numbers, such as $0.1, 0.1, 0.1, 0.05, 0.3$ and 0.35, that fulfil conditions (1.1.1–1.1.3) are a legitimate probability function. Only the choice $1/6$ has the additional property that the events are identical. Every experiment, including the operation of an industrial system, has some probability functions attached to it. In principle, the number of possible functions is unlimited. This is a source of uncertainty regarding the system operation. It is through added assumptions and through observations that we narrow the range of the choices to the point at which useful results can be obtained.

Example 1.1.1: The chaotic organization

An extremely inefficient organization operates n different systems numbered from 1 to n, in n different locations. Once a year these systems must be replaced. A central operator obtains a set of new systems and must transfer them to the appropriate locations. The central operator sends the spare systems in a random manner so that each system may find itself in any location. What is the probability that at least one system will be shipped to the correct place?

The elementary event is a sequence of the form $(j_1, j_2, \ldots, j_n)$, where j_k is the number of the spare system that was sent to location k. There are $n!$ such elementary events. A situation in which the right replacement system reached the appropriate location occurs when at least one of the elements of the sequence fulfils the condition $j_m = m$.

Let A_m be the event – the mth spare system reached location m (i.e. the right place). The event $\mathrm{A}_r \cap \mathrm{A}_s$ contains, for example, all the elementary events for which $j_r = r$ *and* $j_s = s$. Obviously these events are not disjoint. They can all happen if the elementary event $(1, 2, 3, \ldots, n)$ occurs. The event (at least one system reached the correct location) is the union of all these events, namely $\mathrm{B} = \mathrm{A}_1 \cup \mathrm{A}_2 \cup \mathrm{A}_3 \cup \cdots \cup \mathrm{A}_n$. Using equation (1.1.10) yields:

$$P(\mathrm{B}) = \sum_{k=1}^{n} (-1)^{k+1} S_k,$$

where

$$S_k = \sum_{1 \le j_1 < j_2 \cdots < j_k}^{n} P(\mathrm{A}_{j_1} \cap \mathrm{A}_{j_2} \cap \cdots \cap \mathrm{A}_{j_k}).$$

The number of terms in S_k is

$$\binom{n}{k} = \frac{n!}{(n-k)!k!}.$$

This is the number of different groups of k objects that can be drawn from a set of n objects. A specific intersection event $\mathrm{A}_{j_1} \cap \mathrm{A}_{j_2} \cap \cdots \cap \mathrm{A}_{j_k}$ contains all the elementary events in which system j_i is sent to location j_1, system j_2 to location j_2, etc. Thus, replacement systems $j_1, j_2, \ldots, j_k$ are sent to the correct locations. Keeping these points fixed in the elementary events, one can change all other $n-k$ points. This can be done in $(n-k)!$ ways. Therefore, each intersection event is a union of $(n-k)!$ elementary events. The elementary events are symmetric; the probability of each elementary event is $1/n!$, thus:

$$P(\mathrm{A}_{j_1} \cap \mathrm{A}_{j_2} \cap \cdots \cap \mathrm{A}_{j_k}) = \frac{(n-k)!}{n!}$$

Now, S_k is the sum of $\binom{n}{k}$ intersection events, yielding

$$S_k = \binom{n}{k}\frac{(n-k)!}{n!} = \frac{1}{k!}.$$

Substituting this result for $P(\mathrm{B})$ yields:

$$P(\mathrm{B}) = 1 - \frac{1}{2!} + \frac{1}{3!} - \frac{1}{4!} + \cdots + (-1)^{n+1}\frac{1}{n!}.$$

For a set of five systems, the probability that no replacement system will reach the correct place is 0.36666. This probability does not decrease to zero as the number of systems increases. Rather, $1 - P(\mathrm{B})$ approaches the limit $1/e$ as $n \to \infty$.

Example 1.1.2: Sharing failed units

Consider a field containing m identical systems. In each system there are k identical units (engines or radios, or any other type of unit). It is known that n units are failed in the field. What is the probability that exactly $r \leq k$ units are failed in the first system?

The elementary event is an assignment vector of the form:

$$(b_{11}, b_{12}, b_{13}, \ldots, b_{1k}, b_{21}, b_{22}, b_{23}, \ldots, b_{2k}, \ldots, b_{m1}, b_{m2}, b_{m3}, \ldots, b_{mk}),$$

where b_{ij} describes the state of the jth unit in the ith system. This state can be either 1 (operating) or 0 (failed). The assignment vector contains 1's and 0's, and it is given that exactly n of these indicators are zeros. The number of elementary events is the number of different possibilities to chose a set of n elements out from a group of $m \times k$ elements. This is $\binom{m \times k}{n}$. With the assumption that all elementary events are symmetrical, the probability for each event is $1/\binom{m \times k}{n}$. The elementary events in which r units have failed in the first system can be chosen by choosing r elements out of k in the first system (r zeros in the first k elements). This can be done in $\binom{k}{r}$ different ways. For each such choice, the remaining $(n-r)$ elements can be chosen from the remaining $(m-1)k$ units in $\binom{(m-1)k}{(n-r)}$ different ways. Thus, there are $\binom{(m-1)k}{(n-r)} \times \binom{k}{r}$ different elementary events, in which there are r failed units in the first system. The probability for this event is thus given as:

$$P_1(r) = \frac{\binom{(m-1) \times k}{(n-r)} \times \binom{k}{r}}{\binom{m \times k}{n}}.$$

1.1.3 The event space

The above construction of the sample space and the probability leaves a number of vague points, since the sample space includes only the elementary events while the probability function is defined for every event (either elementary or any set theory combination of events such as unions and intersections). The Null event ϕ, for example, is not included in the sample space. A more inclusive space may be defined as follows.

Definition 1.1.2 A probability space $(\boldsymbol{B}, \Omega, P)$ is the triplet including the sample space Ω, the probability function $P(\mathrm{A})$ defined in equation (1.1.1) and the event space $\boldsymbol{B}$ (sometimes also called the events field or Borel space), which is a collection of subsets of Ω with the following properties:

$$\Omega \in \boldsymbol{B} \text{ (the symbol } \in \text{ stands for 'belongs to').} \tag{1.1.12}$$

$$\text{If event } \mathrm{A} \in \boldsymbol{B} \text{ then also } \overline{\mathrm{A}} \in \boldsymbol{B}. \tag{1.1.13}$$

$$\text{If } \mathrm{A} \in \boldsymbol{B} \text{ and } \mathrm{B} \in \boldsymbol{B} \text{ then also } \mathrm{A} \cup \mathrm{B} \in \boldsymbol{B}. \tag{1.1.14}$$

$$\text{If } \mathrm{A}_n \in \boldsymbol{B} \quad (n = 1, 2, 3, \ldots) \Rightarrow \left(\bigcup_{n=1}^{\infty} \mathrm{A}_n \right) \in \boldsymbol{B}, \tag{1.1.15}$$

thus any finite or countable union of events is included in the Borel space.

With the above definition, the probability function is a real valued function defined on the event space.

1.1.4 Conditional probability

Definition 1.1.3: Conditional probability function A and B are events and $P(\mathrm{B}) > 0$, then the conditional probability of A *given* B (the probability that A will occur given that B has occurred) is denoted by $P(\mathrm{A}|\mathrm{B})$, and is defined as

$$P(\mathrm{A}|\mathrm{B}) = \frac{P(\mathrm{A} \cap \mathrm{B})}{P(\mathrm{B})}. \tag{1.1.16}$$

It is straightforward to show that the conditional probability function fulfils all the conditions of a probability function (1.1.1–1.1.3), and is a legitimate probability function. This also implies that properties (1.1.4)–(1.1.10) apply to this function.

The following theorems are brought without proof. They are quite easy to prove, and a proof can be found in any basic text on probability (e.g. see Dudewicz, E. J. (1976). *Introduction to Probability*. Holt, Rhinehart & Winston, New York).

Theorem 1.1.1 (Total probability) *Let* $\mathrm{B}_1, \mathrm{B}_2, \ldots, \mathrm{B}_n$ *be a set (finite or infinite) of mutually exclusive events, i.e.* $\mathrm{B}_i \cap \mathrm{B}_j = \phi$ *for any* $(i \neq j)$, *and also* $P(\mathrm{B}_k) > 0$ *for each* k, *and such that* $\left(\bigcup_{i=1}^{n} P(\mathrm{B}_i)\right) = 1$, *i.e. the union of the events is the universal event.*

Then, for any event A *the probability* $P(\mathrm{A})$ *can be written as*

$$P(\mathrm{A}) = \sum_{i=1}^{n} P(\mathrm{A}|\mathrm{B}_i)P(\mathrm{B}_i). \qquad (1.1.17)$$

Theorem 1.1.2 (Bayes' rule) *Let* $\mathrm{B}_1, \mathrm{B}_2, \ldots, \mathrm{B}_n$ *be a set of events as in Theorem* 1.1.1 *and* A *an event such that* $P(\mathrm{A}) > 0$. *Then, for every* k, *the conditional probability of* B_k *given* A *can be written as*

$$P(\mathrm{B}_k|\mathrm{A}) = \frac{P(\mathrm{A}|\mathrm{B}_k)P(\mathrm{B}_k)}{\sum_{j=1}^{n} P(\mathrm{A}|\mathrm{B}_j)P(\mathrm{B}_j)}; \qquad (1.1.18)$$

recall that n *can be either finite or infinite.*

Definition 1.1.4: Independent events Two events A and B are called independent if and only if $P(\mathrm{A} \cap \mathrm{B}) = P(\mathrm{A})P(\mathrm{B})$.

In many practical instances, the concept of independence simply means that two events are unrelated. The occurrence of any of them has no effect on the occurrence of the other. The assumption of independence relieves much of the complexity of many problems, because it is then allowed to calculate the probability of occurrence of two events (say, a failure of a pump and a protection mechanism during the same time period) as a product of the two probabilities. In realistic systems the independence assumption is often cast in doubt. For example, a failure in a pump may seem unrelated to a failure of a protection mechanism. Yet, protection units are usually inspected periodically rather than continuously. A failure in a pump is often an opportunity to inspect the protection mechanism. Thus, past failures of the pump may affect the probability of future failures of the protection mechanism, and the two events cannot be considered independent.

Example 1.1.3: Energy transmission

A power supply system involves a number of circuit breakers as shown in Figure 1.1.1. The system is operational as long as at least one connection exists between the Entry and the Exit. The probability that a circuit breaker (CRB) functions properly is p. If CRB number 4 fails, the

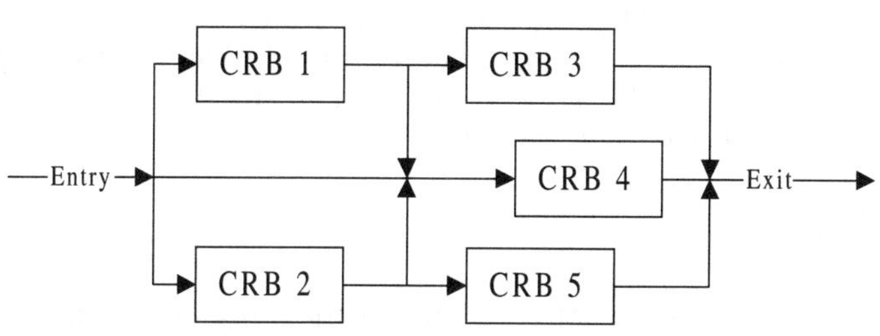

Figure 1.1.1 A system of circuit breakers

load on CRBs numbers 3 and 5 increases and, as a result, their probability to function properly is reduced by a factor $\alpha < 1$. What is the probability that the system functions properly?

Denote by A_i the event {Circuit breaker number i functions}. Let B be the event {the system functions properly}, then according to Theorem 1.1.1:

$$P(\mathrm{B}) = P(\mathrm{B}|\mathrm{A}_4)P(\mathrm{A}_4) + P(\mathrm{B}|\overline{\mathrm{A}}_4)P(\overline{\mathrm{A}}_4).$$

Since $P(\mathrm{A}_4) = p$ and $P(\overline{\mathrm{A}}_4) = 1 - p$, we have

$$P(\mathrm{B}) = p \times P(\mathrm{B}|\mathrm{A}_4) + (1-p) \times P(\mathrm{B}|\overline{\mathrm{A}}_4).$$

When CRB 4 is operating, it is sufficient that either CRB 1 or CRB 2 will operate, thus:

$$P(\mathrm{B}|\mathrm{A}_4) = P[(\mathrm{A}_1 \cup \mathrm{A}_2)|\mathrm{A}_4] = P(\mathrm{A}_1|\mathrm{A}_4) + P(\mathrm{A}_2|\mathrm{A}_4) - P[(\mathrm{A}_1 \cap \mathrm{A}_2)|\mathrm{A}_4] = 2p - p^2.$$

Likewise, if CRB 4 has failed, either CRBs 1 *and* 3 or CRBs 2 *and* 5 must operate, thus

$$\begin{aligned} P(\mathrm{B}|\overline{\mathrm{A}}_4) &= P[(\mathrm{A}_1 \cap \mathrm{A}_3) \cup (\mathrm{A}_2 \cap \mathrm{A}_5)|\overline{\mathrm{A}}_4] \\ &= P(\mathrm{A}_1 \cap \mathrm{A}_3|\overline{\mathrm{A}}_4) + P(\mathrm{A}_2 \cap \mathrm{A}_5)|\overline{\mathrm{A}}_4) - P[(\mathrm{A}_1 \cap \mathrm{A}_3) \cap (\mathrm{A}_2 \cap \mathrm{A}_5)|\overline{\mathrm{A}}_4] \\ &= \alpha p^2 + \alpha p^2 - \alpha^2 p^4 = 2\alpha p^2 - \alpha^2 p^4. \end{aligned}$$

The probability that the system is operational is thus

$$P(\mathrm{B}) = p(2p - p^2) + (1-p)(2\alpha p^2 - \alpha^2 p^4).$$

In the above expressions, use is made of the fact that a conditional probability is a probability function, and the rules of union of events (equation (1.1.8)) and independent events apply to it.

Example 1.1.4: Purchasing considerations

An organization purchased 100 systems from two manufacturers, 30 from manufacturer A and 70 from manufacturer B. The systems of A are known to have a probability of being failed at any point in time equal to $p_a = 0.074$, and those of B, $p_b = 0.034$. Each company declares that as a result of a new maintenance program, the failure probability described above has been reduced by a factor of 2. Yet for the systems of company A, this program was applied only to 50% of the systems, and in the case of company B it was applied to 30% of the systems. What is the probability that a randomly selected system is failed?

Denote the event C = {the system is failed}. Also denote by M the event {the system had improved maintenance} (event $\overline{\mathrm{M}}$ is the complementary event, namely, no improved maintenance). Consider the events: $\mathrm{D}_1 = \mathrm{A} \cap \mathrm{M}$ (a system is from factory A with improved maintenance), and likewise, $\mathrm{D}_2 = \mathrm{A} \cap \overline{\mathrm{M}}$, $\mathrm{D}_3 = \mathrm{B} \cap \mathrm{M}$ and $\mathrm{D}_4 = \mathrm{B} \cap \overline{\mathrm{M}}$. The events $\mathrm{D}_1, \mathrm{D}_2, \mathrm{D}_3, \mathrm{D}_4$ fulfil the conditions of Theorem 1.1.1, and thus

$$\begin{aligned} P(\mathrm{C}) &= \sum_{i=1}^{4} P(\mathrm{C}|\mathrm{D}_i)P(\mathrm{D}_i) \\ &= 0.037 \times 0.7 \times 0.5 + 0.074 \times 0.7 \times 0.5 + 0.017 \times 0.3 \times 0.3 + 0.034 \times 0.3 \times 0.7 \\ &= 0.04752. \end{aligned}$$

Table 1.1.1 Availability range and probabilities for four different manufacturers

Availability range	C_1	C_2	C_3	C_4
1. 0.95–1.0	0.05	0.1	0.15	0.4
2. 0.9–0.95	0.1	0.15	0.25	0.35
3. 0.8–0.9	0.25	0.4	0.4	0.2
4. Below 0.8	0.6	0.35	0.2	0.05

Example 1.1.5: Posterior conclusions

Four different companies, C_1, C_2, C_3 and C_4, are manufacturing the same product with different quality levels. Quality is measured in terms of the relative average time the system is operating, measured over a year. This relative time is also called the *average availability*. The quality is divided into four regions of average availability. Less than 0.8, between 0.8 and 0.9, between 0.9 and 0.95 and between 0.95 and 1. The probabilities that a system will be in any given range for each manufacturer are shown in Table 1.1.1.

A supplier bought a single system for an organization, claiming that he obtains such systems from all four manufacturers in a completely random manner. Following a year of operation, it was verified that the average availability was 0.73. What is the probability that the system was manufactured by company C_4?

Let C_i be the event {the manufacturer is C_i} and A_j the event {the average availability is in range j}. The desired probability is $P(C_4|A_4)$. Using the Bayes' Rule (Theorem 1.1.2), one obtains:

$$P(C_4|A_4) = \frac{P(A_4|C_4)P(C_4)}{\sum_{j=1}^{n} P(A_4|C_j)P(C_j)}.$$

According to the supplier, $P(C_i) = 0.25 \ \ (i = 1, 2, 3, 4)$, thus

$$P(C_4|A_4) = \frac{0.05 \times 0.25}{(0.6 + 0.35 + 0.2 + 0.05) \times 0.25} = 0.0416.$$

Customers observing such performance will find it hard to believe that the *a priori* probability for getting a C_4 system is 0.25, while the above (*a posteriori*) calculation indicates, in view of the actual performance, that this probability is less than 5%. Note that if the supplier claimed he received the system from C_4 with certainty, i.e. $P(C_4) = 1$, and $P(C_i) = 0.0 \ (i = 1, 2, 3)$, then the above equation will yield $P(C_4|A_4) = 1$, i.e. his claim will be verified despite the poor performance. In general, the stronger is the *a priori* claim of the supplier, the higher will be the agreement of the *a posteriori* results with it. This is demonstrated in Table 1.1.2, in which $P(C_4)$ varies in the range 0.25–1, and the prior probabilities for the first three manufacturers are given as $(1 - P(C_4))/3$.

The above discussion suggests that, when purchasing a system, it is recommendable to define the performance requirement directly, rather than settle for a supplier promise of a good manufacturer.

Table 1.1.2 Posterior and *a priori* probabilities

$P(C_4)$	$P(C_4 \mid A_4)$	$P(C_4 \mid A_4)/P(C_4)$
0.25	0.0416	0.1664
0.5	0.115	0.23
0.6	0.163	0.271
0.7	0.23	0.328
0.8	0.34	0.425
0.9	0.54	0.6
0.95	0.712	0.749
0.99	0.928	0.937
1.0	1.0	1.0

1.1.5 Random variables

So far we have considered the outcome of experiments in terms of events. The connection of such events with the mathematical framework was done through the probability function. Yet, to obtain full advantage of the richness of mathematical theories, it is desirable that the events themselves are defined in terms of real variables. This is especially true when continuous events are considered. Such connection is obtained by creating a correspondence between events and real numbers. Such variables are called *random variables*, as they represent a random numerical outcome of statistical experiments.

Definition 1.1.5: Random variable A random variable is a real valued function defined on the sample space. This can be denoted as $X(\omega)\colon \Omega \to R$, where $\omega \in \Omega$ and $X \in R$. The domain of the function is the sample space, and the range is a subset on the real axis.

Random variables can be either discrete or continuous, depending on the nature of the sample space. In many cases, the definition of the random variable comes in a natural way from the definition of the sample space. For example, a parachutist is attempting to land as close as possible to a marker. The sample space is constructed of the events $\omega \in \Omega$, where $\omega = \{\text{The point of landing}\}$. Obviously, these are elementary and disjoint events. The distance X from the point of landing to the marker is a natural definition for a random variable. It is easy to associate probabilities with the random variables, since probabilities are already attached to the elementary events.

1.2 Distributions and Moments

When events are replaced or presented by random variables, the probability of an event is accordingly replaced by a probability of a single real value or an interval. The probability function is then replaced by a function defined on real variables. Such functions are called *distributions*. In the following section, the concepts of distributions will be introduced and discussed.

1.2.1 Distribution functions

Definition 1.2.1: The cumulative distribution function Given a sample space with elementary events ω and a random variable $\{x(\omega), \omega \in \Omega\}$, the Cumulative Distribution Function (CDF) is defined as:

$$F(X) = P_r[x(\omega) \leq X] \quad \text{for all } X \in (-\infty, +\infty). \tag{1.2.1}$$

P_r stands for the probability function defined in Section 1.1. This is the probability of the union of all the events ω for which $x(\omega) \leq X$. In applications related to systems, the most important and naturally occurring random variable is the time. This is usually the time at which an event of some type occurs. In such cases, the cumulative distribution is presented as

$$F(t_0) = P_r(t \leq t_0). \tag{1.2.2}$$

The CDF has a number of properties that can be easily deduced from the definition and the properties of the probability function. These properties will be described using the random variable t, to introduce this variable and without any loss of generality. It should be noted, however, that whereas random variables are defined over the range $(-\infty, +\infty)$, time is defined in the range $[0, +\infty)$.

Theorem 1.2.1: Properties of the cumulative distribution *If t is a random variable (r.v.) then its CDF $F(T)$ has the following properties:*

1. *$F(T)$ is non-decreasing, i.e. if $T_2 \geq T_1$ then $F(T_2) \geq F(T_1)$. This property results from the fact that the union of events that correspond to $t \leq T_2$* includes *the union of events for which $t \leq T_1$.*
2. *$F(-\infty) = \lim_{T \to -\infty} F(T) \equiv 0$ and $F(+\infty) = \lim_{T \to +\infty} F(T) \equiv 1$. These properties result from the fact that the event for which $t \leq -\infty$ is the null event and the union of events corresponding to $t \leq +\infty$ is the universal event Ω.*
3. *$F(T)$ is continuous from the right, i.e.* $\lim_{h \to 0\, h>0} F(T+h) = F(T)$.

To prove the last property, consider the limit of the difference $[F(T+h) - F(T)]$. The definition of the CDF yields

$$\begin{aligned}\lim_{\substack{h \to 0 \\ h>0}} [F(T+h) - F(T)] &= \lim_{\substack{h \to 0 \\ h>0}} [P(t \leq T) + P[T < t \leq (T+h)] - P(t \leq T)] \\ &= \lim_{\substack{h \to 0 \\ h>0}} P[T < t \leq (T+h)].\end{aligned}$$

The above expression is derived by dividing the range $[0, T+h]$ into a union of two disjoint ranges, namely, $[0, T]$ and $(T, T+h]$. As $h \to 0$ the event $[T < t \leq (T+h)]$ approaches the null event, since no value of t fulfils the condition. Accordingly, the probability of the event approaches zero.

Every real valued function that has the above properties is a CDF of some event space. Random variables may be either discrete or continuous. A discrete r.v. assumes only a finite (or denumerable) set of values. The definition and properties of the CDF apply equally to both types of r.v.

From the definition of the CDF, it is clear that the probability of the random variable to take values in a given interval $[t_1, t_2]$ is given by

$$P[t \in [t_1, t_2]] = P(t_1 \leq t \leq t_2) = F(t_2) - F(t_1) \tag{1.2.3}$$

Likewise:

$$P[t \in [0, t_0]] = P(0 \leq t \leq t_0) = F(t_0) - F(0) = F(t_0) \tag{1.2.4}$$

In the context of systems behavior, the last expression means that the probability that an event (failure, repair or anything else) will occur up to time t_0 is $F(t_0)$. Note that (1.2.3) and (1.2.4) give information about an event within an *interval*, rather than about the point at which the event may occur. To relate the occurrence of an event to a point in time, the interval must be reduced to an infinitesimal size. This is addressed in the following definitions.

Definition 1.2.2: The probability density function Let t be a r.v. with a differentiable CDF. The Probability Density Function (PDF) is defined as the derivative of the CDF:

$$f(t) = F'(t) = \frac{dF(t)}{dt} \tag{1.2.5}$$

The PDF is related to the value of the random variable rather than to an interval. $f(t)$ is non-negative since $F(t)$ is non-decreasing. Equation (1.2.5) also implies that the CDF is an integral of the PDF in the form:

$$F(t) = \int_{-\infty}^{t} f(y)\, dy \tag{1.2.6}$$

Using Definition (1.2.2) and equation (1.2.3), it is easy to see that:

$$P[t \in [t_1, t_2]] = P(t_1 \leq t \leq t_2) = \int_{t_1}^{t_2} f(y)\, dy, \tag{1.2.6a}$$

and since $F(\infty) = 1$, also

$$\int_{-\infty}^{\infty} f(y)\, dy = 1, \tag{1.2.6b}$$

which is called the normalization property of the PDF.

One of the most important perceptions of the meaning of the PDF is revealed when one considers a small interval Δt around a point t_c. If the interval is small enough so that the variation of $f(t)$ in it is negligible, then the probability that the r.v. will be realized in that interval, $P(t \in \Delta t)$, can be approximated by

$$P_r(t \in \Delta t) \cong f(t_c)\Delta t. \tag{1.2.7}$$

The interpretation of equation (1.2.7) is that if a unit starts its operation at time zero and $f(t)$ is the PDF for an event to occur at any time t, then the probability that the event will take place in an interval Δt is given by expression (1.2.7).

Consider again an event in Δt. For the event to occur in that interval it need *not happen* up to that interval. If we denote by Γ the event {The event did not happen up to time t_c}, then equation (1.2.7) can be written as

$$P_r(t \in \Delta t) = P_r(\Gamma) \times P_r(t \in \Delta t|\Gamma). \tag{1.2.8}$$

$P_r(t \in \Delta t|\Gamma)$ is the conditional probability of an event to occur in Δt. It can be described as a conditional probability *per unit time*, $z(t)$, multiplied by the size of the interval, namely $P(t \in \Delta t|\Gamma) = z(t)\Delta t$. From the definition of the CDF, and since Γ is the complementary of the event considered in equation (1.2.4), it is obvious that $P(\Gamma) = 1 - F(t_c)$. Substituting these relations back in equation (1.2.8) and comparing with equation (1.2.7) yields

$$f(t) = [1 - F(t)]z(t) \quad \text{or} \tag{1.2.9}$$

$$z(t) = \frac{f(t)}{[1 - F(t)]}. \tag{1.2.10}$$

$z(t)$ is called the 'hazard' function. It is the probability that an event will happen at time t, per unit time, given that the event did not happen up to that time.

The hazard function plays an important role in system engineering. Data collection, often done by observing failures in systems, is naturally related to the hazard function since, at any point in time, it is given that the system did not fail up to that time. The CDF, PDF and the hazard are equivalent in the sense that knowing any one of them is enough to obtain the two others. Equations (1.2.5), (1.2.6) and (1.2.10) provide the necessary relations. Substituting dF/dt for $f(t)$ in equation (1.2.10) yields a simple differential equation for $F(t)$, which can be written as

$$-\frac{d[\ln(1 - F(t))]}{dt} = z(t),$$

and upon integration:

$$F(t) = 1 - e^{-\int_0^t z(x)\,dx} \quad \text{and} \quad f(t) = z(t)\,e^{-\int_0^t z(x)\,dx}, \tag{1.2.11}$$

which completes the relation between the three functions. The above discussion leads to the following observations:

- $F(t)$ – Gives the probability that an event will occur in the interval from 0 to t.
- $f(t)dt$ – Gives the probability that an event will occur in a small interval dt at t.
- $z(t)dt$ – Gives the *conditional* probability that an event will occur in a small interval dt at t, given that it did not occur up to that time.

The name 'hazard function' was adopted for $z(t)$ because it is related to failures. Although failures are important events, they are not the only possible events in considering the life history of a system. There are many other events, such as repairs, inspections, maintenance, change of operating state and many more that may be of a statistical nature. All types of stochastic events (events of a statistical nature) have PDFs, CDFs and hazard functions attached to them, so the name 'hazard' is not always appropriate. In the following, the terms 'local rate function' or 'rate function' will also be used.

The following example can further clarify the role and interpretation of each of the above defined functions.

Example 1.2.1: A ministerial visit

A company introduced a new production line into operation. The properties of the line are known and the PDF for failure is given in the form:

$$f(t) = k\lambda t^{k-1}\,\mathrm{e}^{-\lambda t^k} = 2 \times 6.23 \times 10^{-8} \times t\,\mathrm{e}^{-6.23\times 10^{-8} t^2}. \tag{1.2.12}$$

This function is generally known as a *Weibull distribution.* In the present case, $\lambda = 6.23 \times 10^{-8}$ and $k = 2$. On the day at which the line is operated, it is known that a ministerial visit is planned in about 4.6 months (3400 h). The duration of the visit will be four hours and is intended mainly to observe the operation of the new line. The plant manager is concerned about the possibility of a production line failure during the visit (Event A).

The probability for this event can be obtained from equation (1.2.7) (the duration of the visit represents a small time interval) namely, taking $t_c = 3400$ and $\Delta t = 4$ yields:

$$P(\mathrm{A}) = f\,(t_c)\Delta t = 2 \times 6.23 \times 10^{-8}(3400)\,\mathrm{e}^{-6.23\times 10^{-8}(3400)^2} \times 4 = 0.0008247.$$

A probability of 0.0008247 is good enough for the plant manager. Time passes and the day of the visit arrives. While the minister is on his way to the plant, a sudden anxiety urges the manager to inquire again for the probability of a failure during the visit. The situation is now different. The present time is t_c and it is known that the system has not failed so far. The probability for a failure during the visit is now given by

$$z(t)\Delta t = \frac{f(t)\Delta t}{[1 - F(t)]} = 0.001694$$

(see equation (1.2.10)). The failure probability has increased by about a factor of 2. The state of affairs is shown in Figure 1.2.1.

Note that in this example the approximate forms $f\,(t)\Delta t$ and $z(t)\Delta t$ are used. This is justified because the time interval is small (the variation of the corresponding functions over the interval is negligible). The next example deals with a longer interval.

Example 1.2.2: Failure in larger intervals

Consider the plant discussed in Example 1.2.1. The conditions now are that the first 12 months of operation (8760 hours) are a startup period. Following that, a four month (2960 hours) period of

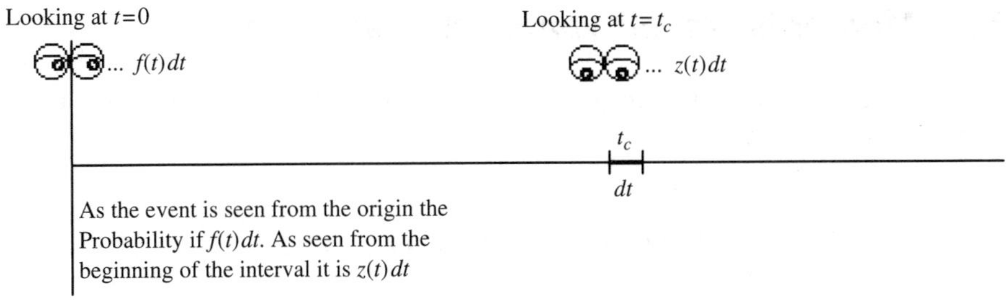

Figure 1.2.1 Probability of event in dt as seen at $t = 0$ and at $t = t_c$

extensive production is expected. A failure during this period will be very costly, and may cause loss of potential markets. Looking from $t = 0$, the probability to have at least one failure in this period (event A) is given by equation (1.2.4) with $t_1 = 8760$ and $t_2 = 11\,720$. Denoting the event by A:

$$P(\mathrm{A}) = \int_{t_1}^{t_2} f(t)\,dt = F(t_2) - F(t_1) = [1 - \mathrm{e}^{-\lambda t_2^k}] - [1 - \mathrm{e}^{-\lambda t_1^k}] = 0.008197. \tag{1.2.13}$$

As t_1 is approached, the event to look at is $(\mathrm{A}|\overline{\mathrm{B}})$, where $\overline{\mathrm{B}}$ is the event {No failure up to t_1}. Now, $P(\overline{\mathrm{B}}) = 1 - F(t_1)$, and $\mathrm{A} \cap \overline{\mathrm{B}} = \mathrm{A}$. Using equation (1.1.16) yields:

$$\begin{aligned} P\{t_{\text{failure}} \in [t_1, t_2] |\text{ No failure upto } t_1\} &= P(\mathrm{A}|\overline{\mathrm{B}}) = \frac{P(\mathrm{A} \cap \overline{\mathrm{B}})}{P(\overline{\mathrm{B}})} \\ &= \frac{P(\mathrm{A})}{P(\overline{\mathrm{B}})} = \frac{F(t_2) - F(t_1)}{1 - F(t_1)} \end{aligned} \tag{1.2.14}$$

which in this case is

$$\frac{\mathrm{e}^{-\lambda t_1^k} - \mathrm{e}^{-\lambda t_2^k}}{\mathrm{e}^{-\lambda t_1^k}} = 0.977.$$

This is a much higher probability than that seen from $t = 0$. The probability of failure in the interval as seen from $t = 0$ is very modest, about 0.8%. Yet if the system reaches the beginning of the interval without a failure, an expected failure becomes almost certain (97.7%).

Equation (1.2.14) is a general expression for the probability of a failure in an interval given that there was no failure up to the beginning of the interval. The expressions used in Example 1.2.1 can be obtained from expressions (1.2.13) and (1.2.14) correspondingly by taking the interval $t_2 - t_1$ to be small enough so that the general approximation $\int_{t_1}^{t_2} f(t)dt \approx f(t_1)[t_2 - t_1]$ applies.

Let t_c be any intermediate time point $0 \le t_c \le t_1$. The probability of failure in the interval $[t_1, t_2]$, given that the system did not fail up to t_c is easily seen to be:

$$P\{t_{\text{failure}} \in [t_1, t_2] |\text{ No failure upto } t_c\} = \frac{F(t_2) - F(t_1)}{1 - F(t_c)} \tag{1.2.15}$$

Expression (1.2.15) is a generalization of (1.2.13) and (1.2.14). If $t_c = 0$ it reduces to (1.2.13). If $t_c = t_1$, (1.2.14) is obtained.

1.2.2 Moments, mean value and variance

Definition 1.2.3: Expected value (mean) The expected value (or mean value) of a r.v. X with PDF $f(x)$ is defined as

$$E(x) = \mu_x = \langle x \rangle = \int_{-\infty}^{+\infty} x f(x)\, dx \qquad (1.2.16)$$

($E(x)$, μ_x and $\langle x \rangle$ are common notations used). In the discrete case, the definition takes the form

$$E(x) = \mu_x = \sum x_i p(x_i), \qquad (1.2.17)$$

where x_i takes all the possible discrete values of the r.v. and $p(x_i)$ is the probability that the r.v. takes a given value x_i.

Definition 1.2.4: The *n*th moment The nth moment of a r.v. is defined as

$$E(x^n) = \langle x^n \rangle = \int_{-\infty}^{+\infty} x^n f(x)\, dx \qquad (1.2.18)$$

($\sum x_i^n p(x_i)$ for the discrete case). The first moment is simply the mean value. Of particular importance is the second moment, which has a special notation

$$S_x^2 = E(x^2) = \langle x^2 \rangle = \int_{-\infty}^{+\infty} x^2 f(x)\, dx \qquad (1.2.19)$$

($\sum x_i^2 p(x_i)$ for the discrete case).

Definition 1.2.5: The *n*th central moment The nth central moment of a r.v. is defined as

$$E[(x-\mu_x)^n] = \langle (x-\mu_x)^n \rangle = \int_{-\infty}^{+\infty} (x-\mu_x)^n f(x)\, dx \qquad (1.2.20)$$

($\sum (x_i - \mu_x)^n p(x_i)$ for the discrete case).

Of particular importance is the second central moment, commonly referred to as the *variance* of the r.v., namely:

$$V_x^2 = E[(x-\mu_x)^2] = \int_{-\infty}^{+\infty} (x-\mu_x)^2 f(x)\, dx. \qquad (1.2.21)$$

The mean value and the variance play an essential role in many statistical applications, and in the application of the Monte Carlo method. The mean gives information about a specific value around which the realizations of the r.v. are scattered, while the variance is a measure of the dispersion of the PDF, or the range over which the r.v. is scattered. If the variance is small, one may expect that the r.v.s will be concentrated around the mean, while a large variance indicates that it will be highly dispersed over a large interval about the mean. These notions will be formally verified in the following sections. Now let $h(x)$ be any function of a r.v., then $h(x)$ is also a random variable. The mean, variance and moments of $h(x)$ are defined as above by replacing x with $h(x)$. Thus, the mean and variance of $h(x)$ will be

$$E(h) = \mu_h = \langle h(x) \rangle = \int_{-\infty}^{+\infty} h(x) f(x)\, dx \tag{1.2.22}$$

($\sum h(x_i)p(x_i)$ for the discrete case)

$$V_h = E[(h(x) - \mu_h)^2] = \int_{-\infty}^{+\infty} [h(x) - \mu_h]^2 f(x)\, dx \tag{1.2.23}$$

($\sum [h(x_i) - \mu_h]^2 p(x_i)$ for the discrete case).

It is easy to see that the variance can be expressed as the difference between the second moment and the square of the mean, i.e.

$$V_h = S_h^2 - {\mu_h}^2. \tag{1.2.23a}$$

The square root of the variance is called the *standard deviation*, and is denoted by $\sigma_h = \sqrt{V_h}$.

The most concise definition of the Monte Carlo method is that it is a method to calculate the mean of functions of random variables, as in equation (1.2.22). These could be elaborate averages of multidimensional PDFs defined on many random variables, but basically, they are all straightforward extensions of the above definition. Furthermore, the method gains its applicability from the ease with which such averages can be calculated, and from the fact that every quantity, being part of a statistical process or not, can be presented as the mean value of a function of one or more random variables.

1.3 Some Important Distributions

Distributions are essential in describing the behavior of systems. Unlike the situation in physics, there are rarely known 'correct' distributions. In most cases, distributions are inferred by fitting them to measured data or by making assumptions on the behavior of the system.

1.3.1 The uniform distribution

Defined on an interval $[a, b]$. This distribution is obtained by assuming that all points in the interval are symmetrical. This assumption is the continuous equivalent of the assumption made in Section 1.1 regarding the symmetry of elementary events. The probability of the r.v. to appear in any interval dt is then proportional to the size of the interval, thus $f(t)\,dt = C\,dt$, where C is a constant. The normalization property (1.2.6b) yields

$$\int_a^b f(t)\,dt = \int_a^b C\,dt = C(b-a) = 1 \Rightarrow C = \frac{1}{b-a},$$

thus:

$$f(t) = \frac{1}{b-a} \quad \text{for } a \le t \le b \tag{1.3.1}$$

and zero outside this range. The CDF and rate function are obtained using equations (1.2.4) and (1.2.8):

$$F(t) = \int_{-\infty}^{t} f(t)\,dt = \int_a^t \frac{1}{b-a}\,dt = \frac{t-a}{b-a} \tag{1.3.2}$$

also $F(t) \equiv 0$ for $t < a$ and $F(t) \equiv 1$ for $t \ge b$. The local rate function is:

$$z(t) = \frac{f(t)}{1-F(t)} = \frac{1}{b-t} \quad \text{for } a \le t \le b \tag{1.3.3}$$

and zero outside this interval.

Note that the rate function is diverging as $t \to b$. This property is common to all distributions defined over a finite range. If a continuous PDF is defined over a finite range $[a, b]$ and $f(b) > 0$ then, since $F(t)$ approaches unity as $t \to b$, $z(t)$ approaches infinity. This is compatible with the intuitive meaning of $z(t)\,dt$. If the event did not occur up to t while $t \to b$, and since the probability of occurring in the interval $[a, b]$ is unity, $z(t)$ must increase indefinitely such that the product $z(t)\,dt$ will approach unity. The uniform distribution in the range $[0, 1]$ is of special importance, and will be denoted by $u(t)$. The mean and variance are given by

$$E(t) = \langle t \rangle = \int_a^b \frac{t}{b-a}\,dt = \frac{b+a}{2} \tag{1.3.4a}$$

$$V = \langle t^2 \rangle - \langle t \rangle^2 = \int_a^b \frac{t^2}{b-a}\,dt - \left[\frac{b+a}{2}\right]^2 = \frac{(b-a)^2}{12}. \tag{1.3.4b}$$

1.3.2 The exponential distribution

This distribution is obtained by assuming that the *past has no effect on the future*. The discussion so far implies that, if one wants to make the future independent of the past, it must be done through looking at $z(t)\,dt$, because this is the probability to have an event at dt given that there was no event in the *past*. So, if $z(t)\,dt$ is to be independent of the past, it has to be independent of t, thus $z(t)$ must be a constant independent of time. This means:

$$z(t) = \alpha\left[\frac{1}{h}\right], \quad 0 \le t \le \infty, \tag{1.3.5}$$

and using equation (1.2.9), $f(t) = \alpha\,\mathrm{e}^{-\alpha t}$ and $F(t) = 1 - \mathrm{e}^{-\alpha t}$ (the units of the rate function are inverse time). The kth moment is given by:

$$E(t^k) = \langle t^k \rangle = \lambda \int_0^\infty t^k\,\mathrm{e}^{-\lambda t}\,dt = \frac{k!}{\lambda^k} \tag{1.3.6}$$

and thus the mean is $1/\lambda$ and the variance is $1/\lambda^2$.

Some of the interesting properties of this distribution are reflected in the next example.

Example 1.3.1

An electronic circuit serves as a control unit for a pump. The circuit has an exponential PDF for failure with $\alpha = 1.23 \times 10^{-4}\,h^{-1}$ and the CDF of the circuit is $F_c(t) = 1 - \mathrm{e}^{-1.23\times 10^{-4} t}$. The pump has a piece-wise uniform PDF for failure, defined as:

$$f_p(t) = \begin{cases} 3.125 \times 10^{-5} & \text{for } 0 \le t \le 8000 \\ 9.375 \times 10^{-5} & \text{for } 8000 < t \le 16\,000 \\ 0.0 & \text{for } t > 16\,000\,. \end{cases}$$

The distribution is normalized in the interval $[0, 16\,000]h$, i.e. $3.125 \times 10^{-5} \times 8000 + 9.375 \times 10^{-5} \times 16\,000 = 1$, and an event must happen up to 16 000 hours. What is the probability that a *first* failure will occur in the first month (740 h), the second, the third, etc., given that there was no failure up to the beginning of the month. This, for the first 20 months of operation for each of the units.

Using equation (1.2.14) with $\Delta t = t_2 - t_1 = 740$ h and t_1 the time up to the beginning of the ith month the probability, $P(i)$, for failure in the ith month can be written as

$$P(i) = \frac{F(t_1 + \Delta t) - F(t_1)}{1 - F(t_1)} \tag{1.3.7}$$

The CDF for the pump is

$$F_p(t) = \begin{cases} 3.125 \times 10^{-5} \times t & \text{for } 0 \le t \le 8000 \\ 0.25 + 9.375 \times 10^{-5} \times (t - 8000) & \text{for } 8000 < t \le 16\,000 \\ 1.0 & \text{for } t > 16\,000 \end{cases}$$

The results are shown in Table 1.3.1.

Table 1.3.1 First failure conditional probabilities in advancing intervals

Month	t_1	$F_p(t_1 + \Delta t)$	Conditional probability of first failure of pump	$F_c(t_1 + \Delta t)$	Conditional probability of first failure of circuit
1	0.00	0.023125	0.023125	0.0870005	0.0870005
2	740	0.04625	0.0236724	0.166432	0.0870005
3	1480	0.069375	0.0242464	0.238953	0.0870005
4	2220	0.0925	0.0248489	0.305164	0.0870005
5	2960	0.115625	0.0254821	0.365615	0.0870005
6	3700	0.13875	0.0261484	0.420807	0.0870005
7	4440	0.161875	0.0268505	0.471197	0.0870005
8	5180	0.185	0.0275913	0.517203	0.0870005
9	5920	0.208125	0.0283742	0.559207	0.0870005
10	6660	0.23125	0.0292028	0.597556	0.0870005
11	7400	0.263125	0.0414634	0.632569	0.0870005
12	8140	0.3325	0.0941476	0.664536	0.0870005
13	8880	0.401875	0.103933	0.693721	0.0870005
14	9620	0.471250	0.11598	0.720368	0.0870005
15	10360	0.540625	0.131206	0.744696	0.0870005
16	11100	0.61	0.151020	0.766908	0.0870005
17	11840	0.679375	0.177885	0.787187	0.0870005
18	12580	0.748750	0.216374	0.805702	0.0870005
19	13320	0.818125	0.276119	0.822606	0.0870005
20	14060	0.8875	0.381443	0.838039	0.0870005

The probability of failing in an interval is steadily increasing for the piece-wise uniform case, but remains *unchanged* for the exponential case. This has a profound meaning in terms of 'aging'. Intuitively, 'aging' reflects the notion that if an event did not happen up to a given time, its probability to happen in a future interval, is increasing. This does not happen for the exponential case. The exponential conditional probability of an event happening in the interval is *independent of the time elapsed up to the beginning of the interval.* This can be seen by substituting the exponential CDFs into expression (1.3.7):

$$P(i) = \frac{F(t_1 + \Delta t) - F(t_1)}{1 - F(t_1)} = \frac{\left(1 - e^{-\alpha(t_1+\Delta t)}\right) - \left(1 - e^{-\alpha t_1}\right)}{1 - \left(1 - e^{-\alpha t_1}\right)} = 1 - e^{-\alpha \Delta t}. \qquad (1.3.8)$$

This property is unique to the exponential distribution, and is a direct result of the initial condition applied to the rate function, namely that it is independent of time. The exponential distribution has no 'memory', and processes following an exponential distribution are not 'aging'. The fact that this property is unique to the exponential distribution can be deduced with the following lemma.

Lemma 1.3.1 *Let $F(t)$ be a differentiable CDF in $[0, \infty]$ with a rate function $z(t)$ defined at each point, and such that the probability for first event occurrence in any interval Δt depends only upon Δt, then $F(t)$ is an exponential distribution.*

Proof The conditions of the lemma suggest that

$$\forall\, t : \frac{F(t+\Delta t) - F(t)}{1 - F(t)} = G(\Delta t).$$

Using the definition of the rate function and the above equation, one may write

$$z(t) = \frac{f(t)}{1 - F(t)} = \lim_{\Delta t \to 0} \frac{F(t+\Delta t) - F(t)}{\Delta t(1 - F(t))} = \lim_{\Delta t \to 0} \frac{G(\Delta t)}{\Delta t}.$$

The conditions of the lemma imply that the limit exists for each t. The right-hand side implies that the limit does not depend upon t, and hence $z(t)$ is constant, yielding the exponential distribution.

The peculiarity of the exponential distribution in 'having no past' is further demonstrated by the following example.

Example 1.3.2: Measuring the mean life time of components

Consider the life of a component. On an intuitive basis it is obvious from daily experience that almost every man-made system has a limited lifetime (by extension, this is true also for the entire universe). Nothing stays forever. To measure the average lifetime of a component (such as a battery), one observes n new identical units until their first failure (assuming that these failures mean end of life). The time of failure t_i for each unit is recorded, and the sample average

$$\tilde{t} = \frac{1}{n}\sum_{i=1}^{n} t_i \tag{1.3.9}$$

is calculated, yielding an estimate for the mean lifetime for the unit (it will be shown later that this is a proper procedure in the sense that as n increases to infinity, the arithmetic average converges to the mean value $\int_0^\infty t \times f(t)dt$ (as defined in (1.2.16)).

If the mean lifetime of the unit is, say, four years, then one expects the failure times to be scattered around this number. If the measurement is done on units that are known to be two years old, one would naturally expect that the measured residual lifetimes will be scattered around two years. This is not the case if the failure distribution is exponential – it cannot be, because the exponential distribution has no recollection of the past. Two year old units do not 'know' that they are older than new units. In fact, they are just as new as new units are, and their measured lifetime will be scattered around four years. To see this, consider the mean time to failure, as measured from t_c, of a unit with a failure PDF $f(t)$ that is known not to have a failure up to t_c.

The PDF, as seen from t_c, is defined in the range $t_c \le t < \infty$ with the condition that up to t_c there was no failure. The CDF for an event up to any time t is given by equation (1.2.14). The PDF can be obtained by taking a derivative of (1.2.14) with respect to t (set $t_c = t_1$ and $t = t_2$ in (1.2.14)) yielding:

$$f(t|t_c) = \frac{f(t)}{1 - F(t_c)}. \tag{1.3.10}$$

The same result can be obtained by noting that for $t \ge t_c$ $f(t|t_c)$ is proportional to $f(t)$, thus $f(t|t_c) = C \times f(t)$, where C is a proportionality constant. Also, $f(t|t_c)$ is normalized in the range $t_c \leq t < \infty$, so that

$$\int_{t_c}^{\infty} f(t|t_c)\,dt = C_x \int_{t_c}^{\infty} f(t)\,dt = C_x[1 - F(t_c)] = 1,$$

thus

$$C = \frac{1}{[1 - F(t_c)]}.$$

Measuring the lifetime from t_c, the mean time is obtained as

$$E(t|t_c) = \int_{t_c}^{\infty} (t - t_c) f(t|t_c)\,dt = \frac{1}{1 - F(t_c)} \int_{t_c}^{\infty} (t - t_c) f(t)\,dt. \tag{1.3.11}$$

Substituting the exponential distribution yields

$$E(t|t_c) = \frac{1}{\mathrm{e}^{-\lambda t_c}} \int_{t_c}^{\infty} (t - t_c)\lambda\,\mathrm{e}^{-\lambda t}\,dt = \int_{t_c}^{\infty} (t - t_c)\lambda\,\mathrm{e}^{-\lambda(t - t_c)}\,dt = \int_{0}^{\infty} x\lambda\,\mathrm{e}^{-\lambda x}\,dt = \frac{1}{\lambda}. \tag{1.3.12}$$

The residual mean lifetime is thus independent of the past.

This surprising result does not agree with daily experience, and the concept of aging. Yet the exponential distribution is, by far, the most frequently and widely used distribution in system engineering applications, mainly because it is the simplest function to work with, not because it is a correct one.

Physical processes that have no memory are called *Markovian processes*. In such models, the future is determined entirely by the state of the system in the present.

1.3.3 The Weibull distribution

The rate function is an exponent of time in the form:

$$Z(t) = \lambda\beta t^{\beta-1}, \quad \lambda > 0 \text{ and } \beta > 0, \tag{1.3.13}$$

where λ and β are called the *scale* and *shape* parameters, respectively. The PDF and CDF are given by

$$f(t) = \lambda\beta t^{\beta-1}\,\mathrm{e}^{-\lambda t^{\beta}} \quad \text{and} \quad F(t) = 1 - \mathrm{e}^{-\lambda t^{\beta}}. \tag{1.3.14}$$

The kth moment of this distribution is given by

$$E(t_k) = \frac{\Gamma(1 + k/\beta)}{\lambda^{k/\beta}}, \tag{1.3.15}$$

where $\Gamma(r)$ is the Gamma function $\Gamma(r) = \int_0^{\infty} x^{r-1}\,\mathrm{e}^{-x}\,dx$. The mean and variance are

$$E(t) = \langle t \rangle = \frac{\Gamma(1 + 1/\beta)}{\lambda^{1/\beta}} \quad \text{and} \quad V = \frac{1}{\lambda^{2/\beta}}\left\{\Gamma\left(1 + \frac{2}{\beta}\right) - \left[\Gamma\left(1 + \frac{k}{\beta}\right)\right]^2\right\}. \tag{1.3.16}$$

1.3.4 The normal distribution

The PDF of the normal distribution is defined as

$$f(x) = \frac{1}{\sqrt{2\pi\sigma^2}} e^{-(1/2)((x-\mu)/\sigma)^2}, \tag{1.3.17}$$

where μ and σ^2 are the mean and variance, respectively. The PDF is commonly denoted as $N(\mu, \sigma)$. This function has no explicit form of the CDF, and thus also no explicit form of the rate function. Of special importance is the normal distribution with $\mu = 0$ and $\sigma^2 = 1$. It is called the *standard normal distribution*, and is denoted by $N(0, 1)$. If x is a normal variant with mean μ and standard deviation σ, then the transformed variable, $y = (x - \mu)/\sigma$ is a standard normal variable.

1.3.5 The lognormal distribution

Let x be a normal r.v. (i.e. a random variable with a normal PDF) with mean μ and standard deviation σ. The random variable $y = e^x$ is called a lognormal variable. It is defined over the range $(0, \infty)$. To obtain the PDF of y note that, in general, if $f_x(x)$ is the PDF of x and y is a single valued function of x, then the PDF of y, $f_y(y)$, is given by

$$f_y(y) = f_x(x)\left|\frac{dx}{dy}\right|. \tag{1.3.18}$$

Using the definition of y and the normal distribution (1.3.17), one obtains

$$f_y(y) = \frac{1}{y\sqrt{2\pi\sigma^2}} e^{-1/2((\ln(y)-\mu)/\sigma)^2}. \tag{1.3.19}$$

The mean μ_L and standard deviation σ_L of the lognormal variable are given by

$$\mu_L = e^{\mu+(1/2)\sigma^2} \quad \text{and} \quad \sigma_L^2 = e^{2\mu+\sigma^2} \times (e^{\sigma^2} - 1) = (e^{\sigma^2} - 1) \times \mu_L^2. \tag{1.3.20}$$

1.3.6 Non-parametric, piecewise constant distributions

This distribution is defined over a finite range $[a, b]$. The range is divided into n intervals $t_0, t_1, \ldots, t_n$ such that $t_0 = a$ and $t_n = b$. In each interval (t_i, t_{i+1}) the PDF has a constant value f_i, and the PDF is zero outside the interval. The values of the PDF must fulfil the normalization condition

$$\sum_{i=1}^{n} f_i \times (t_i - t_{i-1}) = 1. \tag{1.3.21}$$

The CDF of the distribution is given by

$$F(t) = \sum_{i=1}^{k} f_i \times (t_i - t_{i-1}) + f_{k+1} \times (t - t_k)$$

for $k = 0, \ldots, n-1$ and $t_k \leq t \leq t_{k+1}$ with the convention that for $k = 0$ the first term is zero.

Non-parametric distributions are often useful in approximating parametric distributions.

2

Basic Concepts in System Engineering

In this chapter the concepts of a component and a system are formally defined. The behavior of a single component will be examined, and some of the methods used for the analysis of systems are discussed.

2.1 Components and Systems

In general, the concepts of a component and a system are used in a quite free context mainly relying on the intuitive understanding of these terms. Here these concepts will be formally defined to establish a clear framework for the theory of systems behavior.

2.1.1 The concept of a component

Definition 2.1: Component An entity will be called a component if it has the following properties:

1. It has a finite number of discrete known states.
2. The mechanism of transfer of the entity between its states is known.

It is to be noted that the assignment of states to a unit is, to a large extent, an arbitrary process. After all, one has to remember that a component is an element in a model which is an abstract structure intended to describe a given reality. Thus, it is expected that the states defined for a component are those which are relevant to its function in the system, and to its future behavior.

A state is denoted by a 'State indicator' variable, b, taking integer values which identify the states. Consider, for example, an engine in a production line. Two natural states can be defined for it:

- **State 1:** operational state – $b = 1$. The name *operational* will be used throughout this book to describe a unit which is not failed and is active.
- **State 2:** failed state – $b = 0$. In this state the engine is failed.

Some other states may be considered in the reality of the engine, such as:

- **State 3:** failed undetected – $b = -3$. In this state the engine is failed, but the failure is undetected. As a result, no restoration operations take place. This state is common to components that are not in continuous operation.
- **State 4:** passive state – $b = -2$. In this state the engine is not failed but is not operating.
- **State 5:** failed detected, missing resource – $b = -1$. In this state the engine is failed, the failure is detected but cannot be repaired because a resource, such as a spare part, necessary for its repair, is not available.

In view of the definition of States -3 and -1, the description of state '0' should now be rewritten as: *The engine is failed, the failure is detected and all the resources needed for repair are available, active repair is in progress.*

- **State 6:** failed, detected, temporarily non-repairable – $b = -4$. In this state, the engine cannot be repaired because of some external conditions. For example, an aircraft engine fails in mid-flight.
- **State 7:** engine is in periodic maintenance – $b = -5$. The engine is not operating and is not failed. It is being maintained to reduce the probability of future failures.
- **State 8:** operation under heavy load – $b = 2$. This mode of operation may accelerate wear out and fatigue processes which may cause failures.

Note that some of the above-defined states have the same meaning in terms of the present reality of the component. Namely, states 0, -3, -1, -4 are all failed states. In all of them the engine is failed. In fact, for all of them, the appearance of the engine may be exactly the same. Furthermore, states -2 and -5 can also be included in this group, since the engine is not operating in these state. Such states differing from each other are mainly because the future looks different for each of them. States -3, -1 and -4 must first reach state '0' before repair can commence. All of these states depend upon some external elements: detection of the failure, arrival of a spare part, completion of the flight and landing, respectively. Only in state '0' is the repair process actually in progress. Thus, if a repair started at time t_0, a PDF for repair, $g(t - t_0)$, may be assumed for the time of repair. In most cases, it is impossible to adopt a single distribution to states ($-3, -1, -4$), since the transfer from these states to state '0' may involve more then a single PDF and, in any case, the PDFs that describe transfers from these states are not an inherent property of the component because of the external elements involved.

The life history of the component can be described as a sequence of transfers between its states. These transfers can be divided into two categories.

1. *Inherent stochastic processes* – these are processes of a statistical nature that can be viewed as internal properties of the component. A sudden failure of a light bulb is such a process. In general, failure and repair processes fall into this category.
2. *Interactions* – these are state transfers resulting from events happening either elsewhere in the system or outside the system. For example: a transfer from state -2 (passive) into state 1 can be caused by a failure of another component for which the component serves as a back-up. A transfer from state -3 (failed undetected) can occur due to inspection operation initiated by a failure elsewhere in the system. A lightning strike can cause changes and induce a failure in one or more components.

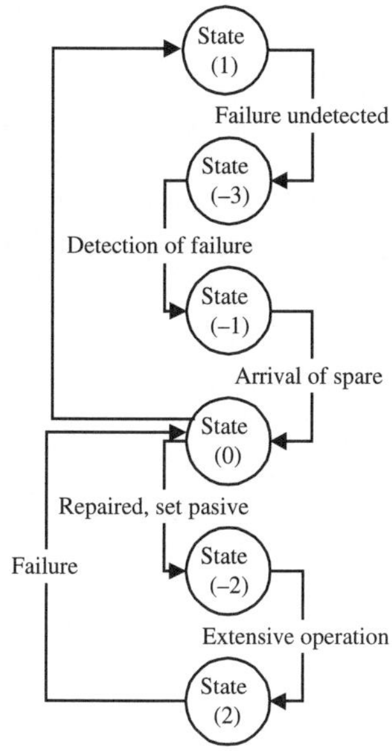

Partial life history

The component fails. Failure is not detected. Following inspection the failure is detected but a spare part is missing. A spare part arrives and repair is commenced. Upon repair the unit is set in a passive state. Then, when needed, the component is set into extensive operation. It then fails. The failure is repaired and the component is set into regular operation.

Figure 2.1.1 A possible life history of a component

Some of the state transfers are not of a statistical nature: pre-planned events such as periodic maintenance or changes in operating conditions may be of a deterministic nature. However, from the point of view of modeling, deterministic events can be described as statistical events. For example, a normal distribution with a very small standard deviation may be used to describe the occurrence of a deterministic event. Figure 2.1.1 demonstrates a possible life history of a component.

The PDF associated with a given state b will be denoted either as $f(b;t)$ or $f_i(t)$, where i is the integer value of the state indicator b. $f_i(t)$ yields the probability density that if a component entered state i at $t = 0$, it will exit from this state at time t (per unit time). If the transfer from any state can be only into a unique possible state, then $f_i(t)$ contains all the information needed to analyze the life history of the component. But, if transfer is possible into more than one state a somewhat more elaborate treatment is needed.

Let $z_{ij}(t - t^i)$ denote the rate function for a transfer from state i into state j. Thus, $z_{ij}(t - t^i)$ is the probability (per unit time) that the component that entered state i at time t^i in the past, will transfer into state j at time t given that it remained in state i up to time t. $z_{ij}(t - t^i)$ is called the *partial rate function* (or *partial hazard*). Assuming that the partial rates are independent, the total rate function is given by

$$z_i(t - t^i) = \sum_{j=1}^{n} z_{ij}(t - t^i). \tag{2.1.1}$$

Namely, the rate function of a component in a given state is the sum of all the independent partial rate functions defined for the state. To see that equation (2.1) is a direct result of the, assumed, independence of the transfer events, consider the CDF of a single transfer. By equation (1.2.11), it is given as:

$$F_{ij}(t) = 1 - e^{-\int_0^t z_{ij}(x)\,dx}. \tag{2.1.2}$$

The probability that a transfer to state j will not occur up to time t is then $1 - F_{ij}(t)$, and the probability that no transfer to any state will occur is given (note the independence of the events) by the product of the probabilities, namely:

$$1 - F_i(t) = \prod_{j=1}^{n}[1 - F_{ij}(t)] = \prod_{j=1}^{n}\left[e^{-\int_0^t z_{ij}(x)\,dx}\right] = e^{-\int_0^t \left\{\sum_{j=1}^{n} z_{ij}(x)\right\} dx} = e^{-\int_0^t z_i(x)\,dx}. \tag{2.1.3}$$

Thus:

$$F_i(t) = 1 - e^{-\int_0^t z_i(x)\,dx} \tag{2.1.4}$$

where $z_i(t)$ is given by expression (2.1.1).

The PDF can be written as (see equation (1.2.11))

$$f_i(t) = \frac{dF_i(t)}{dt} = z_i(t)\,e^{-\int_0^t z_i(x)\,dx} = \left(\sum_{j=1}^{n} z_{ij}(t)\right) e^{-\int_0^t z_i(x)\,dx} = \left(\sum_{j=1}^{n} \gamma_{ij}(t)\right) z_i(t)\,e^{-\int_0^t z_i(x)\,dx}, \tag{2.1.5}$$

where

$$\gamma_{ij}(t) = \frac{z_{ij}(t)}{z_i(t)}. \tag{2.1.6}$$

$\gamma_{ij}(t)$ can be interpreted as the probability that a transfer into state j will take place if it is known that an exit from state i took place at time t.

Example 2.1.1: Multiple failure modes

A jet engine has three independent failure modes. Each failure mode leads to a different failed state because the type of treatment needed for each failure is different. The first mode is called Rapture; it is connected to fractures in pipes inside the engine. The hazard for this mode is known to be $z_{1,-6}(t) = 3.502 \times 10^{-6} \times 2.09(t^{1.09})$. The second failure mode is related to wear out of the turbine blades, and its hazard is given by $z_{1,-7}(t) = 9.87 \times 10^{-10} \times 3.57(t^{2.57})$. The third failure mode is a general failure with a constant hazard $z_{1,-8}(t) = 8.759 \times 10^{-8}$. The three failure modes are indicated by -6, -7 and -8, respectively.

What is the probability that the engine will not fail during the first 400 hours of operation?

Given that there was a failure at time 400 h, what is the probability that the failure is in each of the modes?

Substituting the three rate functions in equation (2.1.3), and denoting $1 - F_i(t)$ by $R_i(t)$ yields:

$$R_1(t) = \left[e^{-3.502\times10^{-6}\times(400)^{2.09}}\right] \times \left[e^{-9.87\times10^{-10}\times(400)^{3.57}}\right] \times \left[e^{-8.759\times10^{-8}\times400}\right] = 0.05599.$$

The component hazard at 400 hours is given by the sum of the partial hazards, namely:

$$z_1(400) = 3.502 \times 10^{-6} \times 2.09(400^{1.09}) + 9.87 \times 10^{-10} \times 3.57(400^{2.57}) + 8.759 \times 10^{-8}$$
$$= 0.021707.$$

By equation (2.1.6), the probability for each failure mode to have occurred (given that a failure occurred) is the ratio of the partial hazard and the component hazard, namely 0.22642, 0.773574 and 3.95×10^{-6}, respectively.

2.1.2 The system function

A system is a collection of $m \geq 1$ components. Denote the state indicator of the ith component by b_i. The vector $\mathbf{B} = (b_1, b_2, \ldots, b_m)$, whose elements are the state indicators of the components, is called the *state vector* of the system. The space of all possible different state vectors is called the *state space* of the system. The difference between a system and a component lies mainly in adding the fact that a system has a measurable performance. Recall that in the definition of a component, states and transfers between states, are essential, but no attention is paid to the function or purpose of the component. A system is created when at least one measure of performance is added to a collection of components. The simplest form of a performance measure is a *network*. A network is a structure such as that shown in the block diagram of Figure 2.1.2. A set of operating components connecting the 'entry' point with the 'exit' point is called a *tie*. The system is operational if at least a single tie exists.

The system has states much like components. In the simplest case, the system will have two possible states, operational and failed. Obviously, the state of the system depends upon the states of its components, i.e. on the state vector $\mathbf{B}$. An expression that defines the state of the system as function of the states of its components is called a *system function*. In the following discussion, we assume that a component may have only two states, 1 – Operational, and 0 – Failed.

Any subset of components in a network that has a single entry point and a single exit point is a subsystem. Obviously, a single component is also a subsystem. If all the possible subsystems in a network contain only serial and parallel subsystems, the network is called a *serial/parallel* network.

In the case of serial/parallel networks, the system function takes a particularly simple form. The system function of two components in series is $b_1 \times b_2$. For two components in parallel (i.e. the system is operational if at least one component operates) it is $(b_1 + b_2 - b_1 \times b_2) = [1 - (1 - b_1)(1 - b_2)]$. It is a simple matter to verify that these basic structures yield a correct state of the system for every possible combination of the states of the components. Furthermore, one may note that these structures follow the

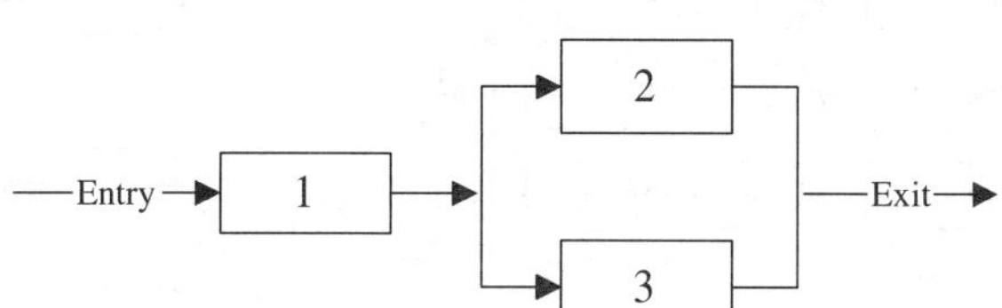

Figure 2.1.2 Network structure of a system

rules of intersection and union of independent events. In the first case, it is the event {both components 1 and 2 are operational} and in the second it is the event {at least one of the components is operational}. Thus, the probability that the structure is in an operational state can be obtained by replacing each b_i by the corresponding probability p_i that $b_i = 1$. Any serial/parallel network can be reduced to a set of parallel and serial substructures. To demonstrate this process, consider the system in Figure 2.1.2. It can be reduced to a serial structure of two subsystems: the first function is denoted by S_1 with $S_1 = b_1$; the second subsystem is denoted by S_2, and its subsystem function is $S_2 = [1 - (1 - b_2) \times (1 - b_3)]$. The system function is denoted by $S(\mathbf{B})$ and

$$S(\mathbf{B}) = S_1 \times S_2 = b_1 \times (b_2 + b_3 - b_2 \times b_3) = b_1 \times [1 - (1 - b_2) \times (1 - b_3)]. \quad (2.1.7)$$

In this form $S(\mathbf{B})$ takes the values 1 (operational) and 0 (failed). Indeed, substituting the probabilities p_i that $b_i = 1$ will yield the probability that each subsystem takes a value 1, and thereby the probability that the system is in state 1. This nesting process is easily extended to any serial/parallel network, and is further demonstrated in the following example.

Example 2.1.2

A water supply system has the block diagram structure shown in Figure 2.1.3. Components 1, 4, 7, 3, 6 and 9 are valves. Components 2, 5, 8 are pumps, component 10 is a pressure regulator and components 11 through 14 are pipes. The probabilities that each type of component is operational are given by $P_v = 0.98$; $P_u = 0.92$; $P_g = 0.975$; $P_e = 0.85$ for the valve, pump, regulator and a pipe, respectively. What is the system function, and what is the probability that the system is operational?

To obtain the system function a nesting process is applied. Going from left to right, the subsystem function of each of the three lines 1, 2, 3; 4, 5, 6 and 7, 8, 9 is simply a product of the state indicators of the components. Denoting these three subsystem functions by S_1, S_2 and S_3, respectively, and the subsystem function of the parallel structure of these three lines in by S_4, results in

$$\begin{aligned} S_4 &= [1 - (1 - S_1) \times (1 - S_2) \times (1 - S_3)] \\ &= [1 - (1 - b_1 b_2 b_3) \times (1 - b_4 b_5 b_6) \times (1 - b_7 b_8 b_9)]. \end{aligned} \quad (2.1.8)$$

Component 10 is included by multiplying $S_4 \times b_{10}$. Denoting the pipes structure by S_5 yields $S_5 = [1 - (1 - b_{11}) \times (1 - b_{12}) \times (1 - b_{13}) \times (1 - b_{14})]$, and the system function takes

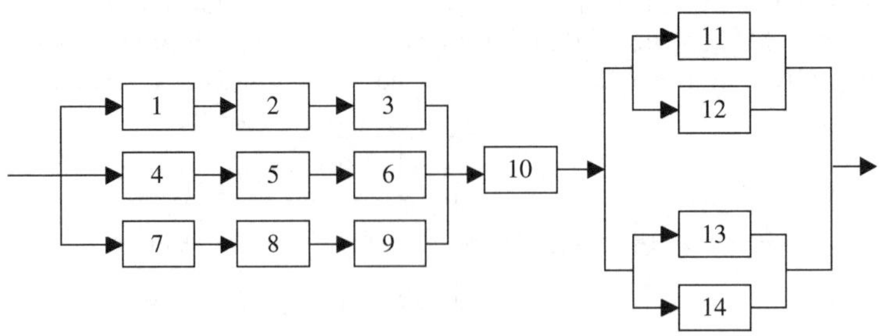

Figure 2.1.3 Block diagram of a water supply system

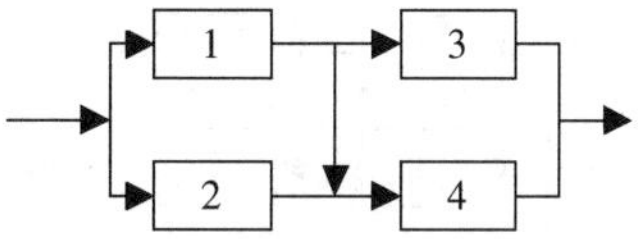

Figure 2.1.4 A non-serial/parallel structure

the form:

$$\begin{aligned}S(\mathbf{B}) &= S_4 \times b_{10} \times S_5 = [1-(1-S_1)\times(1-S_2)\times(1-S_3)]\times b_{10}\times S_5\\ &= [1-(1-b_1b_2b_3)\times(1-b_4b_5b_6)\times(1-b_7b_8b_9)]\\ &\quad \times b_{10}\times[1-(1-b_{11})\times(1-b_{12})\times(1-b_{13})\times(1-b_{14})].\end{aligned} \tag{2.1.9}$$

To obtain the probability that the system is operational, replace each b_i by the corresponding probability p_i. This yields the expression:

$$P_s = P_r[S(\mathbf{B})=1] = [1-(1-P_v^2P_u)^3]\times P_g\times[1-(1-P_e)^4] = 0.9729. \tag{2.1.10}$$

The network shown in Figure 2.1.4 is an example of a structure that is not serial/ parallel. To consider such networks the concepts of a tie set and a minimal tie set are introduced. A tie set is a subset of components such that all of them being operational is a sufficient condition for the system to be operational.

For example, the subsets $\{1, 2, 3\}$ and $\{1, 4\}$ are tie sets. A minimal tie set is one which does not remain a tie set if any component is removed. The tie set $\{1, 2, 3\}$ is not minimal since, upon removing component 2, the remaining set, $\{1, 3\}$, is a tie set. Yet set $\{1, 4\}$ is a minimal tie set. The minimal tie sets of the system in Figure 2.1.4 are: $A_1 = \{1, 3\}; A_2 = \{1, 4\}$ and $A_3 = \{2, 4\}$. A minimal tie set event is the event {All the component in the minimal tie set are operational}. It may be noted that the event {The system state is 1} is the union of the minimal tie sets. Thus the probability that the system is operational is given by $P_s = P_r(A_1 \cup A_2 \cup A_3)$. Using equation (1.1.9) yields

$$\begin{aligned}P_s &= P_r(A_1)+P_r(A_2)+P_r(A_3)-P_r(A_1\cap A_2)-P_r(A_1\cap A_3)-P_r(A_2\cap A_3)\\ &\quad + P_r(A_1\cap A_2\cap A_3).\end{aligned}$$

Since the components are independent, the probability of each tie set event is the product of the probabilities of the components in it. The minimal tie sets, however, are not independent because they share identical components. Thus, the probability $P_r(A_1 \cap A_2)$ is $p_1 \times p_3 \times p_4$, since component 1 appears in both minimal tie sets. Also, $P_r(A_1 \cap A_3) = P_r(A_1 \cap A_2 \cap A_3) = p_1 \times p_2 \times p_3 \times p_4$. With these considerations, the probability that the system is in state 1 can be written as:

$$P_s = p_1\times p_3 + p_1\times p_4 + p_2\times p_4 - p_1\times p_3\times p_4 - p_1\times p_2\times p_4. \tag{2.1.11}$$

To obtain the system function, note that with the notion of the minimal tie set the system block diagram can be replaced by an equivalent diagram constructed of all the minimal tie sets in parallel, as shown in Figure 2.1.5.

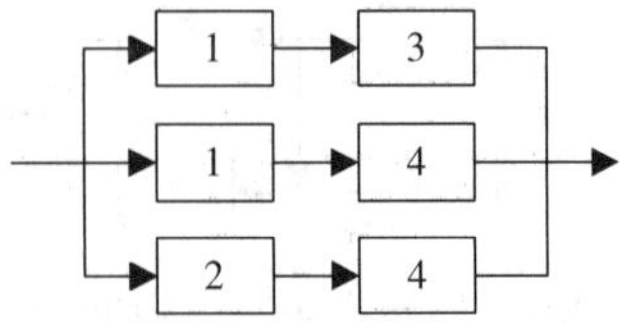

Figure 2.1.5 The equivalent minimal tie set diagram

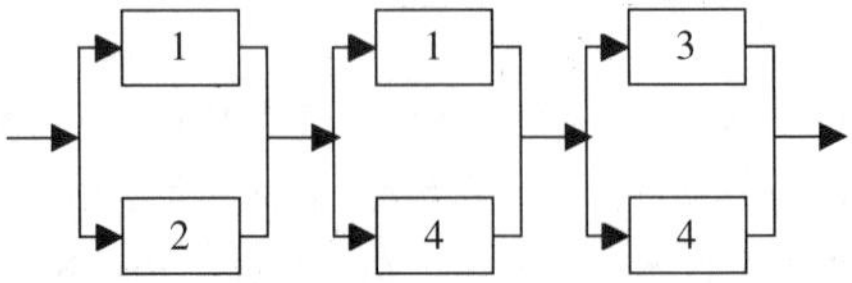

Figure 2.1.6 The equivalent minimal cut set diagram

This diagram is identical in every respect to the diagram in Figure 2.1.3, but it has a serial/parallel form so that the system function can be immediately written as:

$$\begin{aligned} S(\mathbf{B}) &= [1 - (1 - b_1 b_3) \times (1 - b_1 b_4) \times (1 - b_2 b_4)] \\ &= b_1 b_3 + b_1 b_4 + b_2 b_4 - b_1^2 b_3 b_4 - b_1 b_3 b_2 b_4 - b_1 b_2 b_4^2 + b_1^2 b_3 b_2 b_4^2. \end{aligned} \tag{2.1.12}$$

Noting that b_i can be either 0 or 1, one may replace any power of the component state indicator by a power of one, namely, $b_i^n \rightarrow b_i$, so that the system function takes the final form

$$S(\mathbf{B}) = b_1 b_3 + b_1 b_4 + b_2 b_4 - b_1 b_3 b_4 - b_1 b_2 b_4. \tag{2.1.13}$$

The form of equation (2.1.13) is identical to that of (2.1.12), but upon comparison to equation (2.1.11) it is seen that one may replace the state indicators by the operation probabilities to obtain the system operation probability.

In some cases, it is convenient to consider cut sets of a system. A cut set is a set of components such that all of them being failed is a sufficient condition for the system to be in a failed state. For example, the sets $\{1, 2, 3\}$ and $\{3, 4\}$ are cut sets of the above system. The set $\{2, 3\}$ is not a cut set. A minimal cut set is a cut set which, upon removal of any component from it, the remaining set is not a cut set. For example, the cut set $\{1, 2, 3\}$ is not minimal, since removal of component 3 still results in a cut set $\{1, 2\}$. The cut set $\{3, 4\}$ is minimal. A minimal cut set is presented as a parallel structure of all the components in the set, and the system can be presented as a series of the parallel modules of all the minimal cut sets. A minimal cut set presentation of the system in Figure 2.1.4 is shown in Figure 2.1.6.

The system function can now be written using the procedure of a serial/parallel system. This will result an expression identical to (2.1.12) and upon reduction of all powers to unity, to expression (2.1.13). The form of the system function discussed above is called a *normalized form* because for every state vector it yields either zero or unity for the state of the system. The following steps summarize how the system probability may be obtained:

1. Create the normalized system function.
2. Reduce the power of each state indicator to one.
3. Substitute each state indicator by p_i.

When the powers of the state indicators are reduced to unity, the expression obtained is similar in every respect to equation (1.1.10) for a union of events where each event is a product of the components of a particular tie set.

The above discussion is related to a very specific form of a system function. In fact, any function that yields a description of any required aspect of the performance of a system is a legitimate system function. For the system of Figure 2.1.1, the expression

$$S(\mathbf{B}) = \begin{cases} 1 & \text{if } (b_1 \times b_2 + b_1 \times b_3) > 0 \\ 0 & \text{if } (b_1 \times b_2 + b_1 \times b_3) = 0 \end{cases} \tag{2.1.14}$$

may be used instead of expression (2.1.7).

Expression (2.1.14) is a legitimate system function. It yields the state of the system as function of the state of the components, although it is not a normalized form and, therefore, cannot be used to directly calculate the probability that the system is operational. Furthermore, a system function is not necessarily a binary function. Consider, for example, the system of Figure 2.1.2, where each of components 2 and 3 represents a power generator. Let component 2 be a 100 mgW generator and component 3 a 60 mgW generator. In that case, the power produced by the system is a suitable performance measure. An appropriate form of the system function for this case will be:

$$S(\mathbf{B}) = b_1 \times (100b_2 + 60b_3). \tag{2.1.15}$$

The next example demonstrates the use of such a system function and introduces the notion of a 'history' of a system.

Example 2.1.3: Average power production

Consider the system shown in Figure 2.1.2 with the system function (2.1.15). Each component has an exponential failure distribution with failure rates $\lambda_1 = 0.0001, \lambda_2 = 0.0025$ and $\lambda_3 = 0.0035$. The system operates unattended for a month (740 hours). Failed components are not repaired up to the end of the month. What is the average monthly power production?

The answer to the question will be presented through two points of view, each representing a different approach. Note that, although this discussion concerns power output, it is of a quite general nature and will apply to any measure of throughput.

The power production at any given time point, t, is the average value of $S(\mathbf{B})$ at that time. Let the probability that the system is in a specific state $\mathbf{B}_i$ at time t be denoted by $P(\mathbf{B}_i, t)$; then, the average value of the system function is given by

$$\langle S(t)\rangle = \sum_{\mathbf{B}_i} P(\mathbf{B}_i, t) S(\mathbf{B}_i). \tag{2.1.16}$$

This averaging is performed over the state vectors. It is an average over the state space at a point in time. The average production (per hour) is obtained by averaging $\langle S(t)\rangle$ over time, namely:

$$\langle S\rangle = \frac{1}{T_m}\int_0^{T_m} \langle S(t)\rangle\, dt = \frac{1}{T_m}\int_0^{T_m} \sum_{\mathbf{B}_i} P(\mathbf{B}_i, t) S(\mathbf{B}_i)\, dt = \frac{1}{T_m}\sum_{\mathbf{B}_i} S(\mathbf{B}_i)\int_0^{T_m} P(\mathbf{B}_i, t)\, dt, \tag{2.1.17}$$

where T_m is the time span of the averaging.

Looking at expression (2.1.15) we note that $S(\mathbf{B})$ is non-zero only for the states $(1,1,1)$, $(1,0,1)$ and $(1,1,0)$. Using the properties of the CDF of the exponential distribution, the probability of being, for example, in state $(1,1,0)$ is given by $P[(1,1,0),t] = e^{-\lambda_1 t} \times e^{-\lambda_2 t} \times (1 - e^{-\lambda_3 t})$. This is simply the probability that the first and second components did not fail up to time t, and the third component failed. Note that use is made of the fact that the components are independent. With the state probabilities for all other states obtained in a similar way, the expression for the average power production takes the form

$$\begin{aligned}\langle S(t)\rangle &= P[(1,1,1),t] \times S(1,1,1) + P[(1,0,1),t] \times S(1,0,1) + P[(1,1,0),t] \times S(1,1,0)\\ &= S(1,1,1) \times e^{-(\lambda_1+\lambda_2+\lambda_3)t} + S(1,0,1) \times e^{-(\lambda_1+\lambda_3)t} \times \left(1 - e^{-\lambda_2 t}\right)\\ &\quad + S(1,1,0) \times e^{-(\lambda_1+\lambda_2)t} \times \left(1 - e^{-\lambda_3 t}\right).\end{aligned}$$

Substitution in (2.1.17) yields:

$$\langle S\rangle = \frac{1}{T_m}\left[\frac{S(1,0,1)}{\lambda_1+\lambda_3}\left(1 - e^{-(\lambda_1+\lambda_3)T_m}\right) + \frac{S(1,1,0)}{\lambda_1+\lambda_2}\left(1 - e^{-(\lambda_1+\lambda_2)T_m}\right)\right] = 65.33, \quad (2.1.18)$$

where $S(1,0,1) = 60_{\left[\frac{\text{units}}{\text{hour}}\right]}$, $S(1,1,0) = 100_{\left[\frac{\text{units}}{\text{hour}}\right]}$ and $T_m = 740_{\,h}$.

Expression (2.1.17) is general, and will also apply when the system function is explicitly time dependent.

Next consider the behavior of the same system over time. The system starts at $t = 0$ in state $(1,1,1)$, which is denoted by $\mathbf{B}_0$. It may remain in this state up to T_m, or a failure may occur at some time point $t_1 < T_m$. If such a failure occurs, the system will transfer into a new state, $\mathbf{B}_1$. A possible such state is $\mathbf{B}_1 = (1,0,1)$. The system may then remain in this state, or it may transfer at time $t_2 < T_m$ into a new state $\mathbf{B}_2$, such as $(0,0,1)$. The time point at which the system changes its state vector is called an *event point*, and the event point together with the state into which the system enters are called a *phase space point*. A set of phase space points defines a single possible 'history' of the system. A history of the system is then comprised of a sequence of such phase space points. The phase space is the product space of the time domain and the state vectors. The behavior of the system over time may be viewed as if it is a particle 'traveling' in the phase space, 'jumping' from one phase space point to another. A single history such as the one described above is schematically shown in Figure 2.1.7.

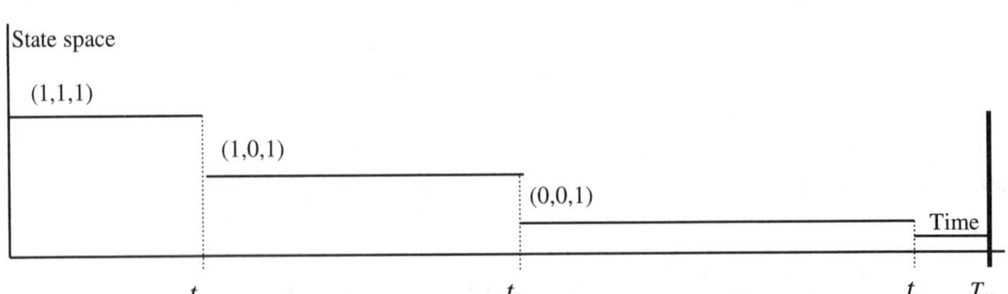

Figure 2.1.7 A possible life history of the system

The total power production for the above history is $S(1,1,1) \times t_1 + S(1,0,1) \times (t_2 - t_1)$. A general life history is denoted by $c_k = (\mathbf{B}_1, t_1; \mathbf{B}_2, t_2; \ldots; \mathbf{B}_k, t_k)$. This means that the system starts at $t = 0$ in state $\mathbf{B}_0$. In the interval dt_1 around t_1 the system transfers into state $\mathbf{B}_1$ and so on, until a transfer is made at t_k into state $\mathbf{B}_k$, and the system remains in this state up to the time T_m. The total power production in this history is

$$\eta(c_k) = \sum_{i=0}^{k} S(\mathbf{B}_i) \times (t_{i+1} - t_i) \quad \text{where } t_{k+1} = T_m. \tag{2.1.19}$$

The 'history' c_k is an elementary event, and $\eta(c_k)$ is a *random variable defined on it.* Averaging $\eta(c_k)$ over all the possible histories of the system will yield the average power produced in the interval $[0, T_m]$. A system's histories are continuous events: k, the number of events in the history, can take any integer value, while the event points are continuous time points. Let $f(c_k)$ denote the probability density of the elementary event c_k, then the average of $\eta(c_k)$ can be formally written as

$$\langle \eta \rangle = \sum_{k=0}^{\infty} \sum_{\mathbf{B}_1,\ldots,\mathbf{B}_k} \int \eta(c_k) f(c_k)\, dc_k. \tag{2.1.20}$$

The averaging process involves summing over all the possible states in the history and integrating over c_k, i.e. integrating all the event points t_1 through t_k over their appropriate range.

The current example presents an opportunity to explicitly demonstrate how this averaging is done. The following types of histories are significant:

1. Histories of type c_0 (zero events): the state $\mathbf{B}_0 = (1,1,1)$ holds throughout time. The probability of this history is $e^{-(\lambda_1+\lambda_2+\lambda_3)\times T_m}$ this is the only history of type c_0, i.e. zero events in the history. $\eta(c_0)$ takes the value $S(1,1,1) \times T_m$.
2. Histories of type c_1 (single event): the state $\mathbf{B}_0 = (1,1,1)$ holds up to time t_1. At this time, the system transfers into state $(1,0,1)$ and remains in this state up to T_m. This history is denoted as $c_1 = ((1,0,1), t_1)$. The PDF for this history is $f(c_1) = e^{-(\lambda_1+\lambda_3)\times T_m}\, \lambda_2\, e^{-\lambda_2 t_1}$, and the contribution to the power generation from this history is $\eta(c_1) = S(1,1,1) \times t_1 + S(1,0,1) \times (T_m - t_1)$.
3. Histories of type c_1: the state $\mathbf{B}_0 = (1,1,1)$ holds up to time t_1. At this time, the system transfers into state $(1,1,0)$ and remains in this state up to T_m. This history is denoted as $c_1 = ((1,1,0), t_1)$. The PDF for this history is $f(c_1) = e^{-(\lambda_1+\lambda_2)\times T_m}\, \lambda_3\, e^{-\lambda_3 t_1}$, and the contribution to the power generation from this history is $\eta(c_1) = S(1,1,1) \times t_1 + S(1,1,0) \times (T_m - t_1)$.
4. Histories of type c_1: the state $\mathbf{B}_0 = (1,1,1)$ holds up to time t_1. At this time, the system transfers into state $(0,1,1)$. At this point, the power production goes to zero and additional possible events are of no consequence. This history is denoted as $c_1 = ((0,1,1), t_1)$. The PDF for this history is $f(c_1) = \lambda_1 e^{-\lambda_1 t_1} e^{-(\lambda_2+\lambda_3)\times t_1}$ and the contribution to the power generation from this history is $\eta(c_1) = S(1,1,1) \times t_1$.
5. Histories of type c_2 or higher: the state $\mathbf{B}_0 = (1,1,1)$ holds up to time t_1. At this time, the system transfers into state $(1,0,1)$. The system stays in this state up to the next event point t_2, at which it transfers into state $(1,0,0)$ or $(0,0,1)$. Once in this state,

only one additional event may occur. Yet this is of no importance, since the production rate is zero. This history is denoted as $c_2 = ((1,0,1), t_1; {(0,0,1) \atop (1,0,0)}, t_2)$. The PDF for this history is $f(c_2) = \lambda_2 \mathrm{e}^{-\lambda_2 t_1}(\lambda_1 + \lambda_3)\ \mathrm{e}^{-(\lambda_1+\lambda_3)\times t_2}$ with $t_2 > t_1$. The contribution to the power generation from this history is $\eta(c_2) = S(1,1,1) \times t_1 + S(1,0,1) \times (t_2 - t_1)$.

6. Histories of type c_2 or higher: the state $\mathbf{B}_0 = (1,1,1)$ holds up to time t_1. At this time, the system transfers into state $(1,1,0)$. The system stays in this state up to a second event point t_2, at which it transfers into state $(1,0,0)$ or $(0,1,0)$. This history is denoted by $c_2 = ((1,1,0), t_1; {(0,1,0) \atop (1,0,0)}, t_2)$. The PDF for this history is $f(c_2) = \lambda_3 \mathrm{e}^{-\lambda_3 t_1}(\lambda_1 + \lambda_2)\mathrm{e}^{-(\lambda_1+\lambda_2)\times t_2}$ with $t_2 > t_1$. The contribution to the power generation from this history is $\eta(c_2) = S(1,1,1) \times t_1 + S(1,1,0) \times (t_2 - t_1)$.

The averaging process over the above histories amounts to multiplying each $\eta(c)$ by the PDF of the corresponding history, and integrating over the event time points:

$$
\begin{aligned}
\langle \eta \rangle = & \sum_{k=0}^{\infty} \sum_{\mathbf{B}_1,\ldots,\mathbf{B}_k} \int \eta(c_k) f(c_k)\, dc_k = S(1,1,1) \times T_m \times \mathrm{e}^{-(\lambda_1+\lambda_2+\lambda_3)\times T_m} \\
& + \int_0^{T_m} [S(1,1,1) \times t_1 + S(1,0,1) \times (T_m - t_1)] \times \mathrm{e}^{-(\lambda_1+\lambda_3)\times T_m} \lambda_2 \mathrm{e}^{-\lambda_2 t_1}\, dt_1 \\
& + \int_0^{T_m} [S(1,1,1) \times t_1 + S(1,1,0) \times (T_m - t_1)] \times \mathrm{e}^{-(\lambda_1+\lambda_2)\times T_m} \lambda_3 \mathrm{e}^{-\lambda_3 t_1}\, dt_1 \\
& + \int_0^{T_m} [S(1,1,1) \times t_1] \times \lambda_1 \mathrm{e}^{-\lambda_1 t_1}\, \mathrm{e}^{-(\lambda_2+\lambda_3)\times t_1}\, dt_1 \\
& + \int_0^{T_m} dt_1 \left[\int_{t_1}^{T_m} [S(1,1,1)\times t_1 + S(1,0,1)\times(t_2 - t_1)] \times \lambda_2 \mathrm{e}^{-\lambda_2 t_1}(\lambda_1 + \lambda_3)\, \mathrm{e}^{-(\lambda_1+\lambda_3)\times t_2}\, dt_2 \right] \\
& + \int_0^{T_m} dt_1 \left[\int_{t_1}^{T_m} [S(1,1,1)\times t_1 + S(1,1,0)\times(t_2 - t_1)] \times \lambda_3 \mathrm{e}^{-\lambda_3 t_1}(\lambda_1 + \lambda_2)\, \mathrm{e}^{-(\lambda_1+\lambda_2)\times t_2}\, dt_2 \right].
\end{aligned}
\tag{2.1.21}
$$

Integration is straightforward, and the result is identical to that of equation (2.1.18).

This example demonstrates the role of the system function in the performance evaluation process and, more importantly, the concept of averaging over the space of a 'system's Histories'. It should be noted that the above example is simple. It has three features that make it simple, all three rarely encountered in realistic cases. First, components are not repaired; secondly, the failure distributions are all exponential; and thirdly, the components are independent of each other. Consider the possibility that components are repaired upon failure and failure distributions are not exponential. This will have the immediate result that the number of events in a history is unlimited (relative to a maximum of three in the present example). An analytic evaluation of the average production will become prohibitively difficult.

It has already been mentioned that the Monte Carlo method is mainly used to obtain averages (see equation (1.2.22)). Equation (2.1.21) is of the same form as (1.2.22). *It is an average of a random variable over elementary events.* It can be seen that every quantity related to the system behavior can be presented in the form of equation (2.1.21), and that such averages are readily calculated by the Monte Carlo method.

2.1.3 The definition of a system

In the above discussion, the term 'system' was used rather loosely. At this point it can be formally defined.

Definition 2.2: A system A system is a collection of components on which at least one real valued function (system function) is defined.

In the context of this discussion, a factory, machine or an organization become systems only if the elements comprising them are components (entities with known states and known properties of transfer between the states), and an explicit measure or measures of performance can be defined as real valued functions. Nothing is a system unless a measure of its performance as a function of the states of its components (and possibly other parameters) is defined.

Consider a single component having three states: operational (1); failed (0); and passive (−2). Let $n_f(t)$ denote the number of failures up to time t. $n_f(t)$ counts the number of transfers from the states $b_1 = 1$ or $b_1 = -2$ into the state $b_1 = 0$. Let $\tau_d(t)$ be the total downtime of the component up to time t. $\tau_d(t)$ is the total amount of time in which the component is either in state 0 or in state −2. Denoting by c_f and c_d the cost of failure and the cost of unit down time, respectively, a system function can be defined as $S(b, t) = c_f n_f(t) + c_d \tau_d(t)$.

This system function measures the performance in terms of the operation cost. It is defined on the phase space of the system. The component with its states and transfer properties between states is not a system. The component becomes a system only when the system function is added.

Considering the behavior of systems in time, we note that as the components change their states the system state vector is changed. As a result, the system function may assume different values. As such, the system function is a random variable indicating the performance of the system. The system can be viewed as a particle moving in its phase space of different states and time. There may be more than one system function as there may be more than one measure of performance. Recalling the discussion in the introduction (see Figure 2), it may now be stated that resources are needed so that the system functions will satisfy given criteria. This may be related to the time average value of the system function (such as the average production in Example 2.1.3), or to any other functional of the system function/s. The goal of system engineering is to calculate such functional/s for a given system design, structure and resources.

2.2 A Single Component System

The simplest possible model is that of a single component having two states. The simplest system function is $S(\mathbf{B}) = b_1$, where b_1 can take only two values; 1 – operational, and 0 – failed. Two generally used performance measures will be introduced, the availability and reliability. The availability will be denoted by $A(t)$, and is defined as:

Definition 2.2.1: Availability $A(t)$ is the probability that at time t, $S(\mathbf{B}) = 1$.

In the above definition, it is assumed that the value 1 is preserved for a state or a union of states for which the system is operational (or available). The reliability is denoted by $R(t)$, and is defined as:

Definition 2.2.2: Reliability $R(t)$ is the probability that up to time t the system never encounters a state in which $S(\mathbf{B}) = 0$.

In the above definition it is assumed that the value 0 is preserved for a state or a union of states for which the system is failed.

The PDFs attached to the states of the component will be denoted as $f_1(t - t_c)$ for the failure distribution and $f_0(t - t_c)$ for the repair (transfer from state 0 to state 1). t_c is the time point at which the system enters the respective state.

The reliability is the probability that the system will not fail up to time t. Noting the definition of the CDF, it is clear that the reliability is simply:

$$R(t) = 1 - F_1(t). \tag{2.2.1}$$

The reliability of a single component system does not involve the repair process because a failure of the component is a failure of the system. The reliability therefore, holds a simple relation to the failure probability of the component. The availability presents a somewhat more elaborate problem. Henceforth, a number of different approaches to its solution will be discussed.

2.2.1 A direct statistical approach

To consider the availability, first define the function $A(t|k)$ as the probability of the system being operational at time t given that it had k failures in the past. The event {The system is available at time t} can be described as a union of the events {The system is available at time t, having k failures in the past}, and since these events are exclusive one may write:

$$A(t) = \sum_{k=0}^{\infty} A(t|k). \tag{2.2.2}$$

It is easy to see that $A(t|0)$ is simply the reliability. It is the probability that the system will have no failures up to time t. The event {system is available at t having one failure in the past} is a union of a set of continuous elementary events in which a failure occurred in an infinitesimal interval dt_1 at time t_1; the probability for this is $f_1(t_1)\,dt_1$. The failure is followed by a repair in an interval dt_2 at t_2, where $t_1 < t_2 < t$. The time for the repair event is measured from t_1, and the probability for this event is $f_0(t_2 - t_1)\,dt_2$. Finally, from the time of repair, t_2, and up to t the system remains in an operational state. The probability for that is $[1 - F_1(t_1 - t_2)]$. This elementary event is made of three independent events, and thus the probability for this event ('history') is $f_1(t_1)\,dt_1 \times f_0(t_2 - t_1)\,dt_2 \times [1 - F_1(t - t_2)]$. In terms of phase space histories, this history can be denoted as $(0, t_1; 1, t_2)$ with $\mathbf{B}_0 = 1$ at $t=0$. The probability of the union

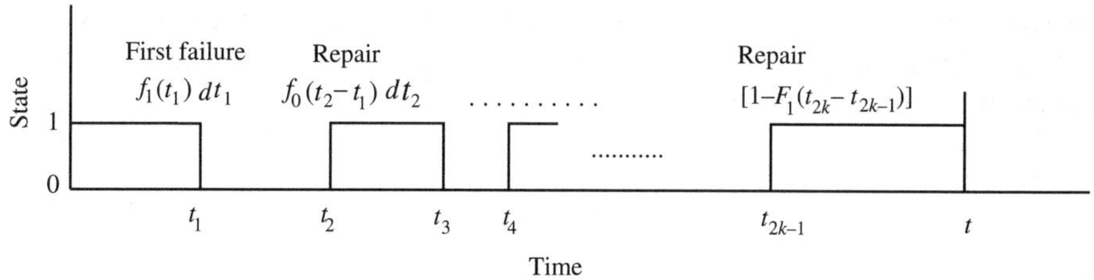

Figure 2.2.1 A sequence of failures and repairs of the kth term in the availability series

event $A(t|1)$ is obtained by integration over t_1 and t_2, namely:

$$A(t|1) = \int_0^t \int_{t_1}^t f_1(t_1) f_0(t_2 - t_1)[1 - F_1(t - t_2)]\, dt_1\, dt_2. \tag{2.2.3}$$

Using the same argument, it is easy to see that the general term can be written as

$$A(t|k) = \int_0^t dt_1 \int_{t_1}^t dt_2 \int_{t_2}^t dt_3 \cdots \int_{t_{2k-1}}^t dt_{2k} f_1(t_1) f_0(t_2 - t_1) f_1(t_3 - t_2) ... f_0(t_{2k} - t_{2k-1})[1 - F_1(t - t_{2k})], \tag{2.2.4}$$

where the event of a single failure is extended to include k failures at points $t_1, t_3 \ldots$ up to t_{2k-1} and each failure is followed by a repair at points $t_2, t_4 \ldots$ up to t_{2k}, respectively. From t_{2k} and up to t the system remains in an operational state (see Figure 2.2.1).

The availability is thus an infinite sum of multi-dimensional integrals.

The exponential case

It was already mentioned that the exponential distribution is often used because of its simplicity. Indeed, one of the few cases in which the sum in equation (2.2.2) can be explicitly obtained is when both $f_1(t - t_c)$ and $f_0(t - t_c)$ are exponential distributions, i.e.

$$f_1(t - t_c) = \lambda\, e^{-\lambda(t - t_c)} \quad \text{and} \quad f_0(t - t_c) = \mu\, e^{-\mu(t - t_c)}. \tag{2.2.5}$$

λ is the hazard for failure, and is often called the *failure rate*, and μ is called the *repair rate*.

To obtain a closed expression for this case, first apply a change in the integration limits of equation (2.2.4). Note that the integration is done on each t_i from t_{i-1} to t. This can be changed so that the integration is done for each t_i form zero to t_{i+1}. This preserves the order of events and covers all possible elementary events. Substituting assumption (2.2.5) only for $F_1(t_{2k} - t_{2k-1})$ and interchanging the integration limits yields

$$A(t|k) = \int_0^{t_2} dt_1 \int_0^{t_3} dt_2 \int_0^{t_4} dt_3 \cdots \int_0^{t_{2k}} dt_{2k-1} \int_0^t dt_{2k} f_1(t_1) f_0(t_2 - t_1) f_1(t_3 - t_2) ... f_0(t_{2k} - t_{2k-1})\, e^{-\lambda(t - t_{2k})}.$$

Upon substitution of this expression into equation (2.2.2), the notation of t_{2k} is changed into t' and integration over it is carried outside the sum. Also, the term $A(t|0)$ is separated from the sum, yielding

$$A(t) = \mathrm{e}^{-\lambda t} + \int_0^t dt' \left[\sum_{k=1}^{\infty} B(k, t') \right] \mathrm{e}^{-\lambda(t-t')}. \tag{2.2.6}$$

$B(k, t')$ is a shorthand notation for the remaining integral over t_1 up to t_{2k-1}, namely:

$$B(k, t') = \int_0^{t_2} dt_1 \int_0^{t_3} dt_2 \int_0^{t_4} dt_3 \cdots \int_0^{t'} dt_{2k-1} f_1(t_1) f_0(t_2 - t_1) f_1(t_3 - t_2) ... f_0(t' - t_{2k-1}). \tag{2.2.6a}$$

Closer examination of $B(k, t')$ reveals that it is a PDF. In fact, it is the probability density function for the time of the last event (the $2k$th event) of k failures followed by k repairs. Furthermore, in considering this process it is clear that an exactly identical process is obtained if one considers k transfers into state 1 followed by k transfers into state 0. This may be viewed as a sequence of k 'down' steps, each having a PDF $f_1(t - t_c)$, followed by a sequence of k 'up' steps each having a PDF $f_0(t - t_c)$. $B(k, t')$ is then the PDF of the time, t' of the last 'up' step. Since the order of the steps is not important, it may be concluded that t' can be viewed as a sum of two r.v.s, namely, the time of k 'down' steps, t_d^k, and the time of k 'up' steps, t_u^k. Denoting the PDF of t_d^k by $g_d(t_d^k)$ and that of t_u^k by $g_u(t_u^k)$, the PDF of the sum $t' = t_d^k + t_u^k$ can be expressed as:

$$B(k, t') = \int_0^{t'} g_d(t' - x) g_u(x)\, dx = \int_0^{t'} g_d(x) g_u(t' - x)\, dx. \tag{2.2.7}$$

Expression (2.2.7) is a general expression for the PDF of a sum of two r.v.s. This type of integral is also called a *convolution integral*. Thus, to obtain $B(k, t')$, it is necessary to obtain $g_d(t_d^k)$ and $g_u(t_u^k)$ explicitly, and substitute them into equation (2.2.7).

Up to this point, everything said about $B(k, t')$ is general, and will hold for any type of failure and repair distributions. Introducing now the exponential assumption, it is clear that $g_d(t_d^k)$ is the PDF of the sum of k independent identical exponential r.v.s. This PDF is known, and is given by the Gamma distribution, $G_k(t)$:

$$g_d(t) = G_k(t) = \frac{\lambda(\lambda t)^{k-1}\,\mathrm{e}^{-\lambda t}}{\Gamma(k)}. \tag{2.2.8}$$

λ is the parameter of the corresponding exponential distribution. A similar expression applies for $g_u(t)$ with μ instead of λ.

To prove this relation, one can show that the moment generating function of the Gamma distribution is identical to the kth power of the moment generating function of the exponential distribution (see, for example, Shooman, M. L. (1968). *Probabilistic Reliability: An Engineering Approach*, McGraw-Hill, New York, p. 59; or Lawless, J. F.

(1982). *Statistical Models and Methods for Lifetime*, John Wiley & Sons). $\Gamma(k)$ is the Gamma function discussed in equation (1.3.15). For integer values of $k, \Gamma(k) = (k-1)!$ The nth moment of the Gamma distribution is given by

$$E(t^n) = \lambda^{-n}[k(k+1)(k+2)\cdots(k+n-1)]. \tag{2.2.9}$$

Substituting the Gamma distributions into equation (2.2.7) yields

$$B(k, t') = \frac{\lambda^k \mu^k}{(k-1)!(k-1)!} \int_0^{t'} x^{k-1} \, \mathrm{e}^{-\lambda x} (t' - x)^{k-1} \, \mathrm{e}^{-\mu(t'-x)}.$$

To calculate this integral, it is convenient to make use of the Laplace transform. The Laplace transform of a convolution of two functions is simply the product of the Laplace transforms of each of them. The Laplace transform of $x^{k-1} \mathrm{e}^{-\lambda x}$ is $(k-1)!/(s+\lambda)^k$, thus, denoting by $H(k, s)$ the Laplace transform of $B(k, t')$ yields

$$H(k, s) = \left[\frac{\lambda}{(s+\lambda)} \frac{\mu}{(s+\mu)}\right]^k.$$

Substituting $H(k, s)$ into the infinite sum of equation (2.2.6) yields the Laplace transform of the infinite sum over $B(k, t')$ in the form of a convergent geometrical series, namely:

$$\sum_{k=1}^{\infty} H(k, s) = \sum_{k=1}^{\infty} \left[\frac{\lambda}{(s+\lambda)} \frac{\mu}{(s+\mu)}\right]^k = \frac{\lambda\mu}{s(s+\lambda+\mu)}.$$

The inverse transform of this is

$$\sum_{k=1}^{\infty} B(k, t') = \frac{\lambda\mu(1 - \mathrm{e}^{-(\lambda+\mu)t'})}{\lambda+\mu}. \tag{2.2.10}$$

Finally, substituting (2.2.10) into (2.2.6) and integrating over t' yields the expression for the availability:

$$A(t) = \mathrm{e}^{-\lambda t} + \int_0^t \frac{\lambda\mu(1 - \mathrm{e}^{-(\lambda+\mu)t'})}{\lambda+\mu} \mathrm{e}^{-\lambda(t-t')} \, dt' = \frac{\mu}{\lambda+\mu} + \frac{\lambda}{\lambda+\mu} \mathrm{e}^{-(\lambda+\mu)t}. \tag{2.2.11}$$

In this section we discuss the solution of the availability of a single component having two states via a direct expression, by considering the elementary events (histories) that comprise the event {The component is available at time t}. The general result appears in equations (2.2.2) and (2.2.4), indicating that the availability is an infinite sum of multidimensional integrals. The exponential case is, in fact, the only case for which an explicit closed solution is available (with a few exceptions, that can be found in Barlow, R. E. and Prochan, F. (1965). *Mathematical Theory of Reliability*, John Wiley & Sons, Chichester or Dubi, A. and Gurvitz N. (1995). A note on the analysis of systems with time dependent transition rates, *Annals of Nuclear Energy*, **22**, no. 3/4).

The availability is one of the basic, most important performance measures. One can hardly think of a problem simpler than that of a single component with two states. The fact that the exponential case is, basically, the only one for which an explicit solution for $A(t)$ is available, and the sharp jump in difficulty arising if any other distribution is considered, could explain the widespread (and in some cases, exclusive) use of the exponential distribution to describe failures and repairs. With the application of Monte Carlo methods, the solution of (2.2.2) becomes a straightforward matter with any distribution. In view of the fact that realistic systems involve more than a single component with more than two states and non-exponential repair and failure distributions, a motivation for the utilization of the Monte Carlo method becomes quite obvious.

It is worth noting that expressions (2.2.2) and (2.2.4) are basically identical to equations (2.1.20) and (2.1.21) i.e. an average of a random variable defined on phase space histories.

To see that the availability is indeed an average over elementary histories, define the general history $c_m = (b_1, t_1; b_2, t_2; \ldots; b_m, t_m)$ with the corresponding PDF for the history $f(c_m) = f_1(t_1)f_0(t_2 - t_1)\cdots f_i(t_m - t_{m-1})[1 - F_i(t - t_m)]$. Note that $f_i(t_m - t_{m-1})$ can be either the failure or the repair distribution, depending on m being odd or even. Define the r.v. on the history $\eta(c_m)$ to be unity if the last state is 1, and zero otherwise. The average of $\eta(c_m)$ is then

$$\langle \eta \rangle = A(t) = \sum_{m=0}^{\infty} \sum_{b_1 \ldots b_m} \int_{\tau} \eta(c_m) f(c_m)\, dc_m, \tag{2.2.12}$$

where τ is the m-dimensional time domain. Noting that the summation over b_1 through b_m yields a single term, and that $\eta(c_m)$ is non-zero only for even values of m, yields expression (2.2.2).

Equations (2.2.12) and (2.1.21) provide a window through which the general framework of system engineering can be viewed. The space in which a system exists is the $\{\mathbf{B}, t\}$ phase space. This is a space of vector states and time. The passage of the system through time (the transport in the phase space) is composed of a sequence of 'jumps' from one phase space point $\{\mathbf{B}_i, t_i\}$ to the next one, $\{\mathbf{B}_{i+1}, t_{i+1}\}$. Through the process of these transfers, a random history of the system is unfolded. In addition to the phase space, there are one or more real functions, defined on the phase space points, that measure the performance of the system. They are the system functions. Every aspect concerning the future of the system is included in the system functions, or can be easily obtained by simple operations on them. Such quantities are, for example, the time dependent availability, the reliability, number of failures, cost of system operation, etc. All these quantities can be obtained by some form of averaging of a particular system function over the space of all possible histories of the system. This is true for any quantity of interest in system engineering. To a large extent, the complexity of the system is defined by the number of the points in the state space. It was shown that even for the most trivial space having only two states, the averaging process may become prohibitively difficult, because of the multidimensional integration over the time domain. An added difficulty appears when the number of points in the phase space increases. This happens as the number of components increases, resulting in an exponential increase in the number of state space points (n components, each having two states, result in 2^n states). Another element to be considered is statistical independence. As long as components are considered to be independent, one may treat each of them as if the others do not exist. This enables a reduction in the number of points in the state space by the separation of variables. Yet in most realistic cases, components are not independent.

It will be demonstrated that the Monte Carlo method is, to a large extent, insensitive to the difficulties described above, and thus enables averaging and calculation in highly complex phase spaces.

The next two examples demonstrate the inclusion of spare parts considerations into the two states component analysis.

Example 2.2.1: Components and spares

Five identical pumps are operating in a chemical plant. The pumps have an exponential failure distribution with $\lambda = 6.23 \times 10^{-5}$. Upon failure, the pump is replaced by a spare pump. Failed pumps are discarded. How many spare pumps should be stored for a period of one year (8760 hours) so as to meet all the needs for spares?

The total failure rate of the five pumps is $\lambda^* = 5\lambda$. Let the number of stored spares be k_S. Let n_f denote the number of pump failures during a year. The probability that the stored spares will be sufficient for all needs is $P_r[n_f \leq k_S]$, which is also the probability that the total number of failures will not exceed k_S. It is called the *sufficiency probability*, or just the *sufficiency*.

Denote the probability to have exactly m failures over a period T_m by $P(m|T_m)$. The probability to have zero failures is simply $P(0|T_m) = \mathrm{e}^{-\lambda^* T_m}$. Assuming that repair is immediate, a recursion relation can easily be established by noting that the event $\{m+1$ failures up to time $T_m\}$ is an intersection of the exclusive events $\{\{m$ failures up to time $t'\}$ and $\{$a single failure in $dt'\}$ and $\{$no failures from t' up to$\}\}$. For the exponential distribution (and only for it), this takes the form $\{P(m|t') \times \lambda^* dt' \times [1 - F(T_m - t')]\}$. By integrating over t' the recursion relation

$$P(m+1|T_m) = \int_0^{T_m} P(m|t')[1 - F(T_m - t')]\lambda^* \, dt' = \int_0^{T_m} P(m|t')\lambda^* \, \mathrm{e}^{-\lambda^*(T_m - t')} \, dt' \quad (2.2.13)$$

is obtained.

Having the expression for $m = 0$, one may easily verify by induction that the solution of (2.2.13) is given by

$$P(m|T_m) = \frac{(\lambda^* T_m)^m \, \mathrm{e}^{-(\lambda^* T_m)}}{m!}. \quad (2.2.14)$$

The probability function of equation (2.2.14) is called the *Poisson distribution*. It gives the probability for exactly m events in an interval of size T_m, given that the rate of occurrence of a single event is a constant λ^*. The average number of events is given by

$$E(m) = \sum_{m=0}^{\infty} mP(m|T_m) = \lambda^* T_m. \tag{2.2.15}$$

The second moment yields

$$S^2(m) = \sum_{m=0}^{\infty} m^2 P(m|T_m) = \lambda^* T_m(1 + \lambda^* T_m), \tag{2.2.15a}$$

so that the variance equals the average $\lambda^* T_m$.

Using the Poisson distribution, the sufficiency takes the form

$$P_r[n_f \leq k_S] = P_s(k_S) = \sum_{n=0}^{k_s} P(m|T_m) = \sum_{n=0}^{k_s} \frac{(\lambda^* T_m)^m \, \mathrm{e}^{-(\lambda^* T_m)}}{m!}. \tag{2.2.16}$$

In Table 2.2.1, values of $P_s(k_S)$ are shown for increasing values of k_S.

From this table one may conclude that the question asked in the example does not have a unique solution. By increasing the number of stored spares, one increases the probability that the spares will be sufficient. To create absolute certainty, an infinite number of spares is needed. Five spares (as many as the number of operating components) yields a sufficiency of 94.09%. Having nine spares will increase the sufficiency to 99.94%. It is not at all clear that increasing the sufficiency by about 6% is worth the additional cost of four spares. It is obvious that the nature of such problems involves an element of uncertainty. One either stores a small number of spares and increases the risk of lacking a spare, or stores a large number of spares at the risk of never using some of them, i.e. wasting money. It should be stressed that in this example, whatever decision will be made, it *does not account for the impact of the spares on the performance*. The sufficiency tells us the probability that spares will be missing, but nothing about the impact of this on the system performance. Note that, since a system function is not

Table 2.1.1 Poisson probability and sufficiency

Number of spares k_S	Probability of exactly k_S failures $P(k_S\|T_m)$	Sufficiency $P_s(k_S)$
0	0.0653	0.0653
1	0.17819	0.2434
2	0.243118	0.4866
3	0.221135	0.7077
4	0.150855	0.8585
5	0.082329	0.9409
6	0.037442	0.9783
7	0.014595	0.9929
8	0.004978	0.9979
9	0.001509	0.9994

defined for the five components, *there is no system* and no system performance is considered. The introduction of a system function may strongly (and correctly) affect the calculation.

Example 2.2.2: Spares for a system

Consider now the exact situation as in Example 2.2.1, with the addition that the five pumps are operating in parallel, i.e. if at least one of them is operational the system is operational. The system function for this structure is simply $S(\mathbf{B}) = 1 - (1 - b_1) \times (1 - b_2) \times (1 - b_3) \times (1 - b_4) \times (1 - b_5)$. In considering the number of spares, one must provide a criterion of performance. The question will now be, how many spare pumps should be stored (for a one year period) so that the average down time of the system over a year will not exceed 40 hours? Replacement of a unit is assumed to be immediate, so that downtime can arise only if no spares are available and all the units are failed.

The availability $A(t)$ can be interpreted as the fraction, per unit time, in which the system is available. Thus, denoting $U(t) = 1 - A(t)$ as the unavailability, the average total down time of the system in a period T_m is simply the integral of $U(t)$, namely:

$$\langle T_d \rangle = \int_0^{T_m} U(t)\, dt. \tag{2.2.17}$$

To obtain the unavailability, one should note that for systems which cannot be repaired once failed, the availability and reliability are identical, i.e. $R(t) \equiv A(t)$. This can be easily proven by considering the fact that the events {System is operational at time t} and {System did not fail up to time t} are identical if the system is not repaired upon failure.

Consider the situation with zero spares. The system is failed at time t if and only if there were exactly five failures, i.e. if all five components failed up to time t'. Thus, the unavailability (and unreliability function) is $U(t) = [1 - e^{-\lambda t}]^5$. In general, for a system of n components in parallel the unavailability will be $U(t) = [1 - e^{-\lambda t}]^n$ (as there are no spares, each failed component is unrepairable). The down time is then

$$\langle T_d \rangle = \int_0^{T_m} [1 - e^{-\lambda t}]^n\, dt = \int_0^{T_m} \sum_{i=0}^{n} \binom{n}{i} (-1)^i e^{-i\lambda t}\, dt = T_m + \sum_{i=1}^{n} \frac{n!}{i!(n-i)!} \frac{(-1)^i}{i\lambda} [1 - e^{-i\lambda T_m}]. \tag{2.2.18}$$

Substituting $n = 5$ and the parameters of the problem, i.e. $\lambda = 6.23 \times 10^{-5}$ and $T_m = 8760$ yields $\langle T_d \rangle = 23.29$ hours, thus the requirement of less then 40 hours of down time over a year is obtained if no spares are bought at all. This can be compared to the value of 5–6 spares that would give a sufficiency of 94–97%, and would seem to be a reasonable decision. The considerations of system performance rather than any spare related quantity (such as the sufficiency) should be the guiding principle. It may be a wild guess, but the key to the $10 billion of waste spares mentioned in the introduction may be hidden in the difference between Examples 2.2.1 and 2.2.2. Spare parts, as well as any other type of resource, are intended to support the system performance and not the sufficiency.

Consider again a system with n components in parallel and with $k > 0$ spares in the store. In this case, the first k failures will be repaired immediately. The next n failures will not be repaired, and upon the $(n + k)$th failure the system will fail. Again, the reliability and availability are identical (because upon a *system failure* there is no repair). The unavailability takes a somewhat more elaborate form now. The probability that the system is failed at time t requires that the kth failure will occur at some intermediate time point t' and there will be n failures in the interval $(t - t')$. The PDF for k failures at t' is the Gamma distribution of equation (2.2.8), namely

$$G_k(t') = \frac{n \times \lambda(n \times \lambda t')^{k-1}\, e^{-n \times \lambda t'}}{\Gamma(k)}.$$

The failure rate for the first k failures is $\lambda^* = n \times \lambda$. Multiplying $G_k(t')$ by the probability of having n additional failures between t' to t, namely $[1 - e^{-\lambda}(t - t')]^n$, yields the PDF for the elementary event in which the system is failed at t (these elementary events are exclusive as t' changes). Thus the unavailability is obtained by integrating over t', or

$$1 - R(t) = U(t) = \int_0^t \frac{\lambda^*(\lambda^* t')^{k-1}\, e^{-\lambda^* t'}}{\Gamma(k)} [1 - e^{-\lambda(t-t')}]^n \, dt', \tag{2.2.19}$$

and the average down time is obtained as in equation (2.2.17) in the form

$$\langle T_d \rangle = \int_0^{T_m} U(t)\, dt = \int_0^{T_m} \left[\int_0^t \frac{\lambda^*(\lambda^* t')^{k-1}\, e^{-\lambda^* t'}}{\Gamma(k)} [1 - e^{-\lambda(t-t')}]^n \, dt' \right] dt. \tag{2.2.20}$$

Equation (2.2.20) is a general expression for the down time of parallel systems with n components and k spares. For $k = 1$ and $n = 5$, one obtains an average down time of 7.56 hours, which is far below the performance demand. The integral of (2.2.20) is rather elaborate to perform, and will be considered when integration by the Monte Carlo method is discussed. The availability's for 0, 1 and 2 spare parts are shown in Figure 2.2.2.

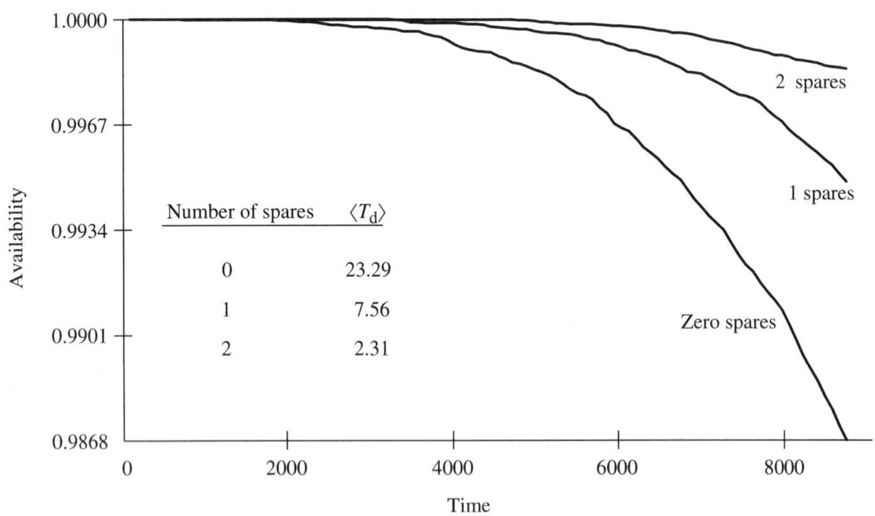

Figure 2.2.2 Availability and average downtime for a one-out-of-five system with 0, 1 and 2 spares. Results obtained using Monte Carlo methods of equations (2.2.19) and (2.2.20)

2.2.2 A transport equation approach

In the preceding section, the availability was derived as an average quantity over the transport of the component in the time and states space. A different approach will be considered next, by which an equation is constructed for the relation between *entries into states*. At this point, a generalization will be introduced such that, instead of having two states, the component may have n possible states.

Consider a single component system with n states. Let ΔT denote a time interval and $N_i(t, \Delta T)$ the number of entries of the component into state i in the interval $[t, t + \Delta T]$. The *event density* $\psi_i(t)$ is defined as:

Definition 2.2.3: The event density $\psi_i(t) = \lim_{\Delta T \to 0} N_i(t, \Delta T)/\Delta T$

The event density is simply the number density of entries of the system into state i per unit time. This definition implies that the probability that the system will enter any state i in an infinitesimal interval dt' around time t' is $\psi_i(t')\,dt'$. Given a time point $t > t'$, the probability that a system will enter state i in dt' at t' and will remain in this state up to time t is $\psi_i(t')\,dt' R_i(t - t')$, where

$$R_i(t) = 1 - F_i(t) = \mathrm{e}^{-\int_0^t z_i(x)\,dx}$$

(see equations (2.1.3)).

Let $P_i(t)$ denote the probability that the system *is in state i at time t*. The above discussion implies that $P_i(t)$ can be expressed in terms of the event density, in the form

$$P_i(t) = \int_0^t \psi_i(t') R_i(t - t')\,dt', \tag{2.2.21}$$

which is simply the probability that the system will enter state i at any time in the past, and will remain in it up to time t. The availability of a two-state single component system is then

$$A(t) = P_1(t) = \int_0^t \psi_1(t') R_1(t - t')\,dt' \tag{2.2.21a}$$

Comparing equation (2.2.21a) with equation (2.2.11) for the exponential case yields

$$\psi_1(t) = \delta(t) + \frac{\lambda\mu}{\lambda + \mu}[1 - \mathrm{e}^{-(\lambda+\mu)t}]. \tag{2.2.22}$$

$\delta(t)$ is the Dirac delta function (see, for example, Marchand, J. P. (1902). *Distributions – An Outline*, North-Holland, Amsterdam, for a detailed discussion of this distribution), accounting for the single entry of the system into state 1 at $t = 0$. The delta function is zero at every point, and has a singularity at the point at which its argument is zero. The most useful property of this function is

$$\int_0^T f(t)\delta(t - t_0)\,dt = f(t_0) \quad \text{for any } 0 \le t_0 \le T.$$

Thus, substituting equation (2.2.22) into equation (2.2.21a) will result upon integration over $\delta(t')R_1(t-t')$, the first term of equation (2.2.11), namely $e^{-\lambda t}$.

The event density fulfills an equation called the transport equation, or the *system transport equation*. The name is derived from the fact that the transport of the system through time and state space is determined by this equation. To derive the equation for the case at hand, consider again two time points t and t' such that $t > t'$. The number of system entries into state j in dt' at t' is $\psi_j(t')\,dt'$. A fraction $\psi_j(t')R_j(t-t')\,dt'$ will remain in state j up to time t, and a fraction $\psi_j(t')R_j(t-t')z_{ji}(t-t')\,dt'$ will enter state i at t, per unit time. This is the contribution to the event density at t, into state i, from preceding entries into state j at dt'. Integrating over time and summing over the states yields the total contribution to $\psi_i(t)$ in the form

$$\psi_i(t) = P_{i0}\delta(t) + \sum_{\substack{j=1\\ j\neq i}}^{n} \int_0^t \psi_j(t')R_j(t-t')z_{ji}(t-t')\,dt'. \tag{2.2.23}$$

The source term $P_{i0}\delta(t)$ accounts for entries into state i at $t=0$, where P_{i0} is the probability that the system starts in state i at $t=0$. Equation (2.2.23) comprises a set of n simultaneous integral equations.

In the case of two states, the equations take the form

$$\psi_1(t) = \delta(t) + \int_0^t \psi_0(t')f_0(t-t')\,dt', \tag{2.2.24a}$$

$$\psi_0(t) = \int_0^t \psi_1(t')f_1(t-t')\,dt'. \tag{2.2.24b}$$

Substituting from (2.2.24b) into (2.2.24a) yields

$$\begin{aligned}\psi_1(t) &= \delta(t) + \int_0^t \left[\int_0^{t'} \psi_1(t'')f_1(t'-t'')\,dt''\right] f_0(t-t')\,dt' \\ &= \delta(t) + \int_0^t \psi_1(t')\left[\int_{t'}^{t} f_1(x-t')f_0(t-x)\,dx\right]dt' = \delta(t) + \int_0^t \psi_1(t')K(t,t')\,dt'.\end{aligned} \tag{2.2.25}$$

The right-hand side (RHS) is obtained by renaming the variables and interchanging the boundaries of integration. $K(t,t')$ is called the *kernel* of the equation.

In the exponential case, the kernel takes the form

$$K(t,t') = \int_{t'}^{t} f_1(x-t')f_0(t-x)\,dx = \int_{t'}^{t} \lambda\,\mathrm{e}^{-\lambda(x-t')}\,\mu\,\mathrm{e}^{-\mu(t-x)}\,dx$$

$$= \frac{\lambda\mu}{\mu-\lambda}[\mathrm{e}^{-\mu(t-t')} - \mathrm{e}^{-\lambda(t-t')}],$$

and the transport equation takes the form

$$\psi_1(t) = \delta(t) + \int_0^t \psi_1(t')\frac{\lambda\mu}{\mu-\lambda}[\mathrm{e}^{-\mu(t-t')} - \mathrm{e}^{-\lambda(t-t')}]\,dt'. \tag{2.2.26}$$

It is easy to verify by substitution that (2.2.22) is a solution of (2.2.26).

One should note that expressions (2.2.21) and (2.2.2) (with (2.2.4)) are identical. This means that expression (2.2.2) must result from a series expansion of the event density. This can be verified with the following iterative approximation process. Let the zero approximation be

$$\psi_1^{(0)}(t) = \delta(t) \tag{2.2.26a}$$

i.e. the zero approximation is the source term, and

$$\psi_0^{(0)}(t) = \int_0^t \psi_1^{(0)}(t')f_1(t-t')\,dt'. \tag{2.2.26b}$$

Likewise, define the first approximation as

$$\psi_1^{(1)}(t) = \delta(t) + \int_0^t \psi_0^{(0)}(t')f_0(t-t')\,dt' \tag{2.2.26c}$$

and

$$\psi_0^{(1)}(t) = \int_0^t \psi_1^{(1)}(t')f_1(t-t')\,dt', \tag{2.2.26d}$$

and in general

$$\psi_1^{(n+1)}(t) = \delta(t) + \int_0^t \psi_0^{(n)}(t')f_0(t-t')\,dt' \tag{2.2.26e}$$

and

$$\psi_0^{(n+1)}(t) = \int_0^t \psi_1^{(n+1)}(t')f_1(t-t')\,dt'. \tag{2.2.26f}$$

Under rather broad conditions, the approximated functions converge to the solution of (2.2.24). Thus

$$\lim_{n\to\infty} \psi_i^n(t) = \psi_i(t).$$

A more explicit form of the expansion terms can be obtained by substituting $\psi_1^{(0)}(t')$ into (2.2.26b). This yields

$$\psi_0^{(0)}(t) = \int_0^t \delta(t')f_1(t-t')\,dt'.$$

It can be easily seen from the property of the delta function given above that $\psi_0^{(0)}(t) = f_1(t)$. Substituting this result in $\psi_1^{(1)}(t)$ yields

$$\psi_1^{(1)}(t) = \delta(t) + \int_0^t f_1(t')f_0(t-t')\,dt',$$

and subsequently

$$\begin{aligned}
\psi_0^{(1)}(t) &= \int_0^t \left[\delta(t') + \int_0^{t'} f_1(t'')f_0(t'-t'')\,dt''\right] f_1(t-t')\,dt' \\
&= f_1(t) + \int_0^{t_2} dt_1 \int_0^{t} f_1(t_1)f_0(t_2-t_1)f_1(t-t_2)\,dt_2 \\
&= f_1(t) + \int_0^{t} dt_1 \int_0^{t_1} f_1(t_1)f_0(t_2-t_1)f_1(t-t_2)\,dt_2,
\end{aligned}$$

and

$$\begin{aligned}
\psi_1^{(2)}(t) &= \delta(t) + \int_0^t \psi_0^{(1)}(t_3)f_0(t-t_3)\,dt_3 \\
&= \delta(t) + \int_0^t \left[f_1(t_3) + \int_0^{t_2} dt_1 \int_0^{t_3} f_1(t_1)f_0(t_2-t_1)f_1(t_3-t_2)\,dt_2\right] f_0(t-t_3)\,dt_3 \\
&= \delta(t) + \int_0^t f_1(t_3)f_0(t-t_3)\,dt_3 + \int_0^t dt_3 \int_0^{t_3} dt_2 \int_0^{t_2} dt_1 f_1(t_1)f_0(t_2-t_1)f_1(t_3-t_2)f_0(t-t_3) \\
&= \delta(t) + \int_0^t f_1(t_1)f_0(t-t_1)\,dt_1 \\
&\quad + \int_0^t dt_1 \int_0^{t_1} dt_2 \int_0^{t_2} dt_3 f_1(t_1)f_0(t_2-t_1)f_1(t_3-t_2)\,dt_2 f_0(t-t_3).
\end{aligned}$$

Proceeding this successive approximation for the nth term, and taking the limit, yields

$$\psi_1(t) = \lim_{n\to\infty} \psi_i^n(t)$$
$$= \sum_{k=0}^{\infty} \int_0^t dt_1 \int_0^{t_1} dt_2 \cdots \int_0^{t_{2k-2}} dt_{2k-1} f_1(t_1) f_0(t_2 - t_1) \ldots f_1(t_{2k-1} - t_{2k-2}) f_0(t - t_{2k-1}). \tag{2.2.26h}$$

To obtain the availability, substitute $\psi_1(t)$ from equation (2.2.26h) into expression (2.2.21) for $P_1(t)$, noting the $R_1(T-t) = 1 - F_1(T-t)$. Comparing the result to expressions (2.2.2) and (2.2.4) shows immediately that the expressions are identical. Thus, expressions (2.2.2) and (2.2.4), which were derived from basic considerations, are in fact a series expansion form of the solution of the transport equation. Equations (2.2.2), (2.2.12) and the above expression are all identical. They represent three different appearances of the same quantity. The quantity (availability in this case) may be derived from direct statistical considerations, it can be presented as an average of a random variable over systems histories in phase space, and this average is also a functional of the solution of the transport equation. The expansion solution of equation (2.2.26h) is known as a Neumann series (see, for example, Kantorovich, L. V. and Akilov, G. P. (1964). *Functional Analysis in Normed Spaces*, Pergamon Press, or any other text on functional analysis (1964)).

The transport equation (2.2.23), presented above, is not the most general form of the transport equation. The system involves only a single component so that a single point in time marks the entry of the system into a state. The kernel is a function of the difference $(t - t')$, and not of each t and t' separately. This means that, upon entry into any state i, the 'future' looks exactly the same, independent of the *time of entry*, and depending only on the elapsed time since entry. Such a model is also called a semi-Markov model. It should be noted that unlike in the semi-Markov model, the above transport equation holds if the kernels depend upon t and t' separately, and not on $(t - t')$.

Another important point concerns the properties of a component just after repair. Equation (2.2.24) implies that if the transfer into state 1 occurs at time t_1, then the failure distribution takes the form $f_1(t - t_1)$. This means that the failure properties of the component are exactly the same as those of a 'new' component starting at $t = 0$ and having the failure PDF $f_1(t)$. The same holds for equations (2.2.4). This limits the model to a situation in which the component is 'as good as new' upon repair. It is quite obvious that in many realistic cases, repair does not renew the component. It remains 'as old as it was before repair' (this would be the case for a human being). Such models will be discussed in forthcoming chapters. Further discussion of the transport equation is deferred until it can be put in the context of a more general theory of the Monte Carlo process.

2.2.3 The Markov equations

The Markov model was mentioned in Section 1.3.2. In this model, knowledge of the state of the system at any given time is sufficient to determine the future. It has no memory and is closely linked with the exponential distribution. (Detailed discussions on Markov and semi-Markov processes can be found in Howard, R. A. (1971). *Dynamic Probabilistic Sytems, Vol. 2*, John Wiley & Sons, New York; Cilnar, E. (1975). *Introduction to Stochastic Processes*, Prentice Hall, Engelwood Cliffs, NJ or Barucha-Reid, A. T. (1960). *Elements of the Theory of Markov Processes and their Applications*, McGraw Hill, New York).

If all the distributions are exponential, then the system is Markovian and the future depends only upon the state in the present. This means that in such a situation, the transport equation (2.2.23) should reduce into an equation relating the probabilities of being in a state $P_i(t)$, rather than between the event densities.

The state reliability function $R_j(t-t')$ takes the form $R_j(t-t') = \mathrm{e}^{-z_j(t-t')}$ and the partial rate function z_{ji} is constant. Equations (2.2.21) and (2.2.23), then take the form

$$P_i(t) = \int_0^t \psi_i(t')\,\mathrm{e}^{-z_i(t-t')}\,dt' \tag{2.2.27}$$

and

$$\psi_i(t) = \sum_{\substack{j=1\\ j\neq i}}^{n} z_{ji} \int_0^t \psi_j(t')\,\mathrm{e}^{-z_j(t-t')}\,dt' = \sum_{\substack{j=1\\ j\neq i}}^{n} z_{ji} P_j(t) \tag{2.2.28}$$

valid for $t > 0$, at which $\delta(t) \equiv 0$.

Taking a time derivative[1] of equation (2.2.27) yields

$$\frac{dP_i(t)}{dt} = \psi_i(t) - z_i \int_0^t \psi_i(t')\,\mathrm{e}^{-z_i(t-t')}\,dt',$$

which upon substitution of (2.2.28) for $\psi_i(t)$ and (2.2.27) for the second term on the right-hand side yields

$$\frac{dP_i(t)}{dt} = -z_i P_i(t) + \sum_{\substack{j=1\\ j\neq i}}^{n} z_{ji} P_j(t). \tag{2.2.29}$$

Equation (2.2.29) is known as the Markov equation. It is a set of simultaneous, first-order differential equations, derived as a special case of the n, single entry, states

---

[1] The derivative of an integral equation follows the formula

$$\frac{d}{dx}\int_{A(x)}^{B(x)} G(x,y)dy = \int_{A(x)}^{B(x)} \frac{\partial G(x,y)}{\partial x}dy + G[x,B(x)]\frac{dB(x)}{dx} - G[x,A(x)]\frac{dA(x)}{dx},$$

provided that $G(x,y)$ and $(\partial G(x,y)/\partial x)$ are continuous functions of x and y and the derivatives of $A(x)$ and $B(x)$ are continuous.

transport equation. The Markov equations have a simple physical interpretation. z_i represents the rate of leaving state i, thus $z_i P_i(t)$ represents the relative reduction of the population in this state. Likewise, units are streaming from other states into state i. The rate of transfer form any state, j, into state i is z_{ji}, resulting in an increase of the population in state i of $z_{ji}P_j(t)$ per unit time. The second term on the right represents the streaming into state i (per unit time). The sum is the net change in the population of the state. An additional equation must be observed, expressing the normalization of probabilities, namely:

$$\sum_{i=1}^{n} P_i(t) = 1 \tag{2.2.29a}$$

(the probability to be in any state, at any time, is 1).

It should be noted that equation (2.2.29a) is also fulfilled by the solutions of the transport equation (2.2.23) when substituted in equation (2.2.21) for $P_i(t)$. This is a property of the transport equation. For the simple case of two states with exponential distributions, equations (2.2.29) take a simple form:

$$\frac{dP_1(t)}{dt} = -\lambda P_1(t) + \mu P_0(t).$$

This, with the normalization equation $P_1(t) + P_0(t) = 1$, yields:

$$\frac{dP_1(t)}{dt} = -(\lambda + \mu)P_1(t) + \mu.$$

It is easy to see that expression (2.2.11), namely

$$A(t) = \frac{\mu}{\lambda + \mu} + \frac{\lambda}{\lambda + \mu} \mathrm{e}^{-(\lambda+\mu)t}$$

is a solution of the equation with the boundary condition $P_1(0) = A(0) = 1$.

The availability of a single component with two states contains two terms. The second term is a transient which goes to zero as time increases. The first term is called the *steady state availability* A_∞, and noting that $1/\lambda$ is the Mean Time To Failure (MTTF) and $1/\mu$ the Mean Time To Repair (MTTR), one obtains:

$$A_\infty = \frac{\mu}{\lambda + \mu} = \frac{MTTF}{MTTF + MTTR}, \tag{2.2.30}$$

which is one of the most commonly used expressions to describe and calculate availabilitys. The main advantage of (2.2.11) and (2.2.30) is their simplicity. Both are based on a simplistic assumption of a single independent component with exponential distributions for failure and repair.

In the following, we consider a number of examples in which the Markov equations are used for modeling spare parts scenarios.

Example 2.2.3: A recycled spare part

In Examples 2.2.1 and 2.2.3 spare parts were considered such that failed units are discarded. Once the stock of spares is deployed, failed units cannot be further repaired (repair is done by replacement). This facilitates the calculation, since life histories of the system have a finite number of possible events.

In many cases, recycling of failed components is possible. The failed unit is replaced by a spare unit and is sent for repair at a remote location. Upon repair it is shipped back and serves as a spare. Three PDFs are involved in the process, $f(t)$ the failure distribution of the component, $g(t)$ – the replacement time distribution, and $h(t)$ – the distribution of recycling time, i.e. the distribution of the time required for the failed unit to be shipped, repaired and returned to the local storage. All distributions are assumed to be exponential with corresponding rates λ, μ and α.

A chemical plant uses a large rotating vessel. A critical part of the system is a large tooth wheel mounted on the rotating vessel. The MTBF of the wheel is in the range of 50 000 hours ($\lambda = 2 \times 10^{-5}$). If the wheel tooth fails, it takes, on average, 12 000 hours to obtain a new wheel. Note that it does not matter if it is the same wheel repaired and returned, or it is a new wheel being ordered. The main point is that it takes 12 000 hours to obtain a wheel ($\alpha = 8.333 \times 10^{-5}$). Although the MTBF is rather large, the recycling time is also large, and one is concerned with the possibility of a long shut down caused by lack of a spare. It is suggested to keep a spare tooth wheel on location. The mean time of replacement is 336 hours ($\mu = 2.976 \times 10^{-3}$). A period of five years is considered, and a decision is required on how many spare wheels (if at all) should be purchased. The cost of a spare wheel is $c_1 = \$190\,000$, and the cost of one hour of down time of the plant is $c_2 = \$627$. At this point, the analysis is based on the steady state availability.

If no spares are bought, the steady state unavailability would, according to equation (2.2.30), be

$$U_\infty = 1 - A_\infty = \frac{MTTR}{MTTR + MTTF} = \frac{12\,000 + 336}{12\,336 + 50\,000} = 0.1978. \qquad (2.2.31)$$

The cost of loss of production (during five years) would be $0.1978 \times 5 \times 8760 \times c_2 = \$5.434\,\text{M}$. The cost of lost production is the only cost considered.

In the case of a single spare it is convenient, in this model, to describe a state by three indices. The first indicator is the state of the system (the operating tooth wheel). This parameter will have three possible values: 1 – Operational, 0 – failed under replacement, and (-1) – Failed waiting for a spare. The second parameter gives the number of spares available on location, and the third parameter gives the number of units being recycled. The initial state is $(1, 1, 0)$, i.e. the system is operational and a single spare is available on location. From this state the system can transfer into the state $(0, 0, 1)$, indicating that the system is under repair and one unit is in recycling. The rate at which the initial state streams into this state is λ. Note that it is assumed that the spare unit cannot fail. From this state, the system can move into state $(1, 0, 1)$ at a rate μ if the unit is replaced before the recycled unit arrives, or into the state $(0, 1, 0)$ (at a rate α) if the recycled unit returns before replacement is completed. From state $(1, 0, 1)$ transfer is possible into state $(-1, 0, 2)$ at a rate of λ, or into the initial state $(1, 1, 0)$ at a rate α. From state $(0, 1, 0)$ transfer is possible only into the initial state at a rate μ. From the state $(-1, 0, 2)$, transfer is possible only into state $(0, 0, 1)$ at a rate 2α. These state transitions are summarized in Figure 2.2.3, where the states are numbered and the possible transitions are indicated along with the corresponding partial rates.

Figure 2.2.3 is called a Markov diagram. In the construction of a Markov diagram, recall equation (2.1.1), and the discussion therein indicating that the rate of exit from a state is the sum of the partial rates. The Markov diagram holds only for Markovian processes in which the time elapsed since entering a state does not effect the partial rates, i.e. the partial rates are constant in time.

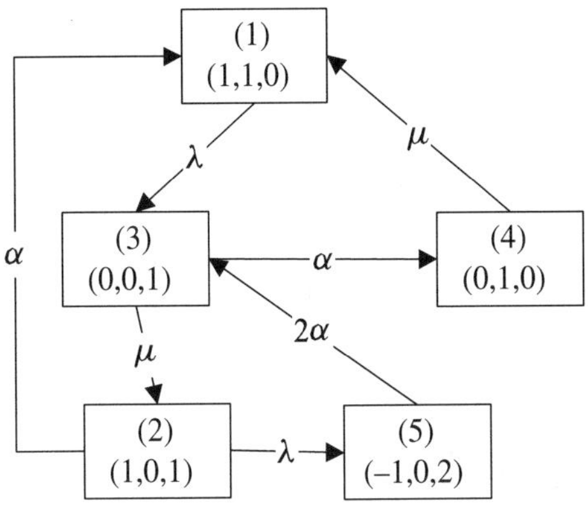

Figure 2.2.3 A Markov diagram for state transitions

The set of Markov equations, equation (2.2.29) corresponding to Figure 2.2.3 is:

$$\frac{dP_1(t)}{dt} = -\lambda P_1(t) + \alpha P_2(t) + \mu P_4(t) \tag{2.2.32a}$$

$$\frac{dP_2(t)}{dt} = -(\alpha + \lambda)P_2(t) + \mu P_3(t) \tag{2.2.32b}$$

$$\frac{dP_3(t)}{dt} = -(\alpha + \mu)P_3(t) + \lambda P_1(t) + 2\alpha P_5(t) \tag{2.2.32c}$$

$$\frac{dP_4(t)}{dt} = -\mu P_4(t) + \alpha P_3(t) \tag{2.2.32d}$$

$$\frac{dP_5(t)}{dt} = -2\alpha P_5(t) + \lambda P_2(t) \tag{2.2.32e}$$

with the normalization condition $P_1(t) + P_2(t) + P_3(t) + P_4(t) + P_5(t) = 1$. This set of equations can be written in a matrix form as:

$$\begin{pmatrix} \dot{P}_1(t) \\ \dot{P}_2(t) \\ \dot{P}_3(t) \\ \dot{P}_4(t) \\ \dot{P}_5(t) \end{pmatrix} = \begin{pmatrix} -\lambda & \alpha & 0 & \mu & 0 \\ 0 & -(\alpha+\lambda) & \mu & 0 & 0 \\ \lambda & 0 & -(\alpha+\mu) & 0 & 2\alpha \\ 0 & 0 & \alpha & -\mu & 0 \\ 0 & \lambda & 0 & 0 & -2\alpha \end{pmatrix} \begin{pmatrix} P_1(t) \\ P_2(t) \\ P_3(t) \\ P_4(t) \\ P_5(t) \end{pmatrix}. \tag{2.2.33}$$

The matrix can be referred to as the *transition matrix* of the Markov process.

A numerical solution of these equations can be easily obtained using a simple iterative procedure:

1. Define a time interval Δt and set the initial conditions $P_1(0) = 1$; $P_2(0) = 0$; $P_3(0) = 0$; $P_4(0) = 0$; $P_5(0) = 0$; set $k = 0$ and $t_k = 0$.
2. Calculate the increment of the discrete equation

$$\Delta P_i = \left[-z_i P_i(t_k) + \sum_{j=1}^{5} z_{ji} P_j(t_k) \right] \times \Delta t. \tag{2.2.34a}$$

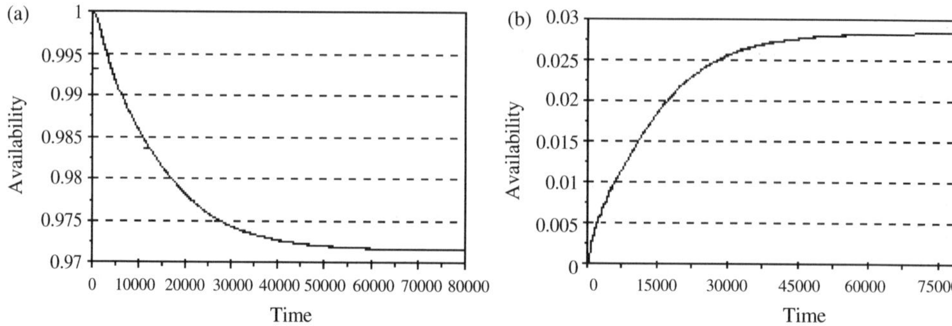

Figure 2.2.4 (a) Availability of the system; (b) unavailability of the system

3. Determine $t_{k+1} = t_k + \Delta t$ and $P_i(t_{k+1}) = P_i(t_k) + \Delta P_i$.
4. Increase $k = k + 1$ and go to step 2.

The availability is the probability of being in any of the states 1 or 2, i.e. $A(t) = P_1(t) + P_2(t)$. The time dependent availability and unavailability, calculated with the above procedure, are shown in Figures 2.2.4(a) and (b).

It can be seen that a steady state in this system is reached only at about 60 000 hours, thus using steady state analysis is conservative in this case, since the availability is underestimated.

To obtain steady state results, note that the time derivatives of the state probabilities are approaching zero as time increases. Substituting zeros for the time derivatives in equations (2.2.32) results in a set of six linear equations (including the normalization equation). One can ignore any of the first five equations, and the results can be obtained as shown below:

$$P_1 = \frac{\alpha(\alpha + \lambda + \mu)}{\lambda(\alpha + \lambda)} P_3; \quad P_2 = \frac{\mu}{(\alpha + \lambda)} P_3; \quad P_4 = \frac{\alpha}{\mu} P_3; \quad P_5 = \frac{\lambda\mu}{2\alpha(\alpha + \lambda)} P_3$$

and

$$P_3 = \left[\frac{\alpha(\alpha + \lambda + \mu)}{\lambda(\alpha + \lambda)} + \frac{\mu}{(\alpha + \lambda)} + 1 + \frac{\alpha}{\mu} + \frac{\lambda\mu}{2\alpha(\alpha + \lambda)}\right]^{-1}.$$

Substituting the partial rates yields:

$$P_1 = 0.78861; \quad P_2 = 0.18291; \quad P_3 = 0.006352; \quad P_4 = 0.000178 \quad \text{and} \quad P_5 = 0.02195.$$

The steady state unavailability is then $U_\infty = P_3 + P_4 + P_5 = 0.02848$, and the total cost of a single spare policy is $c_1 + c_2 T_m U_\infty = 190\,000 + 627 \times 5 \times 8760 \times 0.02848 =$ \$97 2134. It is thus considerably smaller than the cost of a 'no spares' policy.

Example 2.2.4: A recycled spare part with undetected failures

In the above discussion, it was assumed that the spare unit cannot fail while in storage. This is a common assumption. Experience shows that very often, spare units (such as electric engines) may

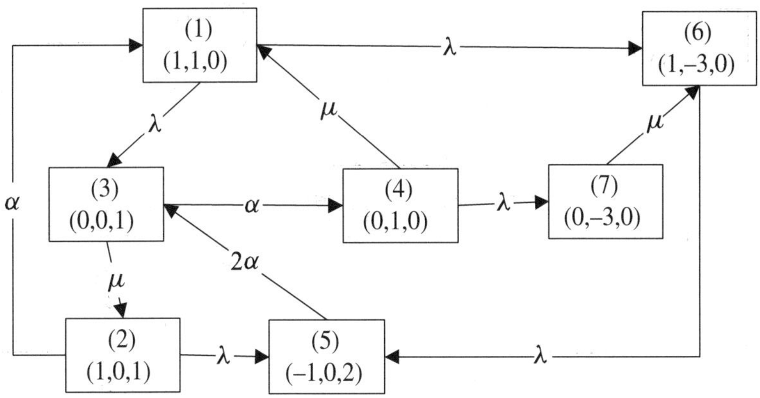

Figure 2.2.5 The Markov diagram with undetected spare unit failures

fail in storage. How will the situation change if the spare unit can fail? When the system unit fails it is immediately detected, since the system stops operating. The spare unit is undetected unless a specific inspection is applied to check its state. Assuming that no such inspection is applied a failed spare will be in state (−3), and its failure will be detected *only* when needed for replacement. Assuming that the spare unit has the same failure rate, λ, as the operational unit, the resulting Markov diagram is shown in Figure 2.2.5.

The transition matrix and the equations for this problem are

$$\begin{pmatrix} \dot{P}_1(t) \\ \dot{P}_2(t) \\ \dot{P}_3(t) \\ \dot{P}_4(t) \\ \dot{P}_5(t) \\ \dot{P}_6(t) \\ \dot{P}_7(t) \end{pmatrix} = \begin{pmatrix} -2\lambda & \alpha & 0 & \mu & 0 & 0 & 0 \\ 0 & -(\alpha+\lambda) & \mu & 0 & 0 & 0 & 0 \\ \lambda & 0 & -(\alpha+\mu) & 0 & 2\alpha & 0 & 0 \\ 0 & 0 & \alpha & -(\mu+\lambda) & 0 & 0 & 0 \\ 0 & \lambda & 0 & 0 & -2\alpha & \lambda & 0 \\ \lambda & 0 & 0 & 0 & 0 & -\lambda & \mu \\ 0 & 0 & 0 & \lambda & 0 & 0 & -\mu \end{pmatrix} \begin{pmatrix} P_1(t) \\ P_2(t) \\ P_3(t) \\ P_4(t) \\ P_5(t) \\ P_6(t) \\ P_7(t) \end{pmatrix} \quad (2.2.35)$$

The resulting steady state linear equations are

$$P_1 = \frac{\alpha}{2\lambda} P_2 + \frac{\mu}{2\lambda} P_4; \quad P_2 = \frac{\mu}{(\alpha+\lambda)} P_3; \quad P_3 = \frac{\lambda}{(\alpha+\mu)} \lambda P_1 + \frac{2\alpha}{(\alpha+\mu)} P_5; \quad P_4 = \frac{\alpha}{(\mu+\lambda)} P_3;$$

$$P_5 = \frac{\lambda}{2\alpha} P_2 + \frac{\lambda}{2\alpha} P_6; \quad P_6 = P_1 + \frac{\mu}{\lambda} P_7; \quad P_7 = \frac{\lambda}{\mu} P_4.$$

The solutions of the above equations yields

$$P_{1\infty} = 0.3764, \quad P_{2\infty} = 0.17465, \quad P_{3\infty} = 6.0638 \times 10^{-3}, \quad P_{4\infty} = 1.6865 \times 10^{-4},$$
$$P_{5\infty} = 6.6146 \times 10^{-2}, \quad P_{6\infty} = 0.37657, \quad P_{7\infty} = 1.1334 \times 10^{-6}.$$

The unavailability is the sum of probabilities of states 3, 4, 5 and 7. Note that the probability of state 5 (failed and waiting for a spare) has increased from about 2% to about 6.6%. The time dependent availability is calculated with the procedure described in Example 2.2.3, and the result for the unavailability is shown in Figure 2.2.6.

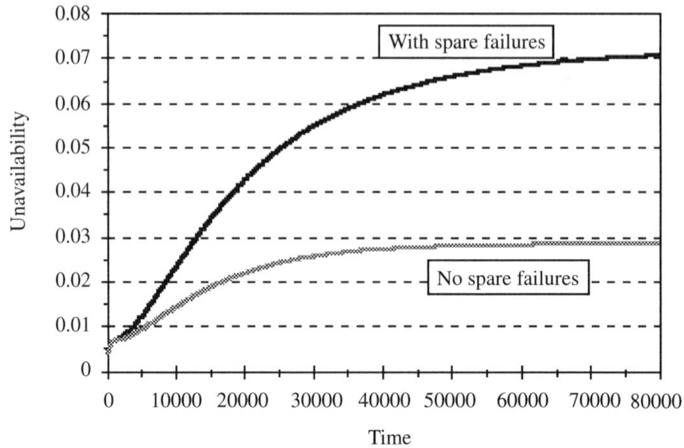

Figure 2.2.6 Unavailability with and without spares failure

The total cost is now

$$c_1 + c_2 T_m U_\infty = 190\,000 + 627 \times 5 \times 8760 \times 0.072379 = \$2\,177\,731.$$

The cost has thus increased by about 1.2 million. This creates a motivation to consider the possibility of inspection.

Example 2.2.5: Recycled spare with failures and inspection

The undetected failures of the spare unit are seen to considerably decrease the efficiency of the very idea of having a spare part. The problem can be eased by inspecting the spare part periodically so as to minimize the probability that it is found failed when needed. Inspection is introduced at equal time intervals, say every 5000 hours. For simplicity, we assume that inspection is immediate and has no effect on the availability of the system itself. The only effect of inspection is the immediate detection of a failed spare unit. This model obviously disrupts the continuous flow of the Markov equations, since at every inspection state number $(6), (1, -3, 0)$, is changed abruptly into state number 2, $(1, 0, 1)$. Also, state number 7, $(0, -3, 0)$, is abruptly changed into state number 3, (0,0,1).

This model can be addressed as follows. First, we solve the time dependent equations in the interval $[0, T_{ip}]$, where T_{ip} is the inspection interval. Looking at the next interval $[T_{ip}, 2T_{ip}]$, one may note that the state probabilities at T_{ip} are the initial conditions for the second interval. With one exception, the probability of state 6 must be added to that of state 2, and that of state 7 to that of state 3. Denoting by $P_i^*(T_{ip})$ the probability of state i obtained in the first interval, the solution in the interval $[T_{ip}, 2T_{ip}]$ can be regarded as a solution of the equations starting from zero with the initial conditions:

$$P_1(0) = P_1^*(T_{ip});\quad P_2(0) = P_2^*(T_{ip}) + P_6^*(T_{ip});\quad P_3(0) = P_3^*(T_{ip}) + P_7^*(T_{ip});\quad P_4(0) = P_4^*(T_{ip});$$
$$P_5(0) = P_5^*(T_{ip});\quad P_6(0) = 0 \quad\text{and}\quad P_7(0) = 0.$$

Once the solution for the range $[T_{ip}, 2T_{ip}]$ is obtained, the same procedure can be used to obtain the solution in the range $[nT_{ip}, (n+1)T_{ip}]$ with $n = 2, 3, \ldots,$ where the solutions at the end of each

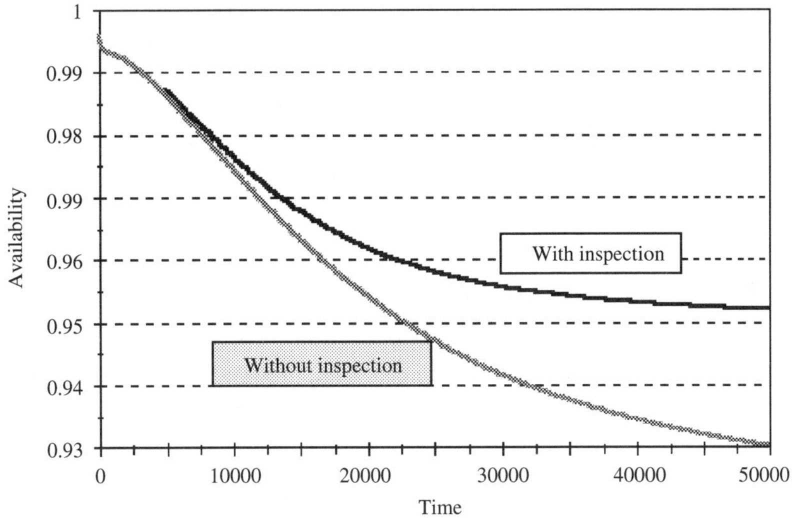

Figure 2.2.7 Availability of the system with and without the inspection of spares failures

interval (after adding state 6 to 2 and 7 to 3) serve as initial conditions for the next interval. The solution within each interval can be obtained using the procedure of equation (2.2.34).

This procedure is enabled by the Markov nature of the model. Once the initial conditions (the present) are given for any time interval, the equations apply in that interval. The results obtained for the time dependent availability (sum of states 1, 2 and 6) for periodic inspection of 5000 hours and those without such inspection are shown in Figure 2.2.7.

In this case, the steady state availabilities are not used. Instead, numerical integration of the availabilities in Figure 2.2.7 is used to obtain the average availability for both the inspected and uninspected cases. The total down time is obtained as (see equation (2.2.17))

$$\langle T_d \rangle = [1 - \langle A(t) \rangle] \times T_m = T_m \times \left[1 - \frac{1}{T_m} \int_0^{T_m} A(t)\, dt \right].$$

This results in 0.964242 and 0.95371, respectively, for the average availabilities. The average unavailability for the uninspected case is 0.0462, considerably smaller than the steady state result (0.072379) of the preceding example. Steady state results do under-estimate the availability. Assuming that the cost of inspection is $c_{ip} = \$2000$, the total cost in the case of inspection will be $9 \times c_{ip} + c_1 + c_2 T_m \langle U \rangle = \$1.19\,\mathrm{M}$ for the case with inspection, and \$1.461 M without inspection. A saving of about \$250 000 can be realized over a period of five years by applying inspection to the system.

The Markov approach proves to be a useful mechanism for modeling. Yet it suffers a number of deficiencies. It applies only to Markov systems, i.e. when all the distributions are exponential. This eliminates the possibility of treating aging and any other model related to non-exponential distributions. Furthermore, even in those cases where one may assume that the distributions are exponential, there is a limit to the number of states for which the set of equations (even the linear steady state equations) can be solved. This is an additional reflection of the dimensional problem discussed in Section 2.2.1. It should be noted that if a system has n components, each having k states, the number of different states (different state vectors) is k^n. A system of 11 components with three states each (operational, failed-undetected and failed-detected) will have 177 147 states. The number of states grows exponentially, making even the simple Markov approach impractical. The Monte Carlo method is insensitive to the explosion of the number of states.

2.2.4 The concept of age

Consider the common situation in which, upon inspection of a failed unit, a technician says "This unit is so old that its not worth repairing, it should be replaced". What is the mathematical interpretation of this statement, and how can it be introduced into the methods discussed so far?

It has already been mentioned that aging has to do with the fact that the future probability of an event increases as a function of elapsed time. Recall that the transport equation (2.2.23) and its Neumann expansion equations (2.2.2) – (2.2.4) assume that repair results a unit which is 'As good as new'. This is also referred to as 'complete renewal' or 'complete repair'. The next example is used to introduce the concept of aging and partial repair.

Example 2.2.6: Logistic options evaluation of a centrifuge – A simple case study

A centrifuge operates as a critical part of a system. The MTTF of the centrifuge is known to be of the order of $\langle T_f \rangle = 1640\,\mathrm{h}$. It is desired to compare two logistic options: one is to *replace* the centrifuge (by a new one) whenever it fails; the other is to *repair* the centrifuge when it fails, and to apply forced replacement after 2500 hours, i.e. after 2500 hours the unit is replaced independently of its state. The cost of repair is \$7000, and the cost of replacement is \$53 000. The value of one hour's production is \$3200. The analysis is required for a period of 2500 hours, and is intended to point to the optimal policy.

What is the difference between repair and replacement? A difference between repair and replacement can arise only if:

(a) The unit is 'Aging'.
(b) Repair, although bringing the unit from state 0 into state 1, does not fully rejuvenate it, but leaves it as old as it was before the failure. Upon such repair the unit will have a higher failure probability in any future time interval than that of a new unit. This will cause the unit to become increasingly 'older' until, eventually, the frequency of failures will become so high that it would be worth replacing it.

Obviously, the state space description based on two states (0 and 1) is not sufficient to describe the above process. An age-related variable must be added.

The 'probabilistic age' or 'age', $\mathrm{Age}_i(t)$, is defined for each state i as:

$$\mathrm{Age}_i(t) = F_i(t - t^b) \quad \text{for state } i, \tag{2.2.36}$$

where $F_i(t)$ is the CDF in state i, t^b is the birth point (the time point from which the distribution time is measured) of the state and t is the current time. It should be noted that 'age' is an attribute of a state. Each state has an age which is accumulated while the components resides in the state.

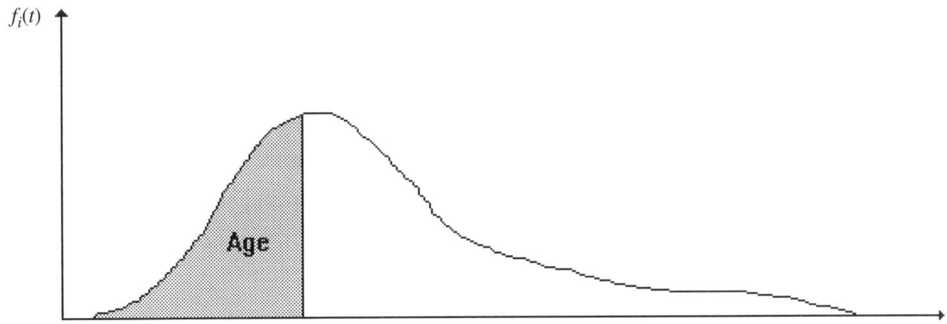

Figure 2.2.8 Age interpretation at time of entry into state i

In this respect, all states are equivalent. One may speak about the age in state 1, which is corresponding to the intuitive notion of age as being accumulated towards failure. However, an age is also accumulated in the failed state and in any other state. The age is the probability accumulated towards exiting from the state. This definition of age has the distinct advantage of naturally incorporating the unique operating conditions of each unit, while yielding information about how 'old' the component is. Consider two identical components starting to operate at the same time, but operating under different scenarios which include different CDFs. Defining age by elapsed time would be ambiguous, while the accumulated CDFs over the life history will yield useful and unambiguous information. If the age of a unit is, for example, 0.04, 0.45 or 0.99, it means that it is very young, in mid 'life', or very old, respectively. In any case, if the CDF is monotonic increasing, the elapsed time can always be calculated as the inverse of the CDF, F^{-1} (age).

Let the time of entry into state i be t^i. The PDF in this state is $f_i(t)$, but where is the elapsed time measured from? If the elapsed time is measured from the entry time t^i, the distribution takes the form $f_i(t - t^i)$, i.e. the origin of the distribution is at t^i. The origin of the distribution does not have to be at t^i. It could be at another time point $t^b \leq t^i$ such that at t^i a certain fraction of the 'life' in the state is already consumed (see Figure 2.2.8). t^b is called the birth point of the distribution or the state, and is the point from which the origin of the distribution is measured.

Indeed, given a birth point t^b and an entry point into the state, t^i, the PDF of the state becomes

$$\frac{f(t - t^b)}{1 - F(t^i - t^b)}$$

(see equations (1.2.14) and (1.3.10)). This is the conditional PDF given that the birth point of the state is at t^b, and there was no event up to t^i. Noting the definition of the age, this can be written as

$$\frac{f_i(t - t^b)}{1 - \mathrm{Age}_i(t^i - t^b)}.$$

If $t^b = t^i$, the age is zero, otherwise the component enters the state with a non-zero age with an impact on its PDF in the state, as described above. The age and the birth point are closely related. The birth point must become an inherent parameter of the transport process.

The differences between repair and replacement may now be considered in view of the above discussion.

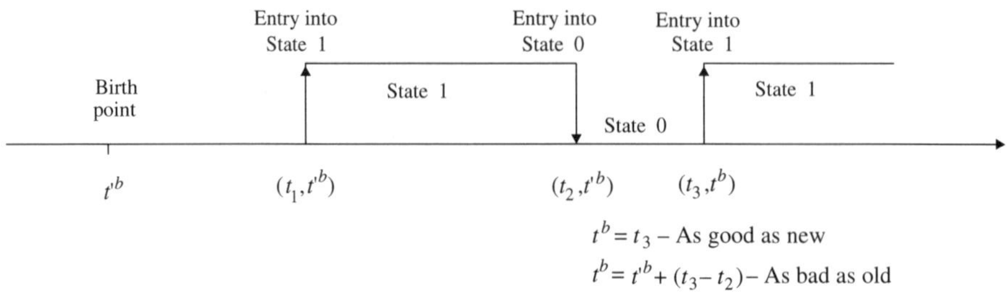

Figure 2.2.9 Repair and birth time

Let $f(t)$ and $g(t)$ denote the failure and repair (or replacement) distributions of the centrifuge, respectively. If the unit enters state 1 at point t_1, the birth point of this state is at some point $t'^b \leq t_1$. In this process, only state 1 has a birth point. The event density of entering into state 1 will thus have to include the birth point. It must then be defined as $\psi_1(t, t^b)$, the number density of entries into state 1 at time t, per unit time, with birth time t^b per unit time.

The next event of the component is a failure at some point t_2 followed by a restoration (either by repair or by replacement) at point t_3. Thus, at t_3 the component goes from state zero back into state 1. What is the birth-point of this state?

Setting $t^b = t_3$ means that the unit is as good as new, setting $t^b = t'^b + (t_3 - t_2)$ means that the state retains the birth point it had in the last entry into that state. This means that the unit is 'as bad as old'. Note that the birth point must be shifted by $(t_3 - t_2)$, since during this time interval the unit was in repair and did not age. This situation is illustrated in Figure 2.2.9.

These are two extremes of the restoration process. Obviously, the birth point can take any value between t'^b and t_3 corresponding to various levels of rejuvenation applied in the restoration process.

The failure PDF can be written in general as

$$f(t, t_c|t^b) = \frac{f(t - t^b)}{1 - F(t_c - t^b)}; \quad t \geq t_c \geq t^b. \tag{2.2.37}$$

With the above definition, the availability $A(t)$ of equation (2.2.37) under minimal repair conditions ('as bad as old') will take the form:

$$A(t) = \sum_{k=0}^{\infty} A(t|k) \quad \text{and}$$

$$A(t|k) = \int_0^t dt_1 \int_{t_1}^t dt_2 \int_{t_2}^t dt_3 \cdots \int_{t_{2k-1}}^t dt_{2k} f(t_1|t^{b1}) g(t_2 - t_1) f(t_3 - t^{b3}|t^{b3}) g(t_4 - t_3)$$
$$\times \cdots f(t_{2k-1} - t^{b(2k-1)}|t^{b(2k-1)}) g(t_{2k} - t_{2k-1}) \times \left[1 - \frac{F(t - t^{b(2k+1)}) - F(t_{2k} - t^{b(2k+1)})}{1 - F(t_{2k} - t^{b(2k+1)})}\right], \tag{2.2.38}$$

where

$$f(t_{2j-1} - t^{b(2j-1)}|t^{b(2j-1)}) = \frac{f(t_{2j-1} - t^{b(2j-1)})}{1 - F(t_{2j-2} - t^{b(2j-1)})}$$

(with the convention $t_0 \equiv 0$), and

$$t^{b1} = 0 \quad \text{and} \quad t^{b(2j+1)} = \sum_{i=1}^{j} (t_{2i} - t_{2i-1}); \quad j = 1, \ldots, k,$$

which is simply the sum of all repair times up to t_{2j+1}. If the repair times are negligible (relative to the time between failures), one may approximate $t^{b(2j+1)} = 0$ in the above expressions.

As was discussed in Section 2.2.2, equation (2.2.38) is a Neumann series expansion of the solution of the transport equation. What is the form of the transport equation with aging and partial repair?

The event density for the 'up' state was already defined as $\psi_1(t, t^b)$. Assuming that only the up state carries a birth time, the event density for state zero is defined as $\psi_0(t, t^b)$, the number density of entries into state 0 at t coming from state 1 with birth time t^b. The total event density is given by

$$\psi_i(t) = \int_0^t \psi_i(t, t^b)\, dt^b. \tag{2.2.39}$$

As the unit changes states from 0 to 1, it may, in principle, enter state 1 with any age or birth point. To incorporate this into the transport equation, a birth point transfer function is defined in the form $U(t', t, t'^b, t^b)$, the probability density that entering state 1 at time t, from state 0 entered at time t' carrying a birth point t'^b, will have a birth point t^b.

The dimension of $U(t', t, t'^b, t^b)$ is 1/time while that of $\psi_i(t, t^b)$ is 1/time^2. With the above definitions, and assuming that at $t = 0$, the unit starts in state 1 with birth time zero the two state transport equation takes the form

$$\psi_1(t, t^b) = \delta(t)\delta(t^b) + \int_0^t \int_0^{t'} dt'^b \psi_0(t', t'^b) g(t - t') U(t', t, t'^b, t^b)\, dt', \tag{2.2.40a}$$

$$\psi_0(t, t^b) = \int_0^t \psi_1(t', t^b) f(t, t'|t^b)\, dt' = \int_0^t \psi_1(t', t^b) \frac{f(t - t^b)}{1 - F(t' - t^b)}\, dt'. \tag{2.2.40b}$$

Substituting $U(t', t, t'^b, t^b) = \delta(t^b - t)$ means that the birth point equals the state entry point, and thus repair is 'as good as new'. It is straightforward to show that substitution of this form into equation (2.2.40) yields equation (2.2.24). Minimal repair ('as bas as old') is obtained when $t^b = t'^b + (t - t')$ (see Figure 2.2.9), thus $U(t', t, t'^b, t^b) = \delta[t^b - (t'^b + t - t')]$.

Substituting this form into the equation, and using the properties of the delta function in integrating over t'^b yields

$$\psi_1(t, t^b) = \delta(t)\delta(t^b) + \int_0^t \psi_0[t', t^b - (t - t')] g(t - t')\, dt'. \tag{2.2.40c}$$

Assuming that the repair time is negligible relative to the time between failures, and integrating over t^b yields equation (2.2.24a).

To precisely evaluate the options described in the current example, equations (2.2.40) must be solved for both cases.

If one assumes that the centrifuge has an exponential failure distribution, the whole case becomes meaningless. There can be no difference between repair and replacement, and the concepts of age and birth point are irrelevant. This can be easily seen by substituting exponential distributions for $f(t)$ and $g(t)$ in equation (2.2.40), and noting that the resulting equations are reduced to equation (2.2.24), and to the Markov equation (2.2.30). If repair differs from replacement, that must mean, that *the failure distribution cannot be exponential.* The very concept of an exponential failure distribution eliminates the concept of aging and thereby the concepts of 'as good as new' and 'as bad as old' altogether. Assuming that the failure distribution is a Weibull distribution, it is noted that the distribution involves two parameters, the scale parameter λ and the shape parameter β. The estimation of these parameters from field data will be discussed in Chapter 4. Here a simple, first order method is described which emphasizes the concept of age.

Using equation (1.3.16), the MTTF provides one equation:

$$\langle T_f \rangle = 1640 = \frac{\Gamma(1 + (1/\beta))}{\lambda^{1/\beta}}. \tag{2.2.41a}$$

A second equation may be obtained by the following consideration. Let T_ℓ denote the maximum possible life time of the component. This is a point beyond which a failure is highly probable. For example, assuming that a car has a Weibull distribution with $\lambda = 1.4 \times 10^{-21}$ and $\beta = 4.3$ (with regard to major failures) and repair is as bad as old, the probability for a failure within a period ΔT at time t is given by $P(t, \Delta T) = 1 - e^{-\{\lambda[(t+\Delta T)^\beta - t^\beta]\}}$. For a one-year old car, the failure probability within a month is 0.0000523. Thus, one car out of about 20 000 will suffer a major failure. After 12 years, the age of the car reaches 0.9958, and the probability for a failure within the next month is 15.46%, the probability that the car will survive the whole year without failing is 10.5%. Only one out of 10 cars will survive the twelfth year without a major failure.

One can often obtain this information from field operators, who have considerable experience with such systems and can make reasonable estimates. Also, for many critical components, an expiration date is set by the manufacturer. This expiration date represents the preception of the manufacturer about T_ℓ. It is a time point beyond which the risk of a critical failure is too high. In the present study, the estimated T_ℓ was 3000 hours of operation. The interpretation of T_ℓ is that the component has consumed a large fraction of its probability pie, i.e. that $F(T_\ell)$ is very close to 1. For the purpose of this estimate, it is assumed that $F(T_\ell) = 0.95$, and substituting this assumption into equation (1.3.14) for the CDF of the Weibull distribution yields a second equation:

$$F(T_\ell) = 1 - e^{-\lambda T_\ell^\beta} = \text{Age}(T_\ell) = 0.95. \tag{2.2.41b}$$

From equation (2.2.41b), one obtains

$$T_\ell = \left[\frac{\text{Ln}(1 - \text{Age}_\ell)}{\lambda}\right]^{1/\beta}. \tag{2.2.41c}$$

Dividing T_f by T_ℓ yields

$$\frac{\langle T_f \rangle}{T_\ell} = \frac{\Gamma(1 + (1/\beta))}{[\text{Ln}(1 - \text{Age}_\ell)]^{1/\beta}}. \tag{2.2.41d}$$

Solving equation (2.2.41d) for β yields $\beta = 1.818$, and substitution in (2.2.41a) yields $\lambda = 1.42624\text{E}{-06}$.

For the case of replacement, the equations to be solved are (2.2.24a, b). The replacement (and repair) time follows a normal distribution $N(36, 12)$; substituting the distributions in (2.2.24) yields

$$\psi_1(t) = \delta(t) + \frac{1}{12\sqrt{2\pi}} \int_0^t \psi_0(t')\, e^{-\frac{(t-t'-36)^2}{2\times 144}}\, dt', \tag{2.2.42a}$$

$$\psi_0(t) = \int_0^t \psi_1(t')\lambda\beta(t-t')^{\beta-1}\, e^{-\lambda(t-t')^\beta}\, dt'. \tag{2.2.42b}$$

For the case of minimal repair, with birth time at $t^b = 0$ and neglecting repair times, the first equation remains unchanged, and in the second equation $f(t-t')$ is replaced by $(f(t))/(1-F(t'))$, yielding

$$\psi_0(t) = \lambda\beta \int_0^t \psi_1(t')(t)^{\beta-1}\, e^{-\lambda[t^\beta - t'^\beta]}\, dt'. \tag{2.2.42c}$$

Both sets of equations are difficult to solve, and no simple known explicit solution exists. However, they are simple to solve using the Monte Carlo method (without the need to neglect the sum of repair times in equation (2.2.42c)). The availability of both cases as a function of time, as obtained by a Monte Carlo solution, are shown in Figure 2.2.10.

The average number of failures can be obtained as $\langle n_f \rangle = \int_0^{2500} \psi_0(t)\, dt$. For the replacement option it is 1.35, and the average down time is 48.29 hours. The total operating cost, for a period of 2500 hours, is \$226 078.

For the 'minimal repair' option, the average number of failures is 2.04 and the down time is 72.4 hours. Adding the cost of replacement applied at the end of a period yields a total of

Figure 2.2.10 Availability with replacement and 'as bad as old' repair. Note the statistical fluctuations, typical for a Monte Carlo calculation

$298 960. The replacement option is, in this case, preferred. It is both cheaper and will provide better performance.

In the above analysis, the Monte Carlo method was used to obtain exact solutions and estimates. For the minimal repair case the following general approximation may be applied. Assuming an immediate repair, the repair distribution in equation (2.2.42a) is replaced by a delta function of the form $\delta(t-t')$, resulting in

$$\psi_1(t) = \delta(t) + \int_0^t \psi_0(t')\delta(t-t')\,dt' = \delta(t) + \psi_0(t),$$

i.e. apart from the delta function corresponding to the initial state the event densities are identical. This can be expected, since each entry into state zero is immediately followed by an entry into state 1. Substituting this result into equation (2.2.42c), for *any failure* PDF yields:

$$\psi_0(t) = \int_0^t \psi_1(t')\frac{f(t)}{1-F(t')}\,dt' = \int_0^t [\delta(t') + \psi_0(t')]\frac{f(t)}{1-F(t')}\,dt' = f(t) + \int_0^t [\psi_0(t')]\frac{f(t)}{1-F(t')}\,dt'.$$

One may guess, now, the solution, $\psi_0(t) = z(t)$, i.e. the hazard of the failure PDF. Substituting this above yields

$$\psi_0(t) = f(t) + f(t)\int_0^t \frac{z(t')}{(1-F(t'))}\,dt.$$

Using the exponential form of the cumulative distribution, it is straightforward to see that the primitive function of the integrand is $1/(1-F(t'))$, since

$$\frac{d[1/(1-F(t'))]}{dt'} = \frac{d\,\mathrm{e}^{+\int_0^{t'} z(x)\,dx}}{dt''} = z(t')\,\mathrm{e}^{+\int_0^{t'} z(x)\,dx} = \frac{z(t')}{1-F(t')},$$

hence the equation takes the form

$$\psi_0(t) = f(t) + f(t)\left[\frac{1}{(1-F(t'))}\right]_0^t = f(t) + \frac{f(t)}{1-F(t)} - f(t) = z(t)$$

which, in the light of the uniqueness of the solution, proves the assumption. This may be summarized in the general useful conclusion: in the case of immediate and minimal repair, the event rates, for a two state component, fulfil the relation

$$\psi_0(t) = \psi_1(t) = z(t), \quad t > 0. \tag{2.2.43}$$

This important result may be used for the present example, the minimal repair, case. The event density will take the form $\psi_1(t) = \lambda\beta t^{\beta-1}$. The expected number of failures up to any time T is then

$$\langle n_f \rangle = \int_0^T \psi_1(t)\, dt = \lambda T^{\beta}. \tag{2.2.44}$$

Substituting the problem parameter yields 2.14 failures (compared with the exact value of 2.04). The down time may now be approximated by multiplying the number of failures by the MTTR, yielding 77.25 hours. The total cost obtained is \$315 180 (compared to the cost of \$298 960 obtained above). For the full repair case a simpler approximation can be applied. Using the steady state availability equation (2.2.30) and the MTTF and MTTR of the problem, an availability of 0.9785 is obtained, resulting in 53.69 hours of downtime. Using equation (2.2.15), the average number of failures is evaluated as

$$\frac{T_m}{MTTF} = \frac{2500}{1640} = 1.524.$$

The estimated cost of this policy would be \$252 580. (Compare with the exact result of \$226 078). Both approximations tend to increase the results by about 10 %. Also, based on the approximated results, correct logistics decisions can be made. Yet these statements, in favor of the approximations, can be made only when the correct results are known. Otherwise, the approximations are uncontrolled, and result from our ignorance and not from a knowledge that they are justified.

3 Basic Concepts in Monte Carlo Methods

The basic concepts of system engineering discussed in the previous chapter indicated that a system is basically a state vector changing its form in time. The components of the system with their states define the state space of the system. The transitions of the system from one state vector to another in time is referred to as 'transport'. Adding the rules by which the system moves from one phase space state to another along with the system function/s completes the model of the system's reality.

A model is 'correct' if the results of the predictions made on it are close to the actual reality measured on the system. The predictions are averages of the system functions. These quantities can always be presented as some form of averaging of a random variable over the space of random system's histories in phase space. It was shown that this averaging process may become extremely elaborate, even for simple cases. The difficulty stems from the multidimensional integration in the time domain done on event points, and from the exponential increase of the number of points in the state space. Age adds an additional complexity as it adds a continuous dimension. The Monte Carlo method is specifically oriented towards the evaluation of multidimensional averages on complex spaces. It therefore bears the potential of providing a feasible tool by which complex, realistic models may be constructed and analyzed.

3.1 The Sample Average

Let $x_1, x_2, \ldots, x_n$ be a set of n identical independent r.v.s, and let $h(x_i),\ i = 1, \ldots, n$, be a function of x. The *sample average* of $h(x)$ is defined as

$$\tilde{h}_n = \frac{1}{n}\sum_{i=1}^{n} h(x_i). \tag{3.1.1}$$

$h(x_i)$ and $\tilde{h}_n$ are also random variables. It will now be shown that, as n increases to infinity, $\tilde{h}_n$ converges to the mean value $\langle h \rangle$.

The mean value of $\tilde{h}_n$ can be expressed as

$$E(\tilde{h}_n) = E\left(\frac{1}{n}\sum_{i=1}^{n} h(x_i)\right) = \frac{1}{n}\sum_{i=1}^{n} E(h(x_i)) = \frac{1}{n}\sum_{i=1}^{n} \langle h\rangle = \langle h\rangle. \qquad (3.1.2)$$

Thus, the mean of the sample average equals the mean of $h(x)$.

The variance of $\tilde{h}_n$ can be expressed, using equation (1.2.23a), as

$$\begin{aligned} V(\tilde{h}_n) &= S_{\tilde{h}}^2 - E^2(\tilde{h}_n) = E(\tilde{h}_n^2) - \langle h\rangle^2 = E\left[\left(\frac{1}{n}\sum_{i=1}^{n} h(x_i)\right)^2\right] - \langle h\rangle^2 \\ &= \frac{1}{n^2}E\left(\sum_{i=1}^{n} h^2(x_i) + 2\sum_{i=1}^{n}\sum_{j=i+1}^{n} h(x_i)h(x_j)\right) - \langle h\rangle^2 \\ &= \frac{1}{n^2}\left(\sum_{i=1}^{n} E[h^2(x_i)] + 2\sum_{i=1}^{n}\sum_{j=i+1}^{n} E[h(x_i)]E[h(x_j)]\right) - \langle h\rangle^2 \\ &= \frac{1}{n^2}\left(nS_h^2 + n(n-1)\langle h\rangle^2 - n^2\langle h\rangle^2\right). \end{aligned}$$

In the last step, use is made of the property that the mean of a product of the independent variable is the product of the means and, since the r.v.s are identical, $E[h(x_i)] = E[h(x_j)] = \langle h\rangle$. Also, S_h^2 is the second moment of h. Thus:

$$V(\tilde{h}_n) = \frac{1}{n}\left(S_h^2 - \langle h\rangle^2\right) = \frac{V(h)}{n}. \qquad (3.1.3)$$

The variance of $\tilde{h}_n$ is, by a factor n, smaller than that of $h(x)$. Obviously, as n increases, the variance of $\tilde{h}_n$ approaches zero. It was already mentioned that the variance is a measure of the 'width' or 'dispersion' of the distribution, in the sense that it is a measure of the 'spread' of the values of the random variable around the mean. This notion is formalized by the Chebyshev inequality, which states that for any random variable h with mean $\langle h\rangle$ and finite variance $V(h)$ and for any positive real number ρ:

$$P_r[|h - \langle h\rangle| \geq \rho] \leq \frac{V(h)}{\rho^2}. \qquad (3.1.4)$$

Consider the interval $[\langle h\rangle - \rho, \langle h\rangle + \rho]$ around the mean. Such an interval is called a ρ interval of $\langle h\rangle$. The inequality $|h - \langle h\rangle| \geq \rho$ implies that h takes a value outside this interval. The Chebyshev inequality states that the probability for this to happen is smaller, then $(V(h))/\rho^2$. Obviously, the smaller the variance the smaller will be the probability for the r.v. to obtain values outside any interval with fixed ρ.

Definition 3.1.1: Convergence in probability We will say that a random variable y_n converges in probability to Y, ($y_n \xrightarrow{P} Y$), if for every $\varepsilon > 0$, the probability that y_n is outside an ε interval of Y converges to zero as $n \to \infty$, namely:

$$\lim_{n\to\infty} P_r[|y_n - Y| > \varepsilon] = 0.$$

Such convergence is also called *stochastic convergence.*

Theorem 3.1.1: Convergence in probability of the sample average *Let $h(x)$ be a random variable (or a function of a random variable) with mean $\langle h \rangle$ and variance $V(h)$. Let $x_1, \ldots, x_n$ be a sample of identical independent random variables, and let*

$$\tilde{h}_n = \frac{1}{n}\sum_{i=1}^{n} h(x_i)$$

be the sample average, then $\tilde{h}_n$ converges in probability to $\langle h \rangle$.

Proof Applying the Chebyshev inequality to $\tilde{h}_n$ and using equations (3.1.2) and (3.1.3) yields

$$P_r[|\tilde{h}_n - \langle h \rangle| > \varepsilon] \le \frac{V(h)}{n\varepsilon^2}, \tag{3.1.5}$$

thus for any $\varepsilon > 0$ as n approaches infinity, the probability converges to zero and the theorem follows.

Theorem 3.1.1 is also called the Weak Low of Large Numbers (WLLN). It is easy to see that the theorem holds also under weaker condition, e.g. the variables $h(x_i)$ are not identical, but have identical first and second moments.

Consider an experiment that has an unknown success probability p. How would one estimate this probability? One may conduct n independent identical experiments and define the random variable:

$$x_i = \begin{cases} 1 - \text{Success} \\ 0 - \text{Failure.} \end{cases}$$

The average of x_i is $\mu_{x_i} = 1 \times p + 0 \times (1 - p) = p$. The variable $Z_n = (1/n)(x_1 + x_2 + \cdots + x_n)$ (which is identical to $\tilde{h}_n$ for $h(x) = x$) seems to be a reasonable estimate of the success probability being the ratio of the number of successes to the number of experiments. It is this quantity that in the initial discussion on the probability function was said to approach p as $n \to \infty$ (equation (1.1.11)). However, this is not and cannot serve as the definition of the probability. It is a proven result of the WLLN. Z_n will converge in probability to the mean value, p because Theorem 3.1.1 guarantees that as n increases, Z_n will be confined to smaller and smaller intervals around p.

In a dice experiment, the frequency of appearances of any number approaches 1/6 because the relative frequency is Z_n and $p = 1/6$. Equation (1.1.11) is not a definition of a probability. It is a direct result of the weak low of large numbers.

The estimation of mean values by the convergent sample average is the corner-stone of the Monte Carlo method.

It should be noted, though, that Theorem 3.1.1 does not reveal much about the error in the estimate. As n increases the sample average gets closer and closer to the mean, but not much is known about the error for any given finite n. This is a critical point, because a numerical estimation will always involve a finite sample size, and not having any information about the size of the error renders the estimate meaningless. The following example shows what can be said about the error at this stage.

Example 3.1.1: Estimating failure probability

Consider a component that has an unknown probability p to be found failed after a given time of operation. It is given that 25 components have been tested and three were found failed. What can one say about p?

From the above discussion it is obvious that in this experiment; $z_{25} = 3/25 = 0.12$. Z_n will converge to the correct probability if the number of components being tested increases indefinitely, but what is the error in z_{25}? Is it 10%, 50%, 5000%, or what?

How far is 0.12 from the correct number?

From Theorem 3.1.1 we know that

$$P_r[|Z_{25} - p| > \varepsilon] \leq \frac{V(x)}{25\varepsilon^2}$$

where $V(x)$ is the variance of the random variable attached to each component. This variance is $V(x) = S_x^2 - \langle x \rangle^2 = [p \times 1^2 + (1-p) \times 0^2] - p^2 = p(1-p)$. Thus,

$$P_r[|Z_{25} - p| > \varepsilon] \leq \frac{p(1-p)}{25\varepsilon^2}.$$

Since p is not known, it is substituted by Z_{25}, yielding

$$P_r[|Z_{25} - p| > \varepsilon] \leq \frac{Z_{25}x(1 - Z_{25})}{25\varepsilon^2} \quad \text{or} \quad P_r[|0.12 - p| > \varepsilon] \leq \frac{0.1056}{25\varepsilon^2}.$$

The right-hand side of the inequality gives an upper bound for the probability that Z_{25} will fall outside an ε interval of p. For a probability of less than 0.2, for example, ε must be 0.1735 or higher. The complementary of the above inequality can be written as

$$P_r[|0.12 - p| \leq \varepsilon] > 1 - \frac{0.1056}{25\varepsilon^2} \quad \text{or} \quad P_r[|0.12 - p| \leq 0.1735] > 0.8.$$

The probability that the estimate lies in an interval of width 0.1735 on both sides of p exceeds 80%. By symmetry, it also means that the correct unknown result has a probability exceeding 80% of being within a 0.1735 interval of 0.12. This state of affairs is shown in Figure 3.1.1. The interval defined around the estimate is called a *confidence interval*.

The probability that the correct value is within this interval is called the *confidence level*. The correct result lies in a confidence interval of 0.1735 (or 144% relative to the estimated mean) with a confidence level of at least 80%. The probable error is 144%, and it is thus possible that the correct result is more than twice the estimate. It will be shown later that estimation of the error can be done with a higher confidence.

In deterministic methods, a single number is sufficient to define the error, because the confidence is always 100%. In statistical estimation there are always two values involved. One for

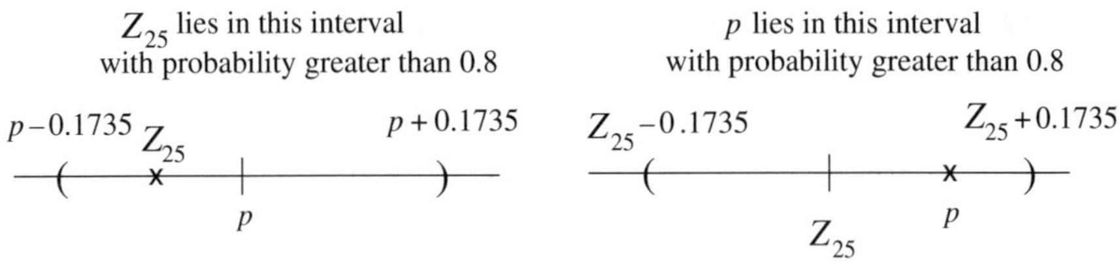

Figure 3.1.1 The confidence interval according to the Chebyshev inequality

the error (the size of the confidence interval), and the other representing the confidence that the correct result is in the interval. The two numbers together comprise the *statistical error*. This uncertainty regarding the estimate accompanies the application of the Monte Carlo method. It is a 'payment' for moving from deterministic methods to statistical methods, and thereby gaining the ability to perform complex multidimensional calculations.

This section can be summarized as follows: let x be a random variable defined on an event space $\omega \in \Omega$ with a *known* PDF, $f(x)$, and mean and variance μ_x and V_x, respectively. Let $h(x)$ be any function of the random variable. The estimation of the average

$$\langle h \rangle = \int_{-\infty}^{+\infty} h(x) f(x)\, dx \tag{3.1.6}$$

can be done by calculating the sample average:

$$\tilde{h}_n = \frac{1}{n}\sum_{i=1}^{n} h(x_i), \tag{3.1.7}$$

where x_i are independent realizations obtained from the distribution $f(x)$.

It should be added that the above discussion and conclusion also hold for multi-variable functions and PDFs. For any finite n, $\tilde{h}_n$ is a consistent estimate of $\langle h \rangle$, in the sense that it will converge to it in probability as $n \to \infty$. The statistical error obtained for any finite calculation is a matter to be further discussed.

How does one obtain realizations of the variable x_i?

One way is to conduct a physical experiment involving the variable. This amounts to field testing or experimental measurement. Yet, if the PDF $f(x)$ is known, one could obtain realization of the r.v. through a mathematical algorithm. This process is called *sampling*. Once sampling of a realization of a random variable from a given PDF is known, equation (3.1.7) can be used to obtain estimates of equation (3.1.6).

3.2 Sampling and Basic Estimation

While the estimation of average quantities by means of evaluating sample averages is the underlying idea of the Monte Carlo method, the process of sampling a realization of a random variable from a given distribution is an essential part of its practical application.

Definition 3.2.1: A proper sampling procedure A mathematical algorithm or process which, for a given PDF $f(x)$ and CDF $F(x)$ generates an outcome x, and such that for every given value x_0 the condition

$$P_r[x \leq x_0] = F(x_0) \tag{3.2.1}$$

is fulfilled, is a proper sampling procedure.

A random variable realization generated by a proper sampling procedure is indistinguishable from an r.v. generated in a process controlled by a PDF $f(x)$.

Consider the equation:

$$\int_{-\infty}^{x} f(x')\,dx' = F(x) = \xi \quad \text{solved for } x, \tag{3.2.2}$$

where ξ is a given random number generated from the uniform distribution $u(x)$ defined on the interval $[0, 1]$. Such uniform random numbers can be easily generated on every computer. Every programming language has functions and routines that generate uniformly distributed variates in $[0, 1]$.

Theorem 3.2.1: The CDF sampling method *A random variable realization, x, obtained from the solution of* (3.2.2) *results in a proper sampling procedure from the CDF $F(x)$.*

Proof Let x_0 be any real number. Let $\xi_0 = F(x_0)$. Since $F(x)$ is a monotonic non-decreasing function, the events $\{x \leq x_0\}$ and $\{\xi \leq \xi_0\}$ are identical, thus

$$P_r[x \leq x_0] = P_r[\xi \leq \xi_0] = \xi_0 = F(x_0),$$

and the conditions of a proper sampling procedure are proven.

It should be noted that the function $F(x)$, viewed as a function of a random variable, is uniformly distributed in the range [0, 1]. Denote a random variable $v = F(x)$. The PDF of v, $g(v)$, maintains the relation $g(v)|dv/dx| = f(x)$. However, $|dv/dx| = f(x)$ so that $g(v) = 1$, which is a uniform PDF.

The CDF method of equation (3.2.2) is the most common sampling method.

- **Sampling the exponential distribution:** for the exponential PDF equation (3.2.2) takes the form $1 - e^{-\lambda t} = \xi$, and the sampling formula is then

$$t = -\frac{\ln(1 - \xi)}{\lambda}. \tag{3.2.3}$$

  Thus, if ξ is a random number generated from a uniform distribution in [0, 1], then t is a realization of an exponentially distributed r.v. Since $(1 - \xi)$ is also an r.v. with a uniform distribution in $[0, 1]$, one may use the expression $t = -(\ln \xi)/\lambda$.
- **The uniform distribution:** to sample from the uniform distribution in the range $[a, b]$, note that $F(t) = (t - a)/(b - a)$. Substituting in (3.2.2) yields

$$t = (b - a) \times \xi + a. \tag{3.2.4}$$

- **The Weibull distribution:** substituting $F(t) = 1 - e^{\lambda t^{\beta}}$ in (3.2.2) yields

$$t = \left[-\frac{\ln(1-\xi)}{\lambda}\right]^{1/\beta}. \tag{3.2.5}$$

- **Sampling uniform points in a rectangle:** consider a rectangle $0 \le x \le a$ and $0 \le y \le b$ in the plane. To sample uniform points in the rectangle, sample the x coordinate uniformly in $[0, a]$ and sample y uniformly in $[0, b]$. Using equation (3.2.4), $x = a\xi$ and $y = b\xi$. A uniform point is then $(a\xi; b\xi)$. Note that, by convention, each ξ means a different random number. Thus, the two values above are two independent values of the uniform random number in $[0, 1]$.
- **Sampling uniform points in a circle:** consider a circle of radius R around the origin. It is desired to sample the circular coordinates (r, φ) for points uniformly distributed in the circular area. Symmetry implies that the PDF for the Azimuthal angle φ is uniform in the range $[0, 2\pi]$. The sampling of the Azimuthal angle is then simply $\varphi = 2\pi\xi$.

To obtain the PDF for the radial coordinate, consider an infinitesimal ring of size dr, as shown in Figure 3.2.1. The probability of a point to be in this ring is proportional to the area of the ring, namely, $2\pi r\,dr$. Thus, the PDF for r must be proportional to this area, or $f_r(r)\,dr \propto 2\pi r\,dr$. This means that $f_r = C \times r$, where C is a normalization constant to be found from the condition $\int_0^R f_r\,dr = 1$. The normalization condition yields

$$f_r(r) = \frac{2r}{R^2} \quad \text{and} \quad F_r(r) = \frac{r^2}{R^2}. \tag{3.2.6}$$

Substitution in (3.2.2) yields

$$r = R\sqrt{\xi}. \tag{3.2.7}$$

Thus, a uniform point in a circle of radius R is obtained as $(r, \varphi) = (R\sqrt{\xi}, 2\pi\xi)$.

In the following a number of simple examples are considered. In these examples, average quantities are evaluated using sample averages to review the initial steps in a Monte Carlo estimation.

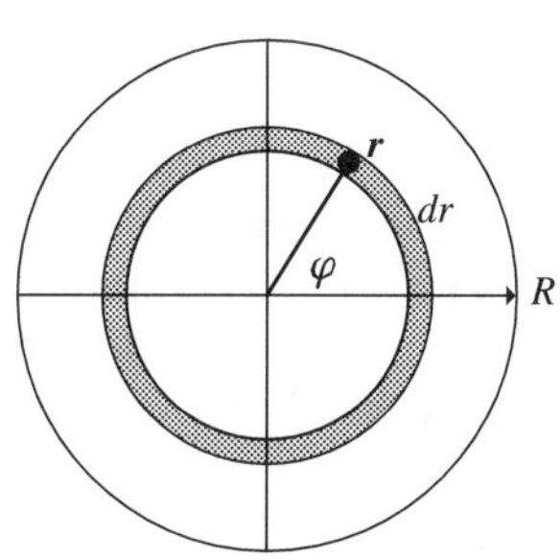

Figure 3.2.1 Sampling uniform points in a circle

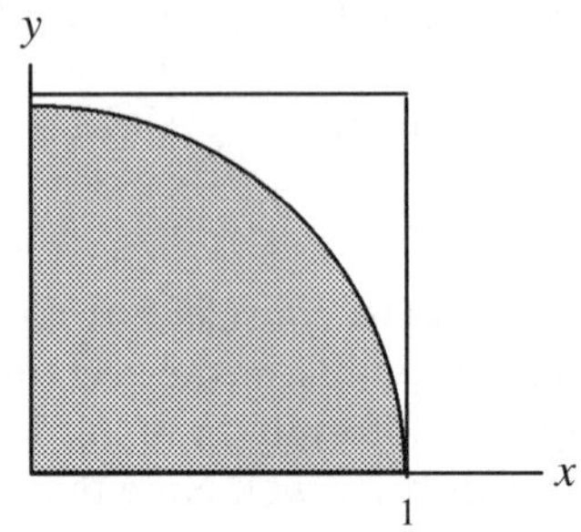

Figure 3.2.2 Sampling in a square of unit sides

Example 3.2.1: Evaluation of π

Consider the quarter circle of unit radius enclosed in a square of unit side, as shown in Figure 3.2.2. The probability of a uniformly sampled point (x, y) in the square being in the quarter circled area is the ratio of the area of the quarter circle to that of the square, namely, $\pi/4$. Define a random variable $h(x, y)$ to be 1 if the point is inside the circle, and zero otherwise. The mean value of $h(x, y)$ is therefore $\pi/4$. By Theorem 3.1.1, the sample average converges to $\pi/4$.

This suggests the following procedure.

1. Sample n independent pairs (x_i, y_i), $i = 1, \ldots, n$ in the unit square.
2. For each i, calculate $h(x_i, y_i)$. $h(x_i, y_i)$ is also called the *estimator*.
3. The sum $\tilde{I}_n = \tilde{h}_n = (4/n)\sum_{i=1}^{n} h(x_i, y_i)$ converges to π.

The notation $\tilde{I}$ or $\tilde{I}_n$ indicates an estimate of the quantity I. The purpose of this example is mainly to see how the estimate $\tilde{I}_n$ converges to the target quantity. This is of essential importance, because the basic behavior observed here is common to almost any Monte Carlo calculations.

The number of samples used in the calculation, n, is called the number of 'histories' or the sample size. In Figure 3.2.3 the values of the estimates, $\tilde{I}_n$, are shown as function of the number of histories up to 40 000 histories. The estimator $h(x, y)$ is a binomial estimator[1] with mean π and variance

$$V_h = 4 \times \frac{\pi}{4}\left(1 - \frac{\pi}{4}\right). \tag{3.2.8}$$

Using equation (3.1.3), the variance of the estimate $\tilde{I}$ is $V(\tilde{I}_n) = V_h/n$, and the Standard Deviation (SD) is

$$\sigma(\tilde{I}_n) = 4 \times \sqrt{\frac{(\pi/4)/(1 - \pi/4)}{n}}. \tag{3.2.9}$$

In the next section it will be shown that the interval $(\pi - 2\sigma(\tilde{I}_n), \pi + 2\sigma(\tilde{I}_n))$ indeed has a confidence level of 0.954. This interval is a function of n, and is shown in Figure 3.2.3 by two gray lines. The confidence interval means that the probability that the correct result lies within this interval is 0.954. The confidence interval as function of the number of histories is called the *convergence envelope* of the estimation.

[1]A binomial estimator takes two possible values: 1 with probability p and 0 with probability $1 - p$. The variance of the estimator is $p(1 - p)$.

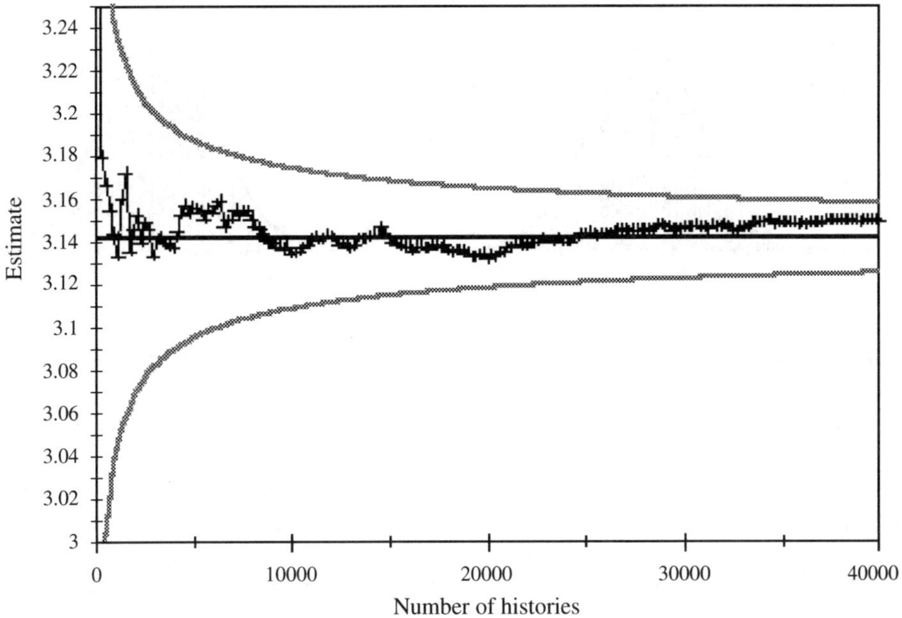

Figure 3.2.3 Estimate of π as function of number of histories. 2σ convergence envelope

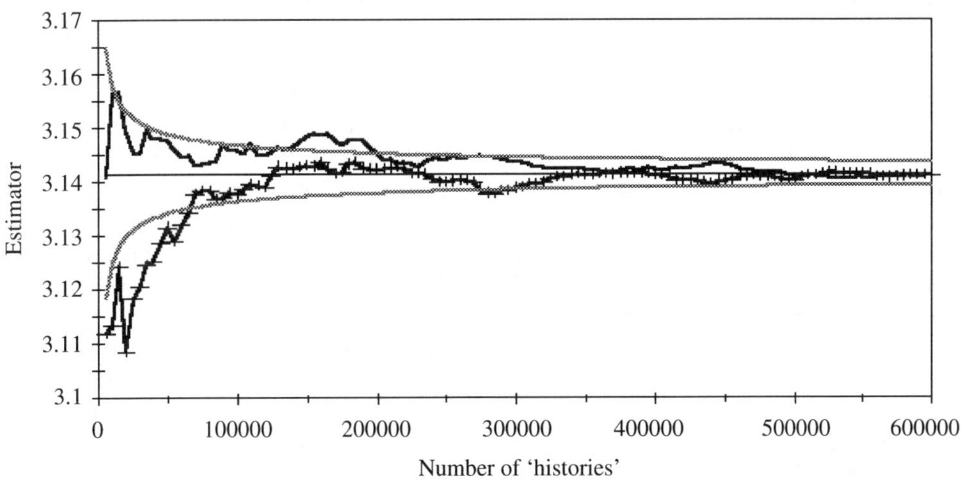

Figure 3.2.4 Estimate of π as function of number of histories. 1σ convergence envelope

It can be seen that all the estimates are confined to the interval. It should be noted that the results, as function of number of histories, in Figure 3.2.3, are not independent. They are all obtained in a calculation of 40 000 histories with intermediate printouts every 200 histories [`Program estim1.for`].[2] In Figure 3.2.4 a similar result is shown

---

[2]Accompanying software can be downloaded from `ftp://ftp.wiley.co.uk/pub/books/dubi/` for all programs cited in this way within this book.

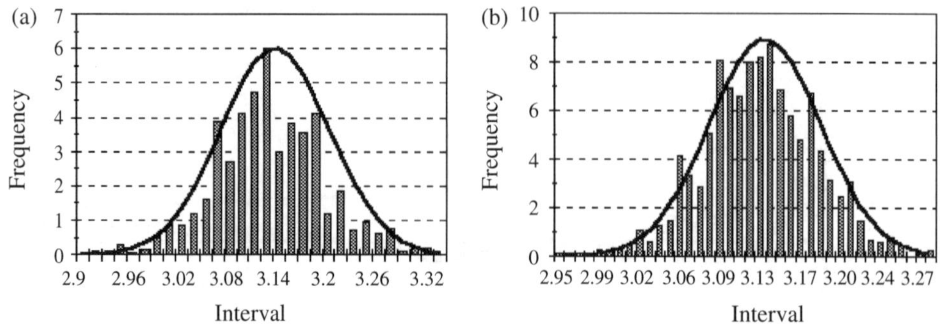

Figure 3.2.5 Histogram and fitted normal distributions for samples of (a) $\tilde{I}_{800}$ and (b) $\tilde{I}_{1200}$

for up to 600 000 histories where two independent calculations are shown with a 1σ convergence envelope.

The confidence level of a 1σ interval is 0.6826, and it can be seen that some of the estimates are outside the convergence envelope. These results are typical of all Monte Carlo calculations. Note that, in this case, the variance could be easily calculated, since the result, π, is known. In general, the result is not known and its estimate $\tilde{I}_n$ is used instead.

Plotting the estimate as a function of the number of histories, along with the convergence envelope, is one solid practice for observing the convergence of a Monte Carlo calculation. A different, though strongly related, form is shown in Figure 3.2.5. Here a set of $m = 1000$ independent calculations of $\tilde{I}_{800}$ (i.e. each calculation is based on 800 histories) were performed. A histogram of the results is shown in Figure 3.2.5(a). The histogram is in the range (2.9–3.32) with intervals of size -0.015. Noting that $\tilde{I}_{800}$ is a random variable, one may wonder about its PDF. The histogram (normalized to the highest value of the normal distribution) represents an approximate PDF. Since, by inspection, this PDF strongly resembles a normal distribution, it might be a good idea to see if it fits any normal distribution. Two parameters are required to define a normal distribution: the mean and the standard deviation. The mean of the variable $\tilde{I}_{800}$ is π; the standard deviation is given by equation (3.2.9), and with $n = 800$ equals 0.058.

This example is much like training on an aircraft while it is on the ground. The quantity being estimated is known *a priori* and the purpose of the calculation is to observe the properties of the method. Assuming that the quantity is not known, the sample mean of the 1000 calculations of $\tilde{I}_{800}$ can be used to estimate the mean. In the present case, this results in

$$\tilde{\mu}_{800} = \frac{1}{m}\sum_{i=1}^{m} \tilde{I}_{800}(i) = 3.1361 \text{ (representing 0.14\% error with } m = 1000). \quad (3.2.10)$$

The second moment of the variable can be estimated as

$$\tilde{S}_{800}^2 = \frac{1}{m}\sum_{i=1}^{m} \tilde{I}_{800}^2(i) = 9.8934. \quad (3.2.11)$$

This sum will converge in probability to the second moment of the variable, just as any other function of the variable converges to its mean value. The standard deviation is then

$$\tilde{\sigma}_{800} = \sqrt{\tilde{S}^2_{800} - \tilde{\mu}^2_{800}} = 0.0654 \tag{3.2.12}$$

(compare to the exact value 0.058 calculated above).

A similar analysis was done for a sample of size $m = 1500$ for $\tilde{I}_{1200}$. The sample average in this case is 3.139 and the standard deviation is 0.04967. The fitted normal is N(3.139, 0.04967). The normal distributions N(3.1361, 0.0654) and N(3.139, 0.04967) are shown in Figure 3.2.5(a,b) in solid lines.

The close fit of the PDF of $\tilde{I}_n$ to a normal distribution $N(\tilde{\mu}_n, \tilde{\sigma}_n)$ is no coincidence. It is an amazing result of a theorem known as the *central limit theorem*. This theorem states that, as $n \to \infty$ the PDF of $\tilde{I}_n$ approaches $N(\mu, \sigma/\sqrt{n})$. The parameters of the distribution are approximated by the estimates $\tilde{\mu}_n$ and $\tilde{\sigma}_n$, respectively. The central limit theorem provides the basis for the statistical error estimation, and defines the rate of convergence of any Monte Carlo calculation. This will be further discussed in Section 3.2.1. In the next example a similar estimation process is applied, but with no *a priori* knowledge of the result.

Example 3.2.2: The area between two functions in the plane

Consider the functions, $f_1(x) = 0.3x^4 + 0.5x^3 - x^2 - x + 2$ and $f_2(x) = -0.8x^2 + 0.5x + 8$. The functions are shown in Figure 3.2.6(a) in the range [0, 3]. The area to be calculated is enclosed by

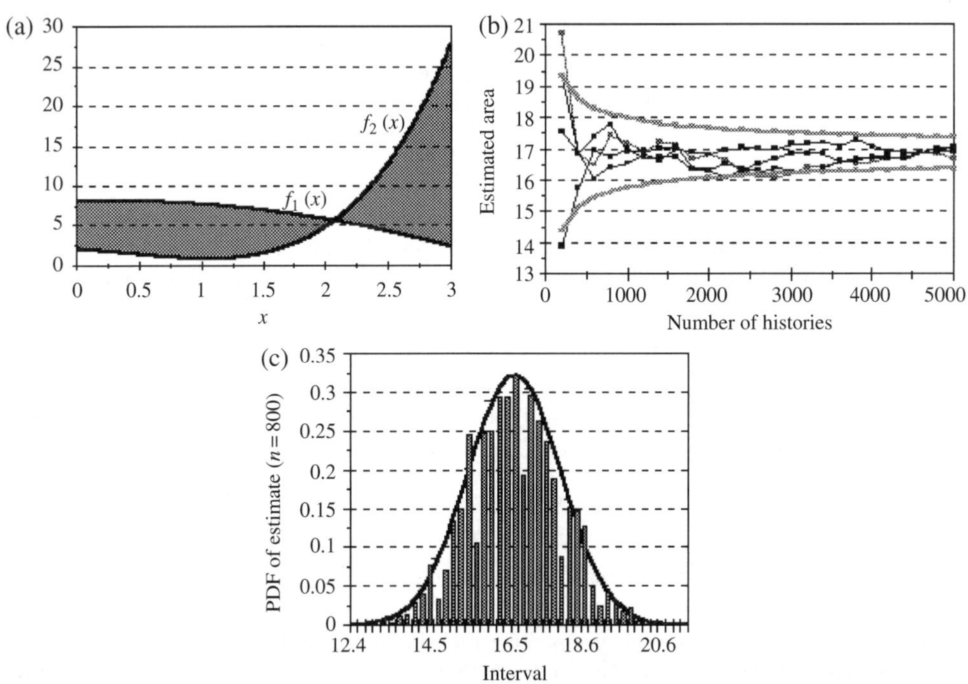

Figure 3.2.6 (a) The area between two functions. (b) $1\tilde{\sigma}$ convergence envelope; (c) normal PDF of $\tilde{h}_{800}$

the functions and is marked in gray in the figure. As in the preceding example, the calculation is based on choosing a large rectangle containing the area such that the probability of a uniformly sampled point in the rectangle to be in the target area equals the ratio of the target area and the area of the rectangle. Such calculations can be easily performed in two, three and higher dimensions.

The steps of the calculation are:

1. Sample N points uniformly: $x_i = 3\xi$; $y_i = 30\xi$.
2. The binomial estimator is

$$h(x_i, y_i) = \begin{cases} 1 & \text{if } \{[y_i \le f_1(x_i)] \cap [y_i \ge f_2(x_i)]\} \cup \{[y_i \ge f_1(x_i)] \cap [y_i \le f_2(x_i)]\} \\ 0 & \text{otherwise.} \end{cases}$$

3. Use the sum

$$\tilde{h}_n = \frac{1}{n}\sum_{i=1}^{n} h(x_i, y_i)$$

to estimate the probability, and multiply this estimate by the area of the rectangle (90).

4. Since $h(x_i, y_i)$ is a binomial estimator, use expression

$$\tilde{\sigma}(\tilde{h}_n) = \sqrt{\frac{\tilde{h}_n(1 - \tilde{h}_n)}{n}} \tag{3.2.13}$$

to estimate the standard deviation of $\tilde{h}_n$.

In Figure 3.2.6(b) the results of four independent calculations as a function of the number of histories (up to 5000) are shown [`Program estim2.for`]. The calculation yields $\tilde{h}_{5000} = 0.1873$, and the area is 16.86. The standard deviation as function of n is then

$$\tilde{\sigma}(\tilde{h}_n) = \frac{90 \times 0.3901}{\sqrt{n}}.$$

The convergence envelope is composed of the lines $\tilde{h}_n + \tilde{\sigma}(\tilde{h}_n)$ and $\tilde{h}_n - \tilde{\sigma}(\tilde{h}_n)$. In Figure 3.2.6(c) a normal approximation to the histogram of a sample of 5000 estimates, each made with 800 histories, is shown. Good agreement is seen between the approximated PDF of $\tilde{h}_n$ and $N[\tilde{h}_n, \tilde{\sigma}(\tilde{h}_n)]$.

In the above example a binomial estimator is used. For such an estimator the calculation of the second moment can be done using the estimate of the first moment. In the next example a similar estimation procedure is used, but the estimator is not binomial – the second moment must be then estimated using a sample average of the square of the estimator.

Example 3.2.3: Average failure probability of a protection mechanism

In many systems protection mechanisms are established to protect the environment against a serious system malfunction. In the system in question, the protection mechanism is a diesel power generator intended to supply power in case of a main power shut down. Power is needed to operate cooling pumps and other emergency units. The diesel generator is regularly in a passive state, and is operated only in case of need. The generator itself has a risk of a 'startup failure', i.e. it may fail to start in case of need. The startup failure probability, P_{sf}, is a function of the time the

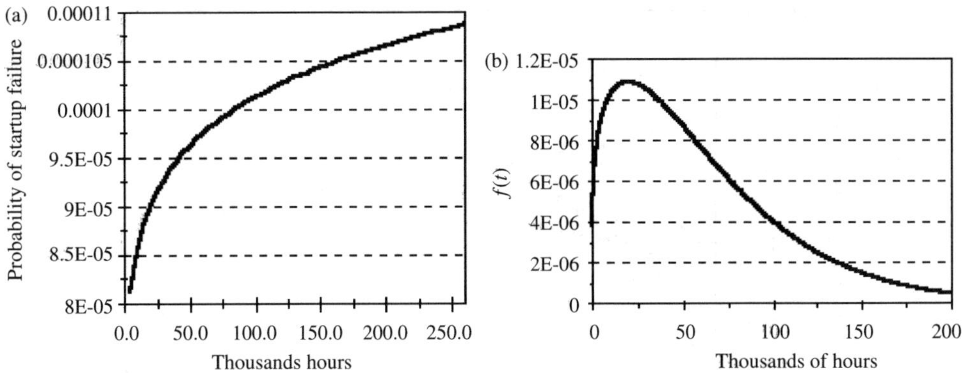

Figure 3.2.7 Startup failure probability as function of (a) time and (b) the Weibull failure distribution

generator is passive, waiting for a demand. The dependence is of the form

$$P_{\mathrm{sf}} = \alpha\left(\frac{t}{t_0}\right)^{\gamma}, \tag{3.2.14}$$

with $\alpha = 1.332\times10^{-6}$, $t_0 = 4.032 \times 10^{+6}$ and $\gamma = 0.074$.

A graph of P_{sf} over time for a period of 260 000 hours (about 30 years) is shown in Figure 3.2.7(a). Consider the situation in which the PDF of failure in the main system is a Weibull distribution, shown in Figure 3.2.7(b): $f(t) = \lambda\beta(t^{\beta-1})\mathrm{e}^{-\lambda t^{\beta}}$ with $\lambda = 7.306 \times 10^{-7}$ and $\beta = 1.27$. The actual startup failure probability depends upon the time of failure. Thus, expression (3.2.14) is a function of a random variable, namely, the time of failure. The *average* startup failure probability is given by:

$$\langle P_{\mathrm{sf}}\rangle = \int_0^{\infty} P_{\mathrm{sf}}(t) f(t)\, dt = \int_0^{\infty} \alpha\left(\frac{t}{t_0}\right)^{\gamma} \lambda\beta(t^{\beta-1})\mathrm{e}^{-\lambda t^{\beta}}\, dt. \tag{3.2.15}$$

Since $P_{\mathrm{sf}}(t)$ is a random variable its second moment is given by

$$S^2 = \int_0^{\infty} P_{\mathrm{sf}}^2(t) f(t)\, dt. \tag{3.2.16}$$

Both $\langle P_{\mathrm{sf}}\rangle$ and S^2 can be evaluated using sample averages.

The estimation procedure follows the previous discussion:

1. Sample t_i $(i = 1, \ldots, N)$ from the Weibull distribution (equation (3.2.5)).
2. Calculate the estimate of the first moment by

$$\tilde{h}_N = \frac{1}{N}\sum_{i=1}^{N} h(t_i) = \frac{1}{N}\sum_{i=1}^{N} P_{\mathrm{sf}}(t_i).$$

3. Estimate the second moment by

$$\tilde{S}_N^2 = \frac{1}{N}\sum_{i=1}^{N} P_{\mathrm{sf}}^2(t_i)$$

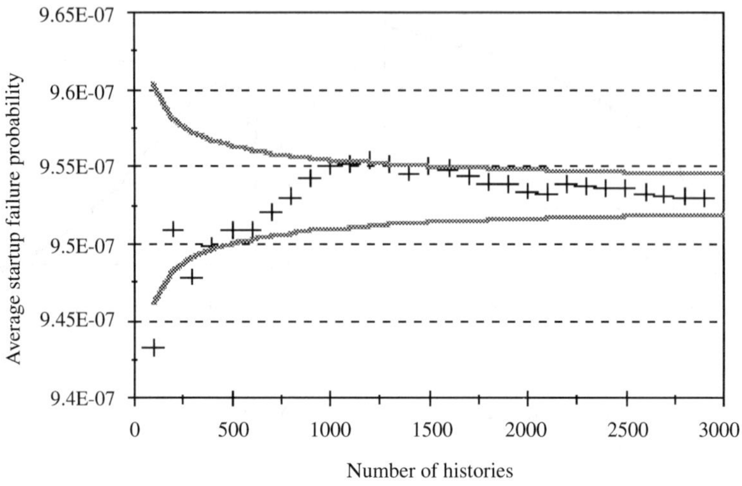

Figure 3.2.8 Convergence of the estimate of $\tilde{h}_N$ to $\langle P_{\text{sf}} \rangle$

and estimate the standard deviation of $\tilde{h}_N$ as:

$$\tilde{\sigma}_N = \sqrt{\frac{\tilde{S}_N^2 - (\tilde{h}_N)^2}{N}} = \frac{\tilde{\sigma}_0}{\sqrt{N}}. \tag{3.2.17}$$

[`Program estim3.for`].

Figure 3.2.8 shows the convergence of the estimator as function of number of histories with a 1σ convergence envelope.

The convergence envelope is composed of the lines

$$\tilde{h}_N \pm \tilde{\sigma}_N \times \sqrt{\frac{N}{n}} = \tilde{h}_N \pm \frac{\tilde{\sigma}_0}{\sqrt{n}}.$$

N is the maximum number of histories used in the calculation.

3.2.1 Error estimation

Monte Carlo estimates are obtained using sample averages. Since the calculation always ends at a finite number of histories, it is essential to have a reliable statistical error estimate for the result. It has already been mentioned that the standard deviation can serve as a measure of a confidence interval. In this section, this notion will be substantiated.

Consider an exact solution $\langle h \rangle$ to be estimated. An exact confidence level α and a confidence interval of the form $[\mu_h - a\sigma_h, \mu_h + a\sigma_h]$ around the mean can be obtained, for any given number of histories n, if the PDF of $\tilde{h}_n$, $f(\tilde{h}_n)$, is known, by noting that

$$\alpha = \int_{\mu - a\sigma}^{\mu + a\sigma} f(\tilde{h}_n)\, d\tilde{h}_n. \tag{3.2.18}$$

The problem is, of course, that one has no idea about the PDF of $\tilde{h}_n$. This key difficulty is remedied by the *central limit* theorem, which guarantees that, under rather weak conditions (the existence of the mean and variance), the PDF of $\tilde{h}_n$ converges asymptotically to the normal distribution.

Theorem 3.2.2: The central limit theorem *Let $h_1, \ldots, h_n$ be a set of independent, identically distributed* r.v.s *with mean μ and standard deviation σ. Let $\tilde{h}_n = (1/n)\sum_{i=1}^{n} h_i$. The PDF of the variable*

$$y_n = \frac{\tilde{h}_n - \mu}{\sigma/\sqrt{n}}$$

converges as $n \rightarrow \infty$ to the standard normal distribution

$$N(0,1) = \frac{1}{\sqrt{2\pi}} e^{-(1/2)y_n^2}.$$

This remarkable theorem states that the limit distribution of the *sample average* is independent of the distribution of the estimator.

The confidence interval of y_n may thus be written as

$$P_r[-a \leq y_n \leq a] = P_r\left[-a \leq \frac{\tilde{h}_n - \mu}{\sigma/\sqrt{n}} \leq a\right] = \int_{-a}^{a} \frac{1}{\sqrt{2\pi}} e^{-(1/2)x^2}\, dx \quad a \geq 0,$$

or

$$P_r\left[-\frac{a\sigma}{\sqrt{n}} \leq \tilde{h}_n - \mu \leq \frac{a\sigma}{\sqrt{n}}\right] = P_r\left[|\tilde{h}_n - \mu| \leq \frac{a\sigma}{\sqrt{n}}\right] = \int_{-a}^{a} \frac{1}{\sqrt{2\pi}} e^{-(1/2)x^2}\, dx. \tag{3.2.19}$$

The right-hand side of (3.2.19) can be found in standard mathematical tables. It has the values 0.6826 for a 1σ interval ($a = 1$) and 0.954 for a 2σ interval ($a = 2$).

The 1σ interval is commonly referred to as the statistical error in $\tilde{h}_n$, with the interpretation that with probability 0.6826 the error does not exceed $\sigma/\sqrt{n}$.

Examples 3.2.1 and 3.2.2 demonstrate the convergence of the statistical error to zero at a rate $(\sqrt{n})^{-1}$, and the convergence of the PDF to a normal distribution. This convergence is in the sense that, for any given confidence level, the measure of the confidence interval converges to zero at this rate. This result is now formally confirmed by expression (3.2.19).

It is to be noted that the convergence rate of numerical integration in one dimension is $1/n$, thus it is faster than that of a Monte Carlo calculation. Yet the central limit theorem and the convergence of the sample average to the mean value are valid for any dimension of the random variable. The convergence of a Monte Carlo calculation follows a $(\sqrt{n})^{-1}$ convergence rate *independently of the dimension of the phase space.* The convergence of numerical integration is critically dependent on this dimension. In a k-dimensional integration the error is inversely proportional to $m = \sqrt[k]{n}$, where n is the

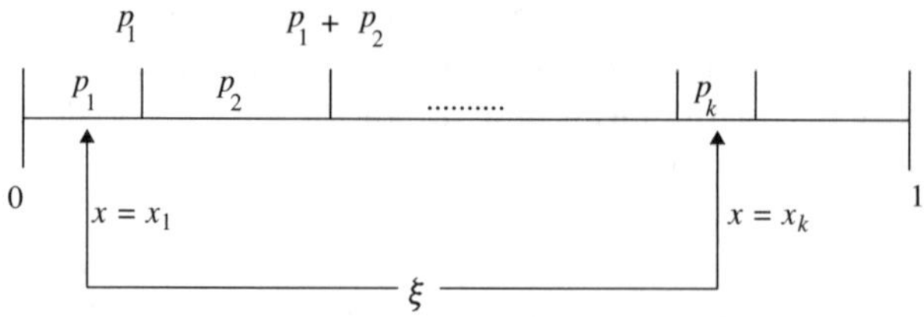

Figure 3.2.9 Sampling a discrete random variable

total number of points used. The insensitivity of the Monte Carlo method to the dimension of the space is the key to its applicability in many areas of science and engineering. System engineering is one particular area in which dimension plays an essential role.

3.2.2 Sampling discrete variables

Consider a discrete random variable X that can take values $\{x_1, \ldots, x_n\}$ with probabilities $\{p_1, \ldots, p_n\}$, such that $p_i = P_r[x = x_i]$. Consider the CDF defined as $G_m = \sum_{i=1}^{m} p_i$. To sample a value x_i for x, let ξ be a random number uniformly distributed in $[0, 1]$ (henceforth a *random number*) and choose k and $x = x_k$ such that

$$G_{k-1} = \sum_{i=1}^{k-1} p_i < \xi \le G_k = \sum_{i=1}^{k} p_i \quad \text{with } G_0 = 0. \tag{3.2.20}$$

The probability that x_k will be sampled is given as $P_r[x = x_k] = G_k - G_{k-1} = p_k$, proving a proper sampling procedure. This sampling technique has a simple graphic interpretation shown in Figure 3.2.9. The probabilities p_i are laid as connected sections on a segment of unit size. The random number is compared to p_1. If ξ is smaller than p_1, then $k = 1$ and $x = x_1$, otherwise ξ is compared to $p_1 + p_2$, etc.

3.2.3 The rejection sampling method

The CDF sampling method (equation (3.2.2)) cannot be used if no explicit expression exists for the CDF. This is, for example, the case with the normal distribution. Sometimes the method is highly inefficient if the CDF equation is complex and 'expensive' in terms of computer time. Sampling is often repeated thousands of times in a Monte Carlo calculation, and it is essential to keep this procedure as efficient as possible.

The rejection method offers a useful option, especially in the first case. Sampling is done from a simple distribution, and a certain proportion of the sample is rejected so that the remaining sample fits the required PDF. The advantage of the rejection method is in its simplicity, and in that it does not require knowledge of the CDF. Its disadvantage lies in the possibility for poor efficiency since part of the sample is rejected.

Let $f(x)$ be a PDF, to be sampled from, defined in the range $[a, b]$. Let $g(x)$ be a PDF defined in $[a, b]$ that is easily sampled by the CDF method.

The rejection procedure

Let

$$r(x) = \frac{f(x)}{g(x)}$$

and let

$$S = \operatorname*{Sup}_{x \in [a,b]} [r(x)]$$

the following steps are taken:

1. Sample x_1 from $g(x)$ using the CDF method.
2. Let ξ_2 be a random number. Accept x_1 if

$$\xi_2 \leq \frac{r(x_1)}{S},$$

otherwise reject x_1 and repeat step 1.

Theorem 3.2.3: The rejection theorem *The rejection procedure is a proper sampling procedure.*

Proof Denote by

$$K\left(x_1 | \xi_2 \leq \frac{r(x_1)}{S}\right)$$

the PDF of the accepted sample. This is the PDF of the sampled points for which the acceptance condition is fulfilled. It is then necessary to show that

$$K\left(x | \xi_2 \leq \frac{r(x)}{S}\right) = f(x).$$

From the definition of conditional probability, the probability that x_1 is sampled in dx_1 given that the acceptance condition is fulfilled is given by

$$K\left(x_1 | \xi_2 \leq \frac{r(x_1)}{S}\right) dx_1 = \frac{g(x_1)\, dx_1 \int_0^{r(x_1)/S} d\xi_2}{\int_a^b \left[g(x_1) \int_0^{r(x_1)/S} d\xi_2\right] dx_1}.$$

The numerator is the probability that x_1 will be sampled in dx_1 *and* the sampled value will be accepted. The denominator is the probability that a sampled value will be accepted.

The above expression can be further reduced to

$$K\left(x_1 | \xi_2 \leq \frac{r(x_1)}{S}\right) dx_1 = \frac{g(x_1)\, dx_1 (r(x_1)/S)}{\int_a^b [g(x_1)(r(x_1)/S)]\, dx_1} = \frac{(f(x_1)/S)\, dx_1}{\int_a^b (f(x_1)/S)\, dx_1} = f(x_1)\, dx_1, \tag{3.2.21}$$

which proves the theorem.

As seen from the above proof, the average probability of acceptance is $1/S$, looking at the definition of S, it is obvious that $S \geq 1$, since for S to be smaller than unity, one must have the condition $g(x) < f(x)\ \forall\ x \in [a, b]$, which is contradicting the normalization condition of the PDFs. To obtain high efficiency, one must have S as close to unity as possible. Indeed, $S = 1$ only if $g(x) = f(x)$.

Uniform rejection

The simplest, though not very efficient, choice of a sampling function $g(x)$ is a uniform distribution, i.e. $g(x) = 1/(b-a)$. The sampling procedure takes a particularly simple form.

Let $M = \mathrm{Sup}_{x \in [a,b]} \{f(x)\}$ so that $S = M(b-a)$:

1. Sample $x_1 = (b-a)\xi_1 + a$, where ξ_1 is a random number.
2. If

$$\xi_2 \leq f(x_1)/M$$

then x_1 is accepted.

The efficiency of the uniform rejection is

$$\frac{1}{S} = \frac{1}{M(b-a)},$$

i.e. the reciprocal of the area of the rectangle comprised by the range of the PDF and the maximum value of $f(x)$.

It should be noted that if the supremum of $r(x)$ cannot be specified, it is possible to take any upper boundary higher than the supremum. This, however, may cause further reduction of the efficiency.

Example 3.2.4: Sampling the normal distribution

The sampling from the normal distribution is a typical example in which the CDF method cannot be applied. The rejection method is then a possible option. In Figure 3.2.10, the standard normal distribution is shown. The range of the distribution is taken from -5 to $+5$. Two forms of the

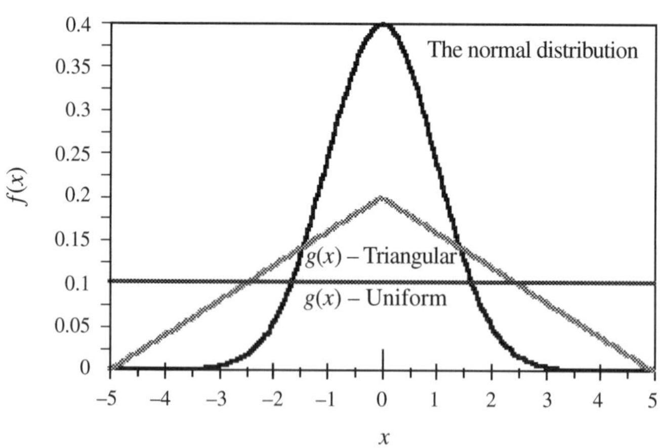

Figure 3.2.10 Sampling the normal distribution by rejection

sampling PDF $g(x)$, are considered. A uniform distribution and a triangular symmetric distribution of the form $f(x) = -0.04|x| + 0.2$. In both cases, the supremum is at $x = 0$.

The value of the normal PDF at that point is $(\sqrt{2\pi})^{-1} = 0.398942$, and that of the uniform distribution is 0.1. The supremum of the ratio is 3.98942 and the efficiency is expected to be 0.2506. Only one out of four samples will be accepted. For the triangular distribution the value at $x = 0$ is 0.2 and the efficiency will be twice as good.

The uniform rejection involes the following steps:

1. Sample a point uniformly in $[-5, +5]$, i.e. $x_1 = 10\xi_1 - 5$.
2. Calculate

$$\frac{f(x_1)}{M} = \mathrm{e}^{-(1/2)x_1^2}.$$

If $\xi_2 \leq \mathrm{e}^{-(1/2)x_1^2}$ accept x_1, otherwise reject it and repeat step 1.

The rejection sampling for the triangular distribution has the following steps:

1. Sample a realization from the triangular distribution. Noting that the distribution is symmetrical, one may first consider the positive section of the distribution. The CDF takes the form $F^a(x) = -cx^2 + dx = -0.04x^2 + 0.4x$. The sampling equation is

$$x_1^a = \frac{d - \sqrt{d^2 - 4c\xi_1}}{2c} = \frac{0.4 - \sqrt{0.4^2 - 0.16\xi_1}}{0.08}.$$

One may examine the sampling formula to verify that it covers the range of the PDF. The above expression yields $x_1^a = 0$ for $\xi_1 = 0$ and $x_1^a = 5$ for $\xi_1 = 5$. Once x_1 is obtained, and since only the positive 'half' of the symmetrical PDF is sampled, with probability 0.5 the number sampled should be negative. This is done by comparing a random number ξ_3 to 0.5. If $\xi_3 \leq 0.5$, then $x_1 = -x_1^a$, otherwise $x_1 = x_1^a$.
2. If

$$\xi_2 \leq \frac{r(x_1)}{S} = \frac{f(x_1) \times g(0)}{g(x_1) \times f(0)} = \frac{0.2 \times \mathrm{e}^{-(1/2)x_1^2}}{[-0.04|x_1| + 0.2]}$$

then x_1 is accepted, otherwise step 1 is repeated.

The results of the sampling are shown in Figure 3.2.11(a,b). A histogram created with one million samples is shown by a solid line, and a histogram of a sample of size 1000 is shown with the markers. Similarly, the solid line in Figure 3.2.11(b) is for 100 000 samples of the triangular distribution, and the markers histogram is for a sample of 1000.

Using uniform rejection, close to four million trials are needed to obtain a sample of one million realizations. Only two million trials are needed for the triangular sampling, yet the longer time required in the triangular sampling offsets the higher efficiency. The sampling time (per accepted realization) for the uniform rejection is 2.7×10^{-4} s and for the triangular sampling it is 2.1×10^{-4} s (time measured on a PC486, 75 MHz machine), thus the real difference is only about 22%. To sample a general Normal variate x, just transform the standard sampled value x_S into $x = \sigma x_S + \mu$. (The listing of the above procedure appears in the program [`sample.for`]. This program generates a given number of realizations from various PDFs. Uniform sampling from the normal distribution appears in the function `SAMNOR1()`, triangular sampling in `SAMNOR2()` and sampling from the triangular PDF in `SAMTRI()`.)

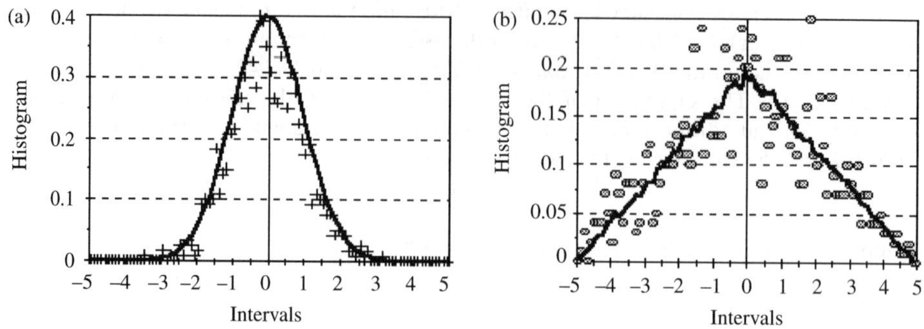

Figure 3.2.11 (a) Histogram of sampled normal variates, (b) histograms of triangular samples

The above discussion indicates that, to obtain efficient rejection, the sampling function $g(x)$ should have two properties. It should be simple to sample from and should be close to $f(x)$, in the sense that S is as close as possible to unity. Example 3.2.4 demonstrates that, although a function, 'closer' to $f(x)$ may be found, any advantage in the acceptance efficiency may be diminished by the increased time spent on the sampling equation.

3.2.4 Additional sampling methods

Relations between random variables can be used to sample a variety of PDFs.

- **Sampling the Gamma distribution**. The Gamma distribution (equation (2.2.8)) can be sampled by recalling that its random variable is a sum of k exponential variables. Accordingly, to sample from a Gamma distribution with parameters k and λ, sample k exponential realizations

$$x_i = -\frac{\ln(\xi)}{\lambda} \quad \text{for } i = 1, \ldots, k$$

  and take the sum. (The function `SAMGAM()` in `sample.for`.)
- **Sampling the Poisson distribution**. The Poisson distribution (equation (2.2.14)) is also closely related to the exponential distribution. Sampling a Poisson variate can be done by calculating the probability for each value of k, $P(k|T)$, and using discrete variates sampling (Section 3.2.2). Yet this may be a lengthy process, since in principle the number of possible values is infinite. One may use the property discussed in Example 2.2.1, namely that k is the number of events (failures) encountered by a unit having an exponential event distribution with parameter λ in an interval of size T with immediate repair. Recall that the average is $\mu = \lambda T$, thus the following process can be used:

1. Consider an arbitrary time T. Set $\lambda = \mu/T$. Set $i = 0$ and $t_i = 0$.
2. Set

$$\Delta t = -\frac{\ln(\xi)}{\lambda}.$$

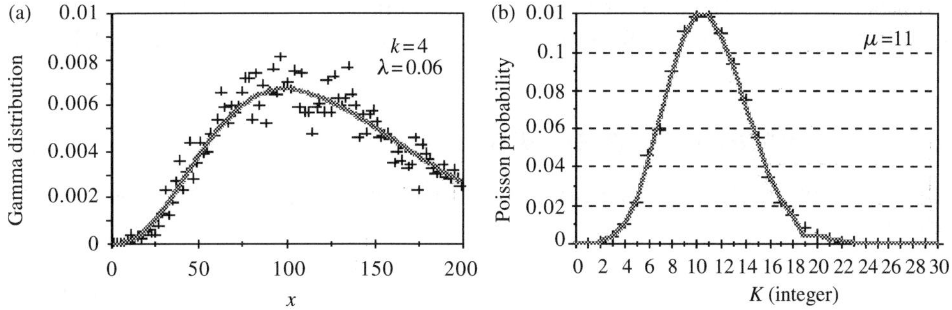

Figure 3.2.12 Sampling (a) the Gamma distribution and (b) the Poisson distribution

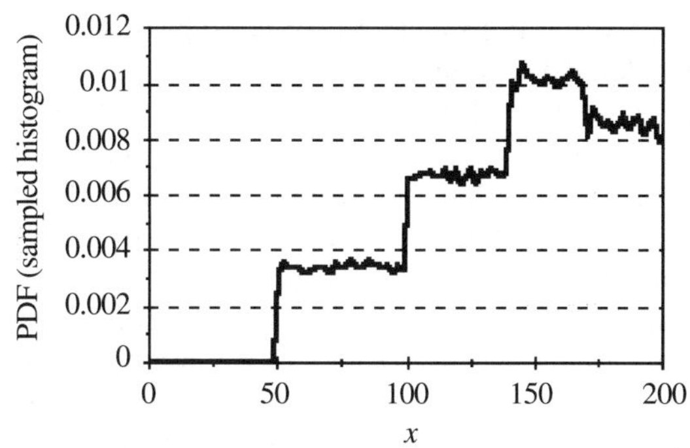

Figure 3.2.13 Sampling histogram of a piecewise constant function

2.1. If $t_i + \Delta t > T$, set $k = i$ and end the process, otherwise;
2.2. Set $t_{i+1} = t_i + \Delta t$, $i = i + 1$ and go to step 2.

(This process is programmed in the function `KSAMPO()` in `sample.for`.)

In Figures 3.2.12(a) and (b), the results of sampling the Gamma distribution and the Poisson distribution are shown. In (a), the gray solid line is the PDF and the markers indicate a histogram of 5000 realizations. In the Poisson distribution, 20 000 samples are used to generate the markers, and the solid gray line is the calculated values.

- **Sampling the lognormal distribution**. A lognormal variate can be sampled by sampling a normal variate x_N and setting $y_L = e^{-x_N}$ for the lognormal variate.
- **Sampling non-parametric piecewise constant distributions** can be done using the CDF method by solving the equation

$$F(t) = \sum_{i=1}^{k} f_i \times (t_i - t_{i-1}) + f_{k+1} \times (t - t_k) = \xi \quad \text{for } k = 1, \ldots, n-1 \text{ and } t_k \leq t \leq t_{k+1}.$$

Alternatively, one may define $P_i = f_i \times (t_i - t_{i-1})$ for $i = 1, \ldots, n-1$. These are the probabilities for the sample to be in the ith interval. Discrete sampling is used to obtain the interval, and the realization is then sampled uniformly within the sampled interval. (The function `SAMUNI()` in `sample.for` samples a variate from a piecewise constant distribution with up to 10 time points.) Figure 3.2.13 shows a histogram of 100 000 samples of a piecewise constant PDF with time points $t_1 = 50, f_1 = 0.0$; $t_2 = 100, f_2 = 0.003389$; $t_3 = 140, f_3 = 0.006779$; $t_4 = 170, f_4 = 0.0101695$; $t_5 = 200, f_5 = 0.0084746$.

3.3 Estimation of Integrals

The framework of the Monte Carlo method can now be presented by the following structure:

$$\left\{\Omega; \omega \in \Omega; \vec{x} : \Omega \to R;\ f(\vec{x}); h(\vec{x}); \mu_h; \sigma_h; \tilde{h}_n = \left[\frac{1}{n}\sum_{i=1}^{n} h(\vec{x}_i)\right] \xrightarrow{P} \mu_h; \mathrm{e} \cong \frac{\tilde{\sigma}_h}{\sqrt{n}}\right\}.$$

A universal event space Ω contains elementary basic events $\{\omega \in \Omega\}$. A random variable (or a multidimensional variable) $\vec{x}(\omega)$ is defined as a real function on the event space with an appropriate joint PDF. $h(\vec{x})$ is a function of the random variable with mean μ_h and standard deviation σ_h. It is desired to calculate that mean. To that end, one takes advantage of the fact that the sample average of the function converges in probability to the mean. The statistical error in the estimate is evaluated using an estimate of the standard deviation. The error converges to zero at a rate proportional to the reciprocal square root of the number of samples (histories) used.

This description suggests that the Monte Carlo method is applied on stochastic processes or models that are statistical in nature. However, the method can be applied to any model provided that the calculated quantity can be presented as an average over an inherent or an artificial stochastic space, with the distinct advantage that the convergence rate is independent of the dimension of the space.

The estimation of integrals serves both as a good example of the application of the Monte Carlo (MC) technique in problems which are not of an inherent statistical nature, and will also provide an opportunity to review some considerations related to the optimization and efficiency of Monte Carlo calculations. Consider an integral of the form:

$$I = \int_a^b g(x)\,dx, \tag{3.3.1}$$

where $g(x)$ is any integrable function defined in $[a, b]$. Let $f(x)$ be any PDF defined in the interval such that $f(x) > 0$ if $g(x) \neq 0$. The integral may then be written as

$$I = \int_a^b \left[\frac{g(x)}{f(x)}\right] f(x)\,dx = \int_a^b h(x) f(x)\,dx = \langle h \rangle. \tag{3.3.2}$$

Thus, the integral is presented in the form of the mean value of a function, $h(x)$, of a random variable x with PDF $f(x)$. The results of Sections 3.1 and 3.2 can be immediately applied so that I can be estimated as

$$\tilde{I}_n = \tilde{h}_n = \frac{1}{n}\sum_{i=1}^{n} h(x_i) \tag{3.3.3}$$

where x_i is a realization of the random variable sampled from $f(x)$, and $h(x)$ is called the estimator of I. The second moment of $h(x)$ is given by

$$S_h^2 = \int_a^b \left[\frac{g(x)}{f(x)}\right]^2 f(x)\,dx = \int_a^b \frac{g^2(x)}{f(x)}\,dx, \tag{3.3.4}$$

and can be estimated by

$$\tilde{S}_h^2 = \frac{1}{n}\sum_{i=1}^{n} h^2(x_i). \tag{3.3.5}$$

The *statistical error* in $\tilde{I}_n$ is given by

$$e_s = \frac{\tilde{\sigma}_h}{\sqrt{n}} = \sqrt{\frac{\tilde{S}_h^2 - (\tilde{I}_n)^2}{n}}. \tag{3.3.6}$$

A general estimator $\eta(x)$ is called an *Unbiased Estimator* (UBE) of a quantity I if

$$\langle \eta(x) \rangle = I. \tag{3.3.7}$$

This property guarantees also that for every value of n, $\langle \tilde{h}_n \rangle = I$. An estimator that converges in probability to I as $n \to \infty$ in the sense of Definition 3.1.1 is called a *consistent estimator*. Note that an unbiased estimator (assuming that the mean and standard deviation of the estimator exist) is also a consistent estimator. Yet a consistent estimator is not always unbiased. Consider the following examples. Let $\tilde{\rho}_n = \tilde{h}_n + a_n/n$, where $\tilde{h}_n$ is an unbiased estimator and a_n is a random variable with a bounded mean. Obviously, the mean value of $\tilde{\rho}_n$ is not I, yet $\tilde{\rho}_n \xrightarrow{P} I$, so that it is a consistent yet not unbiased estimator. Furthermore, in equation (3.3.6) the variance of $\tilde{h}_n$ is estimated as

$$\tilde{V}_n(h) = \tilde{S}_h^2 - \tilde{h}_n^2 = \frac{1}{n}\sum_{i=1}^{n} h^2(x_i) - \left[\frac{1}{n}\sum_{i=1}^{n} h(x_i)\right]^2. \tag{3.3.8}$$

Indeed, as n goes to infinity $\tilde{V}(h)$ converges in probability to the variance of $h(x)$, but is it also unbiased ?

Considering the mean of the left-hand side, one obtains

$$\begin{aligned}\langle \tilde{V}_n(h) \rangle &= \left\langle \frac{1}{n}\sum_{i=1}^{n} h^2(x_i) - \left[\frac{1}{n}\sum_{i=1}^{n} h(x_i)\right]^2 \right\rangle = \frac{1}{n}\sum_{i=1}^{n} S_h^2 - \frac{1}{n^2}\sum_{i=1}^{n} S_h^2 + \frac{2}{n^2}\sum_{i=1}^{n}\sum_{j=i+1}^{n} \mu_h^2 \\ &= S_h^2 - \frac{1}{n} S_h^2 + \frac{n-1}{n}\mu_h^2 = \frac{n-1}{n}(S_h^2 - \mu_h^2) = \frac{n-1}{n} V(h).\end{aligned}$$

Thus, the mean of $\tilde{V}_n(h)$ is not the variance. An unbiased estimator of the variance will be

$$\frac{n}{n-1}\tilde{V}_n(h).$$

For large values of n the difference can be neglected. Since Monte Carlo calculations are usually done with a large sample, one rarely needs to worry about this problem.

The statistical error of a Monte Carlo calculation is determined by $\tilde{\sigma}_h/\sqrt{n}$ of equation (3.3.6). A common error measure is the Percentage Relative Standard Deviation (PRSD), defined as

$$\text{PRSD} = 100 \times \frac{\tilde{\sigma}_h}{\tilde{I}\sqrt{n}}. \tag{3.3.9}$$

This yields the size of a 1σ confidence interval in percentage terms, relative to the estimated value. In many cases, more than one estimator (or even different event spaces) can be used to calculate a given quantity I. It is useful to define a quality measure that will determine the efficiency of a Monte Carlo scheme. As already mentioned, the PRSD or the standard deviation cannot serve this purpose since, although a smaller variance may be achieved for a given number of histories, if the time required to process a history is high, the benefit of a smaller variance may be misleading. An appropriate measure of efficiency would be the computer time required to achieve a given statistical error, or PRSD. Consider a calculation with n histories. Assuming that the processing time of a single history is τ, the time of the calculation is given by $T = n \times \tau$ and the product

$$\text{PRSD}^2 \times T = 100^2 \times \tau \times \left(\frac{\tilde{\sigma}_h}{\tilde{I}}\right)^2$$

is independent of the number of histories. Setting the PRSD to 1%, the product yields the time required for an 1% relative error. The reciprocal of this is called the Figure of Merit (FOM) of the calculation. Thus

$$\text{FOM} = \frac{1}{\text{PRSD}^2 \times T}. \tag{3.3.10}$$

Still, if one may assume that the processing time of two different estimators is similar, one may directly compare the second moments or the PRSDs. The ratio of the FOMs of two calculations of the same quantity is called the *benefit factor*, as it provides the benefit in the calculation time required to obtain the same PRSD.

Consider the integral in (3.3.2). It is obvious that there is an infinite number of possible choices of the sampling function $f(x)$. Any PDF in the range will be appropriate. Is there a best choice? The variance of the estimator can be written as

$$V(h) = S_h^2 - I^2 = \int_a^b \left[\frac{g(x)}{f(x)}\right]^2 f(x)\,dx - I^2$$

$$= \int_a^b \left[\frac{g(x)}{f(x)}\right]^2 f(x)\,dx - I^2 \int_a^b f(x)\,dx = \int_a^b \left[\left(\frac{g(x)}{f(x)}\right)^2 - I^2\right] f(x)\,dx. \tag{3.3.11}$$

Thus, if one takes

$$f(x) = \frac{g(x)}{I}, \tag{3.3.11a}$$

a zero variance results. This is called a 'zero variance Monte Carlo'. Indeed, it means that each value added in the sample average will have the value I. Yet to get such a perfect estimator, one must know the value of I. This means that if one already knows the answer, then it can be exactly calculated with a single history. A zero variance Monte Carlo can not be achieved (although many have tried). But equation (3.3.11a) suggests that a smaller variance is achieved if the sampling PDF follows the integrand. Although this is a trivial conclusion it is of extreme importance, since it also applies to multi-dimensional complex spaces, and is the common guideline of all Variance Reduction Methods (VRM).

From a statistical point of view, this conclusion is consistent with the fact that the UBE $\eta(x)$ (which by equation (3.3.2) equals $g(x)/f(x)$) should have the smallest possible variation in $[a, b]$ to minimize the variance. Obviously, when (3.3.11a) is fulfilled, $\eta(x)$ has zero variation, it is equal to I for every x. If $g(x)$ takes values close to $f(x)$ a small variance can be expected. Conversely, if $g(x)$ is small when $f(x)$ is large, and *vice versa*, then $\eta(x)$ will have large variations resulting in a large variance.

Consider, for example, the function $g(x)$ shown in Figure 3.3.1. The sampling distribution $f(x)$ is strongly mismatched with the integrand. In fact, about 4% of the probability mass (the CDF) covers almost 99% of the area under $g(x)$, while 96% of the probability covers about 1% of the integral. Most of the points will be sampled in zone B, where the values of $g(x)$ are very small and those of $f(x)$ relatively high. This will result in an underestimation of the integral. Eventually, a point will be sampled in zone A. The estimator will take a large value (since $g(x)$ is high and $f(x)$ is small). This will result in a jump in the estimate with a corresponding, and higher, jump in the variance. Subsequently, small contributions will be scored leading to a gradual decline of the estimate until, again, a low probability, high score estimate will be encountered, resulting in a jump in both the estimate and the variance.

This phenomena can be seen in Figure 3.3.2, where the sampling was done from $f(x)$ of Figure 3.3.1 to estimate the integral of $g(x)$. The exact value of the integral is 100. The chainsaw pattern in Figure 3.3.2 is systematic and distinctly different from the

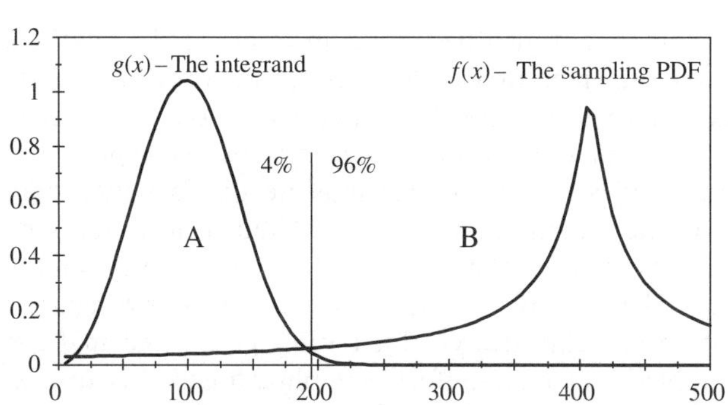

Figure 3.3.1 Improper choice of a sampling distribution

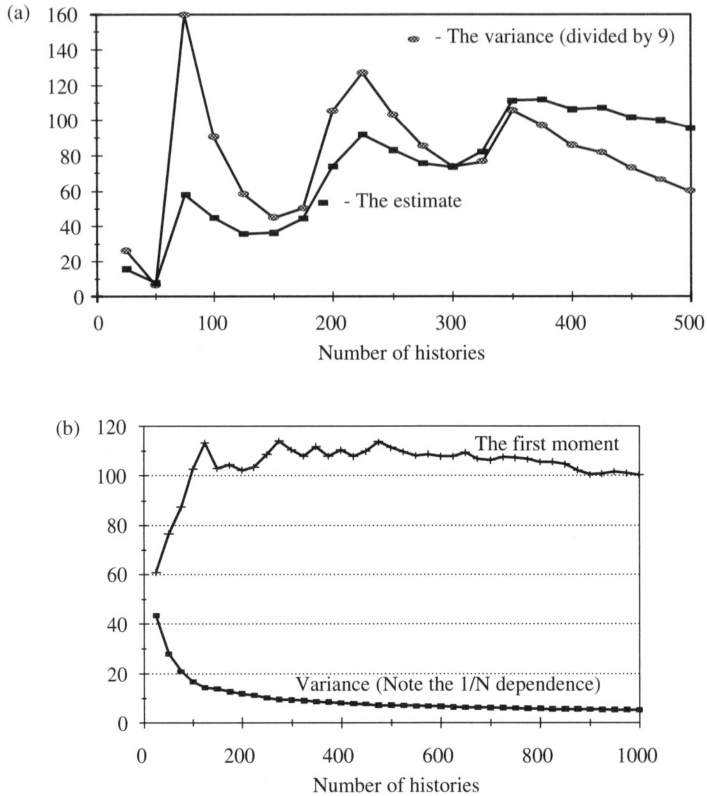

Figure 3.3.2(a) Typical jumps in the estimate and variance resulting from improper sampling. (b) The estimated integral with uniform sampling. A proper convergence

typical statistical fluctuations of proper sample estimates (see Figures 3.2.3–4, 6(b) and 8). This chainsaw pattern is typical of Monte Carlo calculations in which the 'coverage' of the phase space is inappropriate, i.e. when important points are poorly sampled.

For comparison, see also Figure 3.3.2(b), where sampling is done from a uniform distribution.

The above discussion should also be viewed in terms of the relations between the functions and the space on which they are defined. In the present case, the phase space is an interval on the real axis. The function $g(x)$ defines the *true importance* of every point x in the space. $f(x)$ defines the probabilistic, or *sampling importance*, attributed to each point. The results of equation (3.3.11) indicate that a match between the true importance ($g(x)$) and the sampling importance attributed to it guarantees the lowest possible variance. A mismatch, on the other hand, could cause systematic fluctuations and a high variance. The example shown above is, of course, artificial. In a one-dimensional integration one can always plot the integrand and make sure that each region is properly sampled. In fact, a uniform sampling function is always a safe choice. It may attribute high importance to low importance regions, but not the other way around. The problem becomes acute in high dimensional problems, as will be seen in

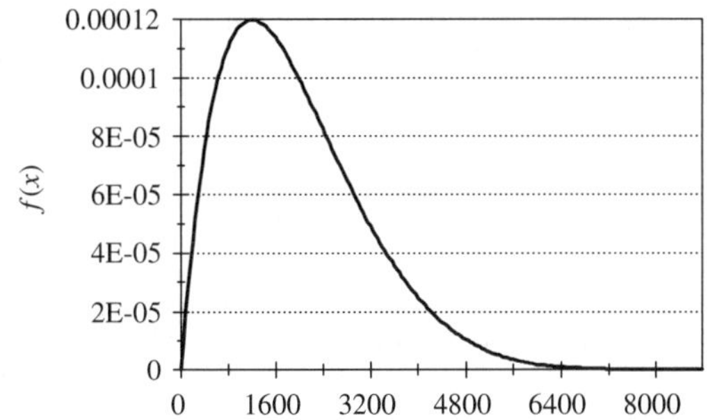

Figure 3.3.3 The importance of the integral in equation (3.3.13)

forthcoming sections. In such problems, identifying the 'true importance' becomes a serious challenge and is very often prohibitively difficult.

The simplest choice of $f(x)$ is a uniform distribution in $[a, b]$, namely $f(x) = 1/(b-a)$. The estimator in (3.3.3) and (3.3.5) takes the form

$$\eta(x_i) = h(x_i) = (b-a) \times g(x_i). \tag{3.3.12}$$

The notation $h(x)$ is used for this specific topic, while the notation $\eta(x)$ is introduced to specify an unbiased estimator in general.

Example 3.3.1: The integral in expression 2.2.19 of Example 2.2.2

$$I = U(t) = \int_0^t \frac{\lambda^*(\lambda^* t')^{k-1} e^{-\lambda^* t'}}{\Gamma(k)} [1 - e^{-\lambda(t-t')}]^n \, dt'. \tag{3.3.13}$$

The integrand of (3.3.13) is shown in Figure 3.3.3. Three possible natural choices of the sampling function can be tested. One is the uniform PDF in the range $[0, t]$, namely $f_1(x) = 1/t$. The estimator is then

$$\eta_1(x) = \frac{\lambda^*(\lambda^* x)^{k-1} e^{-\lambda^* x}[1 - e^{-\lambda(t-x)}]^n}{\Gamma(k)} t.$$

The second option is to take advantage of the fact that the integrand already contains a PDF, i.e. the Gamma distribution $G_k(\lambda^*, t)$. However, the distribution is not normalized in $[0, t]$. From the definition of the Gamma distribution, it is clear that the CDF of the Gamma distribution is the probability that at least k events will occur up to time t. Denoting the CDF by $C_{Gk}(\lambda^*, t)$, it results that

$$C_{Gk}(\lambda^*, t) = 1 - \sum_{i=0}^{k-1} P(i, |\lambda^* t), \tag{3.3.14}$$

where $P(i, |\lambda^* t)$ is the Poisson distribution with mean $\lambda^* t$ (equation (2.2.14)). A Gamma PDF normalized in the range is given by

$$G_k^N(\lambda^*, x) = \frac{G_k(\lambda^*, x)}{C_{Gk}(\lambda^*, t)} = \frac{G_k(\lambda^*, x)}{1 - \sum_{i=0}^{k-1} P(i, |\lambda^* t)}. \tag{3.3.15}$$

To sample from (3.3.15), one can sample from the regular Gamma distribution and reject values of x if $x > t$. Multiplying and dividing (3.3.13) by the denominator of (3.3.15), the estimator takes the form:

$$\eta_2(x) = \left[1 - \sum_{i=0}^{k-1} P(i|\lambda^* t)\right] [1 - e^{-\lambda(t-x)}]^n. \tag{3.3.16}$$

In the example, $n = 5$ and $k = 2$ (five systems and two spares in storage). Also, $\lambda = 6.23 \times 10^{-5}$ and $\lambda^* = 5 \times 6.23 \times 10^{-5} = 3.115 \times 10^{-4}$. Taking $t = 8760$ h, the normalization factor is

$$\left[1 - \sum_{i=0}^{1} P(i|2.7287)\right] = 0.7565077.$$

The estimates and PRSDs as function of the number of histories are shown in Figure 3.3.4.

The exact result is 0.00164. The third estimator shown is obtained by adopting a piecewise constant PDF as shown in Figure 3.3.5.

Table 3.3.1 summarizes the results of the three calculations. It can be seen that the uniform and Gamma sampling are comparable and have a relatively low FOM. This is consistent with the

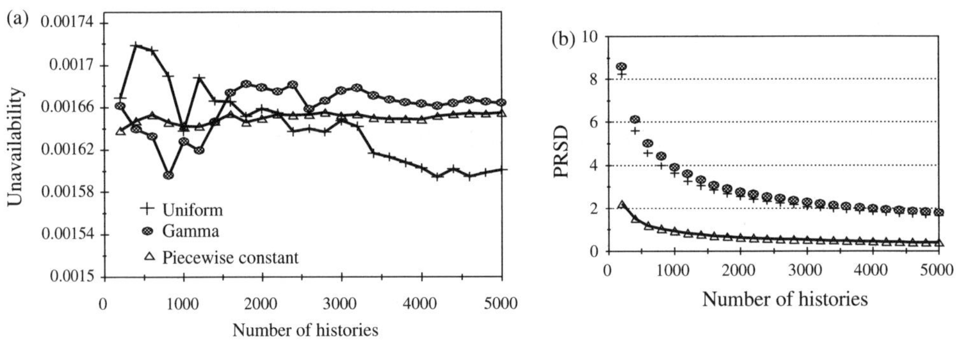

Figure 3.3.4 Three estimators for the integral (2.2.19) and the corresponding PRSDs

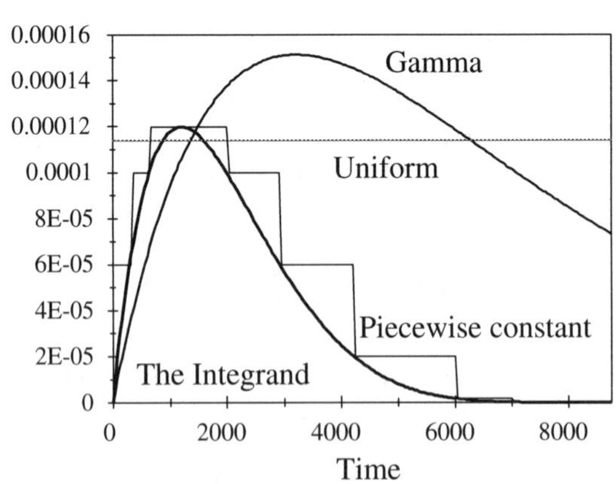

Figure 3.3.5 Choice of sampling functions relative to the integrand (PDFs are not normalized)

Table 3.3.1 Results for three estimators

Sampling	$\tilde{I}$	PRSD	Time (sec)	FOM
Uniform	0.001601	1.63%	0.55	0.684
Gamma	0.001664	1.78%	0.55	0.573
Piecewise	0.001654	0.40%	0.44	14.2

mismatch that can be observed between both the Uniform and the Gamma PDFs and the integrand in Figure 3.3.5 in the range above 4000 hours.

The FOM of the piecewise constant estimator is by a factor ≈ 20 higher than the other two methods. This means that the same statistical error can be obtained 20 times faster.

Uniform sampling is an 'ignorant', safe approach. It is appropriate when no prior knowledge of the structure and importance of the phase space is available. It assigns equal importance to every point. The Gamma sampling presents an alternative by which part of the importance is used for sampling. This does not guarantee any advantage. Sampling from part of the integrand would be useful only if the remainder has a small variation in the range. The use of a piecewise constant distribution in this case is both simple and efficient. The values of the constants can be determined approximately by inspection, as is done above. (`Program integ.for.` allows integration by uniform sampling and piecewise constant sampling.)

Optimal values for a piecewise constant PDF can be obtained with the following considerations: let the range $[a, b]$ be divided in to n intervals, $t_0, t_1, \ldots, t_n$ such that $t_0 = a$ and $t_n = b$. The piecewise constant distribution is defined by a set of probabilities p_i, where $p_i = P_r[t_{i-1} \le t < t_i]$ is the probability of the sample to be in the ith interval. The PDF is then

$$f_i = \frac{p_i}{t_i - t_{i-1}}, \qquad i = 1, \ldots, n.$$

Multiplying and dividing the integral by the PDF, one obtains

$$I = \int_{t_0}^{t_n} g(x)\,dx = \sum_{i=1}^{n} (t_i - t_{i-1}) \int_{t_{i-1}}^{t_i} \frac{g(x)}{p_i} f_i\,dx. \tag{3.3.17}$$

The second moment is given by

$$S^2 = \sum_{i=1}^{n} (t_i - t_{i-1}) \int_{t_{i-1}}^{t_i} \frac{g^2(x)}{p_i}\,dx = \sum_{i=1}^{n} \frac{\Gamma_i}{p_i}, \quad \text{where} \quad \Gamma_i = (t_i - t_{i-1}) \int_{t_{i-1}}^{t_i} g^2(x)\,dx. \tag{3.3.18}$$

Assuming that the sampling time is of no concern, maximal FOM will be obtained with a minimal second moment. This under the constraint $\sum_{i=1}^{n} p_i - 1 = 0$. To obtain a minimum under constraint, the Lagrange multipliers method is utilized (see, for example: Press, W. H., Teukolsky, S. A., Vetterling, W. T. and Flannery, B. P. (1992). *Numerical Recipes in Fortran*, Cambridge University Press or Rade, L. and

Westergren, B. (1995). *Handbook for Science and Engineering*, Studentlitteratur, 3rd ed., p. 220 or any other textbook on mathematics of variations). The constraint equation, which is identically zero, is multiplied by an unknown parameter λ and added to the second moment. A derivative with respect to the parameters to be optimized is taken, yielding

$$\frac{d}{dp_j}\left[S^2+\lambda\left(\sum_{i=1}^{n}p_i-1\right)\right]=0, \tag{3.3.19}$$

yielding

$$-\frac{\Gamma_j}{p_j^2}+\lambda=0 \Rightarrow p_j=\sqrt{\frac{\Gamma_j}{\lambda}}$$

from which one obtains using the constraint

$$\frac{1}{\sqrt{\lambda}}=\frac{1}{\sum_{j=1}^{n}\sqrt{\Gamma_j}}.$$

Substituting this back into the relation for p_j yields the optimal values

$$p_{j\text{opt}}=\frac{\sqrt{\Gamma_j}}{\sum_{i=1}^{n}\sqrt{\Gamma_i}}. \tag{3.3.20}$$

Equation (3.3.20) suggests a method for 'self-optimizing' MC. Initially, uniform sampling is used to obtain the estimate of I and estimates for Γ_i. These estimates can be used to obtain approximate values of the optimal parameters, which in turn are used to improve the estimates.

Example 3.3.2: An ordered sequence of failures, early warning system in chemical corrosion

A protection unit is attached to the internal wall of a chemical process unit grinding corrosive materials. The purpose of the protection is to provide early warning of the corrosive thinning of the wall. The PDF for failure of the early warning mechanism is exponential $f_1(x)=\lambda_1 e^{-\lambda_1 t}$ with $\lambda_1=9.09\times10^{-5}$. The corrosion failure of the wall follows a more complex distribution. The hazard is zero in the first 3000 hours of operation, and then follows a Weibull distribution:

$$z_2(t)=\begin{cases}0 & t\le 3000\\ \lambda_2(k-1)(t-3000)^k, & \end{cases}$$

where $\lambda_2=4.24\times10^{-9}$ and $k=2.1$.

This distribution is often called a three parameter Weibull distribution, and the PDF is given in general as

$$f_2(t)=\lambda_2(k-1)(t-\gamma)^{k-1}e^{-\lambda_2(t-\gamma)^k} \quad \text{and} \quad f_2(t)=0 \quad \text{for } t\le\gamma. \tag{3.3.21}$$

The CDF is accordingly

$$F_2(t) = 1 - e^{-\lambda_2(t-\gamma)^k} \quad \text{and} \quad F_2(t) = 0 \quad \text{for } t \leq \gamma. \tag{3.3.21a}$$

The sampling equation from this CDF is

$$t = \gamma + \left[-\frac{\ln(1-\xi)}{\lambda} \right]^{1/k}. \tag{3.3.22}$$

In this example, a failure of the reactor wall while the early warning system is operational is an acceptable consequence, since necessary preventive operations can take place to minimize the damage. However, if the warning system fails first and a failure of the reactor wall occurs without a warning, the consequences are severe. It is thus desired to minimize the probability of this event. The system is inspected on a periodic basis, and it is desired to set up the periodic interval so as to obtain a low probability for a severe event. At inspection, both the warning system and the chemical reactor are checked and all failures are detected and repaired. The purpose of the example is to calculate the probability of the event $A = \{$The warning system fails *and then* the main unit fails before inspection$\}$. The inspection interval is denoted by T_s. The probability density that the warning system will fail at time t' and the main system will not fail up to this time is $f_1(t') \times [1 - F_2(t')]$. The probability that, given that the main system did not fail up to t', it will fail in the interval $[t', T_s]$ is

$$\frac{F_2(T_s) - F_2(t')}{[1 - F_2(t')]}.$$

The probability density of a severe event is thus the product of the above two probabilities, namely $f_1(t') \times [F_2(T_s) - F_2(t')]$. The probability for a severe event is obtained by integrating this probability density from zero to T_s, or:

$$P_r[A(T_s)] = \int_0^{T_s} f_1(t') \times [F_2(T_s) - F_2(t')]\, dt'.$$

(Note that $F_2(t') \equiv 0$ if $t' \leq \gamma$.) Substituting the explicit distributions yields

$$P_r[A(T_s)] = \int_0^{T_s} \lambda_1 e^{-\lambda_1 t'} \times [r(t', \gamma) - e^{-\lambda_2 (T_s - \gamma)^k}]\, dt' \tag{3.3.23}$$

for $T_s > \gamma$ (otherwise the probability is zero). Also, $r(t', \gamma) = 1$ for $t' \leq \gamma$ and $r(t', \gamma) = e^{-\lambda_2(t'-\gamma)^k}$ for $t' > \gamma$.

Two methods will be used to obtain the probability as function of T_s. First, integration is considered with two estimators. A uniform sampling PDF in the range $[0, T_s]$ of the form $g(t') = 1/T_s$ which results in an estimator of the form

$$\eta_1(t') = T_s \times \left\{ \lambda_1 e^{-\lambda_1 t'} \times \left[r(t', \gamma) - e^{-\lambda_2 (T_s - \gamma)^k} \right] \right\}.$$

This is simply the integrand multiplied by the reciprocal of the sampling PDF.

A second estimator is obtained when t' is sampled from the natural exponential PDF $f_1(t')$, which is part of the integrand. However, this PDF is not normalized in the range of integration.

It is thus necessary to sample from the normalized PDF:

$$f_1^*(t') = \frac{f_1(t')}{1 - e^{-\lambda_1 T_s}}.$$

The sampling equation is then

$$t' = -\frac{\ln\left[1 - \xi \times \left(1 - e^{-\lambda_1 T_s}\right)\right]}{\lambda_1}. \tag{3.3.23a}$$

Dividing and multiplying the integral by the normalization factor yields the estimator

$$\eta_2(t') = [1 - e^{-\lambda_1 T_s}][r(t', \gamma) - e^{-\lambda_2 (T_s - \gamma)^k}].$$

It should be noted that in the above, a history in which the failure in the chemical system occurs before the failure of the warning system is not considered a severe failure, and is not accounted for.

The above two estimators represents variations of the same method. One looks at the integral with no reference to its physical meaning. A sampling function is chosen and the estimation follows according to equations (3.3.3)–(3.3.6). In the next approach, the notion that every quantity of interest can be viewed as an average over the phase space histories of the system is invoked. In the present case, a possible history of the system can be presented as a set of two time points $(t_1; t_2)$, where t_1 is the failure time of the early warning system, sampled from $f_1(t_1)$, and t_2 is the failure time of the main system, sampled from $f_2(t_2)$. Noticeably, these two events are independent. The system function is, in fact, the probability of a subset Φ of all the possible histories such that $\Phi = \{(t_1; t_2) : [t_1 < t_2 < T_s]\}$, i.e. those histories in which both failures occur before the inspection time and the warning system fails before the main system. Let the r.v. $\eta(t_1; t_2)$ be defined on Φ such that

$$\eta_a(t_1; t_2) = \begin{cases} 1 & \text{if } (t_1, t_2) \in \Phi \\ 0 & \text{otherwise.} \end{cases}$$

This estimator singles out those histories in which a severe failure has occurred. The PDF for a history $(t_1; t_2)$ is simply $f_1(t_1) \times f_2(t_2)$ since the components are independent. The probability for a severe event is simply the average of $\eta_a(t_1; t_2)$, namely

$$I = \int_0^\infty dt_1 \int_0^\infty dt_2 \eta_a(t_1; t_2) f_1(t_1) \times f_2(t_2).$$

To see that this expression is identical to (3.3.24), note that

$$F_2(T_s) - F_2(t') = \int_{t'}^{T_s} f_2(t_2)\, dt_2.$$

Substituting this in (3.3.23) yields

$$P_r[A(T_s)] = \int_0^{T_s} f_1(t') \times [F_2(T_s) - F_2(t')]\, dt' = \int_0^{T_s} f_1(t') \times \left[\int_{t'}^{T_s} f_2(t_2)\, dt_2\right].$$

Multiplying the integrand by $\eta_a(t_1; t_2)$ allows the extension of the integration limits to infinity (since $\eta_a(t_1; t_2)$ is zero for t' or t_2 larger than T_s and the integration on t' can be taken from zero, as $\eta_a(t_1; t_2)$ is zero in the range $t_2 < t'$) proving the required identity.

The subscript 'a' in η_a stands for 'analog', since the process simulated is analog to the statistical reality of the system. It should be noted that sometimes the term 'analog' is used for integrals estimation when a uniform PDF is used as the sampling distribution.

Expression (3.3.27) is a two-dimensional integral; it is no different than a one-dimensional integral, and it suggests the second estimation method. Sample a pair $(t_1; t_2)$, calculate $\eta(t_1; t_2)$ and accumulate the sample average. t_1 is sampled from the exponential distribution (equation (3.2.3)) and t_2 is sampled using equation (3.3.22). The estimator $\eta_a(t_1; t_2)$ is a binomial estimator. Each system's history is an 'experiment', and the estimator takes the values 1 or 0 according to the 'success' or 'failure' of the experiment.

The results for the three estimators are summarized in Table 3.3.2. In the table, the estimate, its PRSD and the FOM are presented for each of the estimators for values of T_s ranging from 3250 h up to 9000 h. The calculation time (in seconds) appears in brackets in the FOM section. For the analog estimation the calculation time is 1 s for all cases. The table shows that the values obtained for the first two estimators are almost identical, and have low PRSDs and relatively high FOMs. The analog results have high PRSDs and relatively poor FOMs. Noticeably, the FOM of the first two estimators is better for low probability values, whereas for the analog estimator the results are opposite. The FOM increases by a factor of about 2500 as the probability increases from 0.00019 to 0.338. This illuminates one of the most important elements in MC calculations, that of 'rare events'. The reciprocal of the FOM stands for the computation time required for a 1% relative statistical error. For the case of $T_s = 3250$ hours, it will be 0.0146 seconds with the

Table 3.3.2 Probability for a severe event calculated with three estimators (20 000 histories for all points)

Inspection Time T_s	$P_r[A(T_s)]$			PRSD (%)			FOM * Time measured in seconds		
	η_1	η_2	η_a	η_1	η_2	η_a	η_1	η_2	η_a
3000	0.0	0.0	0.0	0.0	0.0	0.0	–	–	–
3250	0.000115	0.000115	0.0001	0.11	0.081	70.7	68.3 (1.21)	89.1 (1.6)	0.00019
3500	0.000515	0.000515	0.0004	0.144	0.11	35.3	36.8 (1.31)	50.0 (1.65)	0.0008
4000	0.00239	0.002389	0.0025	0.188	0.144	14.2	20.6 (1.37)	29.2 (1.65)	0.005
4500	0.00598	0.00598	0.00645	0.22	0.165	8.77	15.0 (1.37)	21.7 (1.7)	0.013
5000	0.0115	0.0115	0.0108	0.244	0.18	6.78	11.8 (1.42)	18.0 (1.71)	0.0217
5500	0.0194	0.0194	0.0191	0.265	0.192	5.06	9.62 (1.48)	14.9 (1.81)	0.039
6000	0.0296	0.0296	0.0299	0.283	0.201	4.02	8.79 (1.42)	14.0 (1.76)	0.0618
6500	0.0423	0.0423	0.0428	0.299	0.208	3.34	7.26 (1.54)	12.7 (1.81)	0.0896
7000	0.0575	0.0575	0.059	0.314	0.213	2.82	6.85 (1.48)	12.5 (1.76)	0.125
7500	0.0751	0.0751	0.0767	0.328	0.218	2.45	6.23 (1.49)	10.1 (2.07)	0.166
8000	0.095	0.0950	0.0952	0.341	0.222	2.17	5.62 (1.53)	10.5 (1.92)	0.212
8500	0.117	0.117	0.119	0.354	0.226	1.91	5.35 (1.49)	10.4 (1.87)	0.274
9000	0.1408	0.1408	0.143	0.367	0.229	1.72	4.82 (1.54)	9.88 (1.93)	0.338

first estimator, 0.0112 seconds with the second and 5263 seconds with the analog estimation. The ratio of the FOMs is about 460 000.

This is the 'Achilles heel' of the Monte Carlo method. When a binomial estimator is used to calculate a probability P, the variance of the estimate is $(P(1-P))/n$ (where n is the number of histories), and consequently the PRSD is

$$\text{PRSD} = 100 \times \sqrt{\frac{(1-P)}{P \times n}}. \tag{3.3.24}$$

As P decreases the PRSD increases indefinitely. For the above case at 3250 hours, $P = 0.000115$ and $n = 20\,000$, resulting a PRSD $= 65.93$ (note the estimate of 70.7% obtained in the calculation). This error renders the result highly unreliable. A different way to look at the same problem is by estimating the number of histories required to obtain a given PRSD, ε. Isolating n from (3.3.24) yields

$$n_\varepsilon = 10^4 \times \frac{1-p}{p\varepsilon^2}. \tag{3.3.25}$$

Denoting the time to process a single history by τ, the calculation time for a given statistical error becomes

$$T_\varepsilon = 10^4 \times \frac{\tau(1-p)}{p\varepsilon^2}. \tag{3.3.25a}$$

A typical value of the time per history is 10^{-5}_{sec}, the time required to calculate the probability of an event in the range of $p = 10^{-9}$ with a statistical error of 10% would be $\approx 10^{8}_{\text{sec}}$ or *3.17 years*. The calculation becomes impractical. This is the 'rare events' phenomenon. Rare events are a reflection of the simple fact that when the probability of occurrence of an event is very small, a huge number of experiments are required for it to happen. In the current example the probability is of the order of 10^{-4}, and in 20 000 histories about 19 998 yield no information; only in two histories does the event occur.

The Monte Carlo method is insensitive to the complexity and dimension of the problem. This is its basic advantage. It is, however, sensitive to the 'rareness' of the event under consideration. Variance Reduction Methods (VRM) are intended to overcome this obstacle, to some extent. In fact, the first two estimators can be viewed as a VRM for the current example. They indeed improve the calculation considerably by a factor of over 450 000. Unfortunately, the above VRM cannot easily be extended to improve by similar factors general complex problems. VRM is a difficult problem which becomes quite frustrating as the phase space becomes complex. Fortunately enough, most problems in system engineering are in the range of 10^{-2}–10^{-6}, which can be handled satisfactorily with modern computers. The only problems outside this range are those related to risk assessment in catastrophic events, i.e. rare events with catastrophic consequences. Variance reduction methods applicable for such cases will be further discussed in Chapters 6 and 7.

The mean time to failure is a fairly simple concept. It is given by equation (1.2.16) in the form $\langle T \rangle = \int_0^\infty tf(t)\,dt$. In some cases, when the PDF for failure is complex or when a failure is a result of a sequence of events, it may become somewhat difficult to calculate. The next example deals with such a process.

Example 3.3.3: Estimating the MTTF of a set of rollers. A two stages process

Consider two coated, heavy concrete rollers, used in the process of thinning stainless steel slabs to a desired width. Such rollers operate under severe conditions of high temperature and stress, and must maintain a polished surface. Failure is caused in two stages. Initially, small micro cracks are formed on the surface. The cracks are accumulated and increase in number and size until at some point a fast process appears, causing large pieces of the coating layer to brake. The first process is believed to follow a Weibull distribution rule with a linearly increasing hazard ($\beta = 2$). This is the initial failure or triggering failure. The second process starts when the triggering failure occurs, and leads to a distraction of large coating pieces. This constitutes a failure of the roller. The second process follows a lognormal distribution. Denoting the PDF for the first process by $f_{in}(t_i)$ and the PDF of the second process by $f_s(t_s)$, and noting that the second process begins at t_i, the PDF for a failure of a roller, $f_r(t)$, can be expressed as

$$f_r(t) = \int_0^t f_{in}(t') f_s(t - t')\, dt',$$

namely a convolution of the two consecutive processes. The mean time to failure is the sum of the mean times of the two consecutive processes. This can be shown as follows:

$$\langle T \rangle = \int_0^\infty t f_r(t)\, dt = \int_0^\infty \int_0^t t f_{in}(t') f_s(t - t')\, dt'\, dt = \int_0^\infty dt' \int_0^\infty t f_{in}(t') f_s(t - t')\, dt'\, dt.$$

The limits of integration have been interchanged above. Using the transformation $t - t' = x$ yields

$$\begin{aligned}\langle T \rangle &= \int_0^\infty dt' \int_0^\infty (x + t') f_{in}(t') f_s(x)\, dt'\, dx \\ &= \int_0^\infty dt' \int_0^\infty x f_{in}(t') f_s(x)\, dt'\, dx + \int_0^\infty dt' \int_0^\infty t' f_{in}(t') f_s(x)\, dt'\, dx,\end{aligned}$$

or

$$\langle T \rangle = \int_0^\infty f_{in}(t')\, dt' \int_0^\infty x f_s(x)\, dx + \int_0^\infty t' f_{in}(t')\, dt' \int_0^\infty f_s(x)\, dx = \langle t' \rangle + \langle x \rangle. \tag{3.3.26}$$

This is a general result which can be extended to any number of distributions. It simply states that the mean time for a sequence of events is the sum of the mean times of the events.

The explicit PDF for the initial event is a Weibull distribution with $\lambda = 2.22217 \times 10^{-5}$ and $\beta = 2$, namely, $f_{in}(t) = 2\lambda t\, \mathrm{e}^{-\lambda t^2}$. The mean time of this distribution is 188 hours. The PDF for the second event is the lognormal distribution

$$f_s(t) = \frac{1}{t\sqrt{2\pi\sigma^2}} \mathrm{e}^{-(1/2)((\ln(y) - \mu)/\sigma)^2},$$

where $\mu = 3.6984$ and $\sigma = 0.280127$ are the mean and standard deviation of the corresponding normal distribution (see Section 1.3.5). The mean and standard deviation of the lognormal distribution are 42 hours and 12 hours, respectively. The mean time to failure of a single roller is then 230 hours.

The first question concerns the probability that a failure will occur up to a given time T. This is the CDF of $f_r(t)$, namely

$$F_r(T) = \int_0^T f_r(t)\,dt = \int_0^T\int_0^t f_{in}(t')f_s(t-t')\,dt'\,dt. \tag{3.3.27}$$

This is a typical two-dimensional integral. To calculate it by uniform sampling, multiply and divide the external integral by $U_1(t) = 1/T$ and the inner integral by $U_2(t') = 1/t$, obtaining

$$F_r(T) = T\int_0^T tU_1(t)\int_0^t f_{in}(t')f_s(t-t')U_2(t')\,dt'\,dt. \tag{3.3.28}$$

Sample t uniformly in $[0, T]$. Then, for a given sampled t, sample t' in $[0, t]$ and use the estimator $\eta_1(t', t) = T \times t \times f_{in}(t')f_s(t-t')$. (The integration of (3.3.28) is done with the program `integ2d.for`.)

Again, the integral of (3.3.27) can be written as an average over a system's histories. First exchange the limits of integration, writing:

$$F_r(T) = \int_0^T f_r(t)\,dt = \int_0^T\int_0^t f_{in}(t')f_s(t-t')\,dt'\,dt = \int_0^T dt'\int_{t'}^{T'} f_{in}(t')f_s(t-t')\,dt.$$

Now define the variable $x = t - t'$ such that the integral takes the form

$$F_r(T) = \int_0^T dt'\int_0^{T-t'} f_{in}(t')f_s(x)\,dt.$$

Denote $t' = t_1$ and $x = t_2$. Multiply the integrand by the binomial estimator

$$\eta_a(t_1; t_2) = \begin{cases} 1 & \text{if } t_1 + t_2 < T \\ 0 & \text{otherwise,} \end{cases}$$

which sets the integrand to zero beyond the original integration limits, and obtain the expression

$$F_r(T) = \int_0^\infty\int_0^\infty \eta_a(t_1, t_2)f_{in}(t_1)f_s(t_2)\,dt_1\,dt_2. \tag{3.3.29}$$

The desired quantity is an average of $\eta_a(t_1; t_2)$ over the phase space. A single history consists of the pair (t_1, t_2) designating the time of the first failure and that of the second failure. If the second failure occurs before T, it is considered a 'success' in the binomial experiment. To estimate the integral, sample t_1 from the Weibull distribution. Sample t_2 from the lognormal distribution (Section 3.2.4), and evaluate the estimator. Note that, in so doing, the natural histories of the system are simulated. The initial failure point is sampled, then the second failure point, and if the second point lies within the interval $[t_1, T_s]$, the estimator is taken to be unity.

Results of the two estimation methods are shown in Figures 3.3.6(a), (b) and (c) in the time range $[0, 600]$ hours.

It can be seen that when the probability for failure is small, the analog estimator has a high PRSD and a low FOM. The uniform estimator is by far superior in this region, yet for high probabilities the analog estimator is considerably better. The region $[350, 600]$ hours is shown

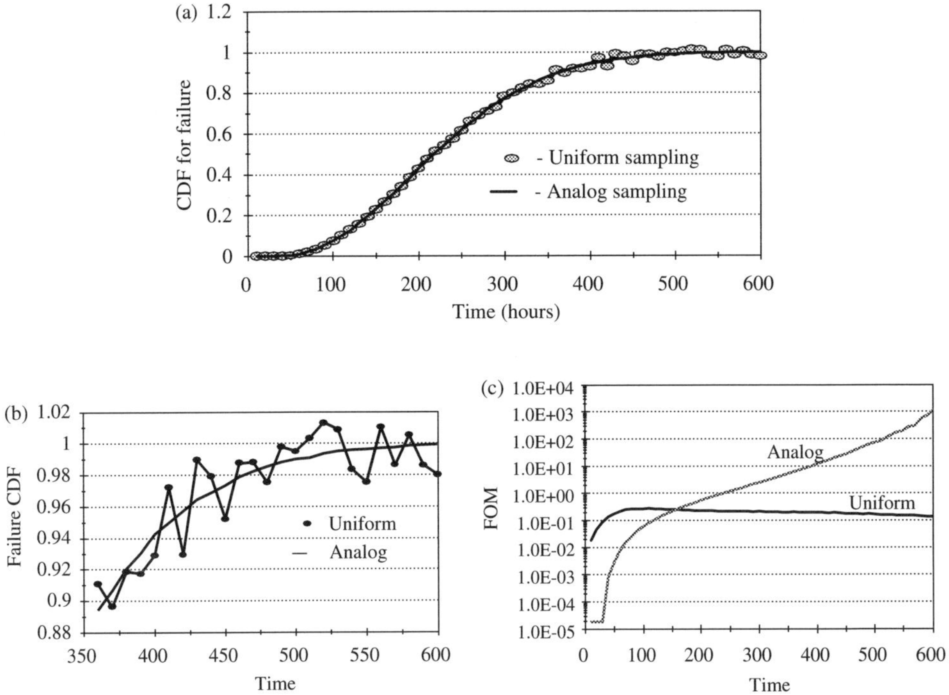

Figure 3.3.6 (a) CDF for a failure of a roller, estimated uniformly and by an analog estimator; (b) CDF for failure in the range [350,600]; (c) the FOM of the estimators

separately in Figure 3.3.6(b), where for the same number of histories (40 000) the fluctuations of the uniform estimator are evident. The FOM as function of the time point T is shown in Figure 3.3.6(c).

The computer time needed for each point is independent of time, and was 3.4 seconds for the uniform case and 5.35 seconds for the analog (for 40 000 histories). The difference between the two methods is rather striking. The FOM of the uniform method is, to a large extent, independent of the time range. It is an integral calculation, and the variance depends only upon the variation of the integrand in the range. The FOM of the analog method varies over eight orders of magnitude. From being a factor of 10 000 worst than the uniform method, in the 'rare event' range, to being by a factor of about 10 000 better.

Next, consider the operation of the two rollers. It is known that as long as a single roller operates, the system can still function. It is desired to find the MTTF of the system of the two rollers. Now let t_1 denote the failure time of the first roller and t_2 the failure time of the second. Let $f_1(t_1)$ denote the failure PDF of the first roller and $f_2(t_2)$ that of the second. Since no repair is applied, the failure time of the system will be $t_s = \max(t_1, t_2) = h(t_1, t_2)$, i.e. the largest of the two. This can also be viewed as a function $h(t_1, t_2)$ of the two r.v.s. As such, the joint PDF of the pair (t_1, t_2) is the product $f_1(t_1) \times f_2(t_2)$, and the average value of the function is given by

$$\text{MTTF} = \langle t_s \rangle = \int_0^\infty \int_0^\infty h(t_1, t_2) f_1(t_1) f_2(t_2)\, dt_1\, dt_2. \tag{3.3.30}$$

Equation (3.3.30) implies that t_1 can be sampled from $f_1(t_1)$, t_2 is sampled from $f_2(t_2)$ and $h(t_1, t_2)$ is the estimator. In this case, the simple integral estimation coincides with the analog averaging over the system's histories, because the sampling functions occur naturally in the statistical process. Using this method, an estimate of 285.71 hours is obtained for the MTTF with 20 000 histories and a statistical error of 0.23% (calculation time 5.3 seconds on a PC 486, 75 MHz). Note that the above is an analog estimation method with a non-binomial estimator.

The same problem can be viewed somewhat differently by noting that the probability that t_s is smaller than any given t is given by

$$P_r(t_s \leq t) = F(t) = F_1(t) \times F_2(t). \tag{3.3.31}$$

namely, it is the probability that both t_1 and t_2 are smaller than t. $F(t)$ is the CDF of the failure time of the system. Taking the derivative with respect to t, the PDF for the system is obtained as

$$f(t) = \frac{d[F_1(t) \times F_2(t)]}{dt} = F_1(t) \times f_2(t) + F_2(t) \times f_1(t). \tag{3.3.32}$$

This expression can be generalized such that the PDF of a variable t defined as the maximum of k r.v.s (i.e. $t = \max\{t_1, t_2, \ldots, t_k\}$ with PDFs $f_1(t_1), \ldots, f_k(t_k)$ is given by

$$f(t) = \sum_{i=1}^{k} \left[\prod_{\substack{j=1 \\ j \neq i}}^{k} F_j(t) \right] f_i(t). \tag{3.3.33}$$

This is the failure PDF of any system with k independent components in parallel. The mean time to failure is then given by

$$\text{MTTF} = \langle t \rangle = \int_0^\infty t f(t)\, dt = \int_0^\infty t \times [F_1(t) \times f_2(t) + F_2(t) \times f_1(t)]\, dt. \tag{3.3.34}$$

Using integration by part, it is seen that the MTTF can also be written as

$$\begin{aligned} \text{MTTF} = \langle t \rangle &= \int_0^\infty t f(t)\, dt = \int_0^\infty t \frac{dF}{dt}\, dt = -\int_0^\infty t \frac{dR}{dt}\, dt \\ &= -R(t) \times t|_0^\infty + \int_0^\infty R(t)\, dt = \int_0^\infty R(t)\, dt. \end{aligned} \tag{3.3.35}$$

$R(t)$ is the reliability, $1 - F(t)$, of the system (equation (2.2.1)). This is true provided $\lim_{t \to \infty}[t \times R(t)] = 0$.

These expressions are not convenient for a uniform Monte Carlo integration for two reasons. First, the infinite range of the integration creates a real problem in sampling from a uniform distribution. Secondly, the integrand contains the CDF of a lognormal distribution which has no explicit form. Still, an approximated evaluation can be applied using the CDF, $F_r(t)$, of a single roller which was calculated in the first part of this example. Substituting this for both $F_1(t)$ and

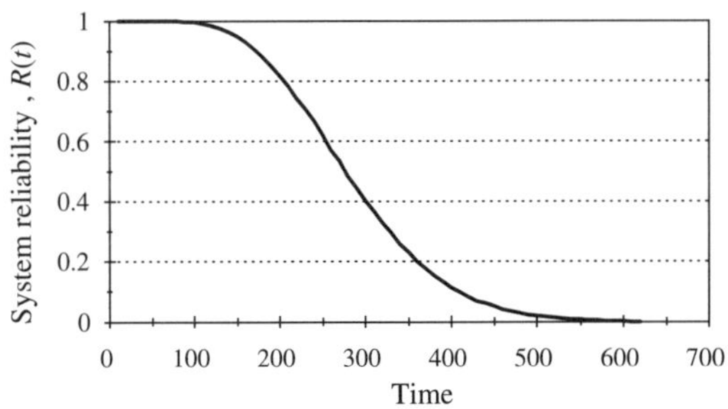

Figure 3.3.7 Reliability of the two rollers system

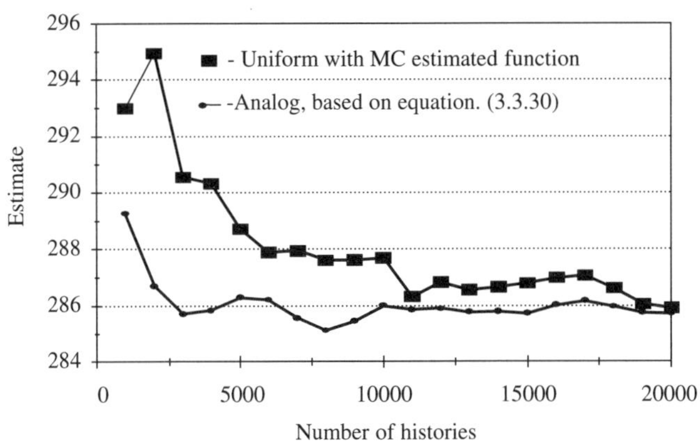

Figure 3.3.8 Analog and uniform Monte Carlo integration of the MTTF of the rollers system

$F_2(t)$ in (3.3.31) and (3.3.35) yields

$$\mathrm{MTTF} = \langle t \rangle = \int_0^\infty [1 - F_r^2(t)]\, dt.$$

A graph of the integrand is shown in Figure 3.3.7.

The figure indicates that the reliability is negligible beyond 700 hours. Thus, the integral can be reduced to the range $[0, 700]$. The integrand can be tabulated and a simple Monte Carlo integration can be applied (the function `roller` can be embedded in `integ.for` to perform this integration). The result of 20 000 histories is 285.9 with a standard deviation of 0.62%, which is in good agreement with the analog calculation described above. The estimates resulting from the analog and uniform methods, as a function of the number of histories, are presented in Figure 3.3.8. The superior convergence of the analog method is evident.

Calculating integrals using the Monte Carlo method is one of the basic applications of this method.

4 Additional Applications

4.1 Availability of Systems with Independent Components

Consider a system with mutually independent components, two states for each component and a system function $S(\vec{\mathbf{B}})$. The availability of the ith component is denoted by $p_i(t)$. The availability of such systems was discussed in Section 2.1.2, where it was shown that it can be obtained by substituting $p_i(t)$ into the normalized system function in place of the state indicators. For example, the availability of the system shown in Figure 4.1.1 is given by

$$A(t) = [p_1(t) \times p_3(t) + p_2(t) \times p_4(t) - p_1(t) \times p_3(t) \times p_2(t) \times p_4(t)] \times p_5(t).$$

If each component has a failure rate λ_i and repair rate μ_i, expression (2.2.11) can be used, yielding

$$p_i(t) = \frac{\mu_i}{\mu_i + \lambda_i} + \frac{\lambda_i}{\mu_i + \lambda_i} e^{-(\mu_i+\lambda_i)t},$$

which can be inserted into the above expression to obtain the time dependent availability of the system.

The above approach is inadequate if a general computerized solution is required, since a different normalized system function must be derived for each system. It is desired to have a general solution by which, for any non-normalized $S(\vec{\mathbf{B}})$, a general solution can be obtained. It should be noted that for large systems, such as that shown

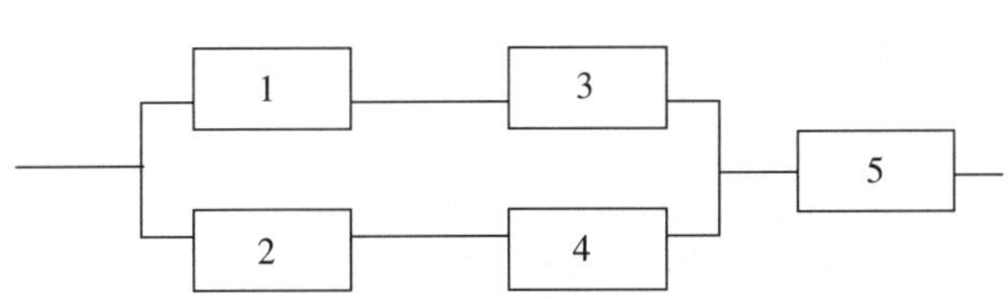

Figure 4.1.1 A system block diagram

in Figure 4.1.8, the derivation of a normalized system function can become prohibitively difficult, whereas a non-normalized system function is relatively easy to obtain.

A general algorithm, based on a non-normalized system function, can be realized with the following consideration: the state space contains 2^n different state vectors of the form $\vec{\mathbf{B}}_j = (b_1^j, b_2^j, \ldots, b_n^j)$ with $j = 1, \ldots, 2^n$; b_i can take the values 1 and 0 and $p_i = P_r[b_i = 1]$. Denoting

$$\beta_i = \begin{cases} p_i & \text{if } b_i = 1 \\ 1 - p_i & \text{if } b_i = 0. \end{cases} \tag{4.1.1}$$

the probability that the state vector is $\vec{\mathbf{B}} = (b_1, b_2, \ldots, b_n)$ at time t (since p_i may be time dependent), is given by

$$P_r(\vec{\mathbf{B}}, t) = \prod_{i=1}^{n} \beta_i. \tag{4.1.2}$$

Now let 0_s denote the set of state vectors $\{\vec{\mathbf{B}}\}$ for which the system is operational, i.e. $0_s = \{\vec{\mathbf{B}}: S(\vec{\mathbf{B}}) > 0\}$. The complementary of 0_s, 0_f, is defined as $0_f = \{\vec{\mathbf{B}}: S(\vec{\mathbf{B}}) = 0\}$. Since the state vectors are mutually exclusive, the probability of the event 0_f is the sum of the probabilities of the state vectors included in it. Thus, the unavailability can be expressed in the form

$$U(t) = 1 - A(t) = P_r[S(\vec{\mathbf{B}}, t) = 0] = P_r[\Gamma_f] = \sum_{\substack{j=1 \\ \vec{\mathbf{B}}_j \in \Gamma_f}}^{2^n} P_r(\vec{\mathbf{B}}_j, t) = \sum_{\substack{j=1 \\ S(\vec{B}_j)=0}}^{2^n} \left(\prod_{i=1}^{n} \beta_i^j(t) \right). \tag{4.1.3}$$

Equation (4.1.3) suggests a very simple algorithm for the calculation of the unavailability, namely:

1. Scan through all 2^n state vectors.
   1.1. For each state vector $\vec{\mathbf{B}}_j$, check if $S(\vec{\mathbf{B}}_j) = 0$.
   1.2. If condition 1.1 is fulfilled, calculate the state probability according to expression (4.1.2) and add it to the unavailability, otherwise proceed to the next state vector.

This procedure is displayed in Figure 4.1.2 and realized in the program `anasys.for`.

Algorithm 4.1.1 requires a scan over all 2^n vector states. Table 4.1.1 shows the results of calculations for a serial system with n components, where n varies from 15 to 21. The availability of each component is $p_i = p = 0.998$, $i = 1, \ldots, n$. Indeed, this is a trivial case. The unavailability could be calculated as $U = 1 - p^n$, but the algorithm is intended for any system function. The number of different state vectors, and therefore the calculation time, depends only upon n (the dimension of the problem), and is independent of the particular system function used.

It is easy to verify from the table that the calculation time increases exponentially. In fact, it increases slightly faster then exponentially, because of the linear increase in the

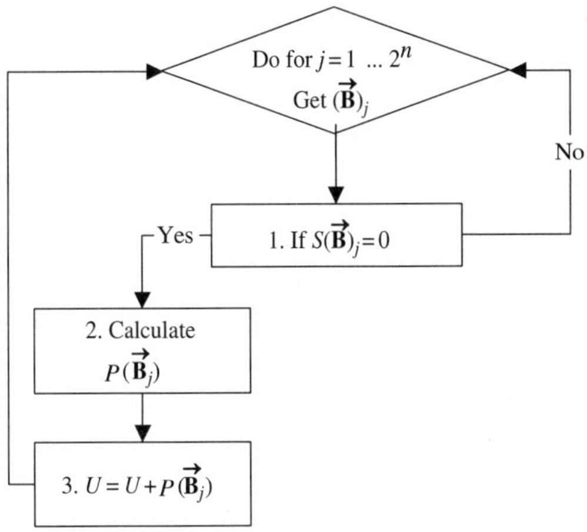

Figure 4.1.2 Flow of a general analytic algorithm

calculation time of equation (4.1.2). The calculation time τ_c can be expressed as $\tau_c = (n\tau_0) \times 2^n$, where τ_0 is a parameter. Fitting τ_c to the results in the table yields $\tau_0 = 5.61 \times 10^{-6}$. The estimated values of τ_c appear in the last column of the table. By extrapolation, the calculation time for 22 components will be 517 s, for 30 components 180 710 seconds or about 50 hours, and for 50 components the calculation time exceeds 1000 years! Calculation of systems with more then 30–40 components is practically impossible. At first glance this seems an odd conclusion; after all, it is such a simple algorithm. A more serious observation may reveal that the obstacle is quite insurmountable. Consider a 50-dimensional integral solved numerically. Dividing every coordinate into only two intervals will result a sum of 2^{50} points. The two phenomena are identical.

Recall Example 2.1.3, where the process of averaging over the phase space was considered. The need to 'scan' through the entire space is a profound feature of all analytic methods. Similarly, in equation (2.2.2) and (2.1.2) every history in the phase space must be accounted for, and in the Markov analysis every state must be considered explicitly. These are all different appearances of the same phenomenon.

When a simple integral of the form $\int_a^b f(x)\,dx$ is considered, every point in the space (the interval $[a, b]$) must be accounted for. One cannot omit any section of a non-zero measure without causing an error. In deterministic calculations, the space may be

Table 4.1.1 Calculation time as function of number of components

Number of components	Unavailability	Number of vector states	Computer time (sec)	Estimated time – τ_c
15	0.029583	32 768	2.59	2.7
16	0.031523	65 536	5.49	5.8
17	0.03346	131 072	11.98	12.5
18	0.03539	262 144	25.5	26.4
19	0.03732	524 288	54.32	56.3
20	0.03924	1 048 576	116.0	117.6
21	0.04116	2 097 152	247.1	247.06

viewed as a 'forest', where each phase space point is represented by a tree. To perform a deterministic calculation, one must enter the forest and 'touch' every tree. In the case of Figure 4.1.8, the number of trees becomes so large that it is impossible to perform this process in any reasonable amount of time. The need to 'touch' every tree does not ever disappear. In performing integration the scanning of the continuum can be done very quickly if the primitive function is known. Indeed, there are some cases in which additional knowledge about the connections or structure of the continuum of the forest allows a very fast process (this will also be the case if a 50 component system is known to be serial). However, if this is not the case one must resort to numerical integration. In such integration, intervals Δx_i are used instead of points of zero measure, and the integral over the continuous region is turned into a sum of the form

$$\sum_{i=1}^{k} f(x_i)\Delta x_i.$$

In terms of the 'forest picture' this means that one groups a large set of trees and assumes that they are the same tree, i.e. the value of the function (which is the only property of that tree) is identical for all the trees in the group. Thus, 'touching' one chosen tree in the group (and multiplying its property by the size of the group) is the same as touching each and every tree. This certainly saves time, but introduces an error which depends on $1/k$. Consider now an n-dimensional forest. To retain the same error, each dimension must be divided into k intervals, and the number of terms in the sum is $m = k^n$. Since the error is proportional to $1/k$, it will converge as $1/\sqrt[k]{m}$. This means, for example that, in a ten-dimensional integration, reducing the error by a factor of two requires an increase of the number of points, and thus the required computer time, by a factor of 2^{10}.

As already discussed in Section 3.2.1, the error in a Monte Carlo calculation is converging as $1/\sqrt{m}$ *independent of the dimension.* As in numerical integration, each of the possible k states of a component is equivalent to an interval, and the number of components n is the dimension of the problem, so that the total number of points in the space is k^n. Furthermore, consider two components having two events, both entering into state 1 at points t_1 and t_2 such that $t_1 < t_2$. The probability density for a next event in any of the components at time $t > t_2$ can be expressed as

$$[1 - F_2(t - t_2)] \times \left[1 - \frac{F_1(t - t_1) - F_1(t_2 - t_1)}{1 - F_1(t_2 - t_1)}\right][z_1(t - t_1) + z_2(t - t_2)].$$

This expression takes into account the fact that, upon the event in component two at time t_2 it is known that the first component did not change its state up to that time. Likewise, if $t_2 < t_1$ the expression is changed into

$$[1 - F_1(t - t_1)] \times \left[1 - \frac{F_2(t - t_2) - F_2(t_1 - t_2)}{1 - F_2(t_1 - t_2)}\right][z_1(t - t_1) + z_2(t - t_2)].$$

The above two expressions are different for any distribution, except the exponential distribution. In both cases, the two components enter into a state $(1, 1)$ at the time at which the last component enters its respective state. It is clear that the state is dependent on the order in which the components enter their respective states. Thus, the number of states in a general system of n components would be of the order of $n! \times K^n$. This provides a glimpse into the source of complexity of reality. This means, for example, that for 20 components there are 8.48×10^{27} states, and it is very easy to reach a number of states larger then the number of molecules in the universe with less then 100 components. Reality is so infinitely complex, not because of the number of its constituents, but because of the number of their states and interactions.

The exploding complexity due to dimension stands as a permanent border for the application of deterministic solutions. This is sometimes referred to as the 'curse of dimension'. The Monte Carlo method overcomes this barrier as it is independent of the dimensions of the problem. This amazing feature is attributed to the Central Limit Theorem (it is one additional corollary of this theorem). Solving the above problem using the Monte Carlo method demonstrates this quality.

4.1.1 Solution by Monte Carlo

The phase space is defined by the vectors $\vec{\mathbf{B}}_j = (b_1^j, b_2^j, \dots, b_n^j)$. Define the estimator as the random variable

$$\eta(\vec{\mathbf{B}}_j) = \begin{cases} 1 & S(\vec{\mathbf{B}}_j) = 0 \\ 0 & \text{otherwise.} \end{cases} \tag{4.1.4}$$

The mean value of $\eta(\vec{\mathbf{B}}_j)$ is given by

$$E(\eta) = \sum_{j=1}^{2^n} P_r(\vec{\mathbf{B}}_j, t)\eta(\vec{\mathbf{B}}_j) = \sum_{\substack{j=1 \\ S(\vec{\mathbf{B}}_j)=0}}^{2^n} P_r(\vec{\mathbf{B}}_j, t) = U(t). \tag{4.1.5}$$

Thus, $\eta(\vec{\mathbf{B}}_j)$ is an unbiased estimator of $U(t)$.

To sample a state vector, discrete sampling (Section 3.2.2) is used. For $i = 1, \dots, n$, a random number ξ_i is compared to p_i. If $\xi_i \le p_i(t)$, then $b_i = 1$, otherwise $b_i = 0$. The estimation of $U(t)$ using N histories takes the form:

1. Use N histories, $j = 1, \dots, N$.
   1.1. In each history sample a state vector $\vec{\mathbf{B}}_j$.
   1.2. If $S(\vec{\mathbf{B}}_j) = 0$ set $\eta(\vec{\mathbf{B}}_j) = 1$, otherwise set $\eta(\vec{\mathbf{B}}_j) = 0$.
2. Upon completion of N histories, the estimate of the unavailability is

$$\tilde{U} = \frac{1}{N}\sum_{j=1}^{N} \eta(\vec{\mathbf{B}}_j),$$

with the PRSD given by

$$\varepsilon = 100 \times \sqrt{\frac{1-\tilde{U}}{N \times \tilde{U}}}.$$

This procedure is shown in Figure 4.1.3.

In Table 4.1.2, results for the same problem as that described in Table 4.1.1 are displayed for systems with up to 300 components.

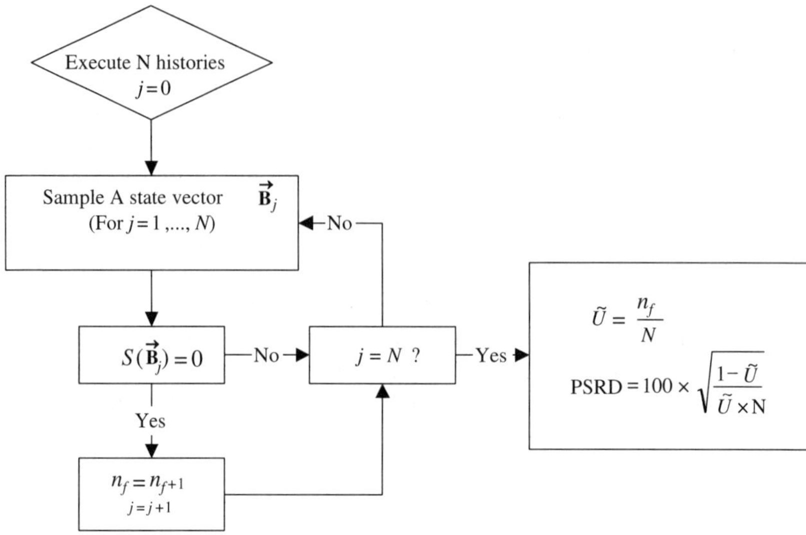

Figure 4.1.3 Flowchart of a Monte Carlo estimation of the unavailability

For each calculation in the table, 100 000 case histories are used [`Program mcsis.for`]. The table indicates that the calculation time depends linearly upon the number of components. The calculation time for 300 components is about 112 seconds (on a 486 PC machine). For the analytic algorithm the same calculation will require the imaginary time of $1.087 \times 10^{80}_{\text{years}}$, by far longer than the time since the 'big bang'.

In this particular problem the unavailability increases with the number of components, therefore the PRSD decreases. The PRSD of the first case is 3.13%, while that of the latter is 0.45%. To obtain a PRSD of 0.45% for the first case, the number of histories must be increased by a factor of

$$\left(\frac{3.13}{0.45}\right)^2 = 48.37,$$

Table 4.1.2 Results of Monte Carlo calculations on a serial system

Number of components	Unavailability estimate $\tilde{U}$	Exact result U	PRSD (%)	Calculation time (sec)
5	0.00983	0.00996	3.13	3.02
15	0.0291	0.0295	1.82	6.65
30	0.0581	0.0582	1.27	12.2
40	0.0774	0.0769	1.1	16.1
50	0.0946	0.0952	0.98	19.5
70	0.1297	0.1307	0.82	26.8
90	0.1655	0.1648	0.71	34.0
100	0.1821	0.1814	0.67	37.0
300	0.3293	0.3299	0.45	111.45

which will increase the calculation time from 3.02 to 145.6 seconds. Thus, for the same statistical error the 300 components case is faster to calculate than that of the five components. The problem of 'rare events', discussed in Section 3.3 is reflected here. This problem arises in every MC calculation, since the number of histories required to obtain a given PRSD is inversely proportional to the root of the probability of occurrence of the event. Thus, whereas the MC method is insensitive to the complexity of the phase space, it is sensitive to the 'rarity' of the phenomena studied. To partially overcome this problem, methods by which the variance of the calculation may be reduced are considered. These methods are called, in general, Variance Reduction Methods (VRM).

4.1.2 Variance reduction for the independent components availability problem

Variance reduction methods are often based on seeking a different set of probabilities for the occurrence of the events. Noting that the reason for the high variance stems from the fact that the event of interest (in this case a state vector for which the system is failed) is rarely sampled, one looks for a set of probabilities that will enhance the sampling of such events. Since the new set of probabilities differs from the natural probabilities (the analog probabilities), the process is also referred to as 'biasing'. Consider expression (4.1.5) for the unavailability

$$E(\eta) = \sum_{j=1}^{2^n} P_r(\vec{\mathbf{B}}_j, t)\eta(\vec{\mathbf{B}}_j).$$

Let k denote the number of zeros in the vector $\vec{\mathbf{B}}$, i.e. k is the number of failed components. Let Λ_k be the subset of all state vectors with exactly k failed components. The number of state vectors in Λ_k is $\binom{n}{k}$. The averaging of expression (4.1.5) can be written as

$$E(\eta) = \sum_{j=1}^{2^n} \eta(\vec{\mathbf{B}}_j)P_r(\vec{\mathbf{B}}_j, t) = \sum_{k=0}^{n} \sum_{\vec{\mathbf{B}}\in\Lambda_k} \eta(\vec{\mathbf{B}}_j)P_r(\vec{\mathbf{B}}_j, t). \tag{4.1.6}$$

The second moment, $S^2(\eta) = E(\eta)$, because $\eta(\vec{\mathbf{B}})$ is a binomial estimator.

Biasing is applied by changing the state probabilities using a new set of probabilities, $P_r^*(\vec{\mathbf{B}}_j, t)$, instead of $P_r(\vec{\mathbf{B}}_j, t)$. To keep the estimator (the first moment) unchanged (unbiased), the expression is written as

$$\begin{aligned} E(\eta) &= \sum_{k=0}^{n} \sum_{\vec{\mathbf{B}}\in\Lambda_k} \left(\eta(\vec{\mathbf{B}}_j) \frac{P_r(\vec{\mathbf{B}}_j, t)}{P_r^*(\vec{\mathbf{B}}_j, t)} \right) P_r^*(\vec{\mathbf{B}}_j, t) = \sum_{k=0}^{n} \sum_{\vec{\mathbf{B}}\in\Lambda_k} (\eta(\vec{\mathbf{B}}_j)W(\vec{\mathbf{B}}_j, t))P_r^*(\vec{\mathbf{B}}_j, t) \\ &= \sum_{k=0}^{n} \sum_{\vec{\mathbf{B}}\in\Lambda_k} \eta^*(\vec{\mathbf{B}}_j)P_r^*(\vec{\mathbf{B}}_j, t). \end{aligned} \tag{4.1.6a}$$

This can be compared to the multiplication and division by a new PDF done for integrals estimation.

$$W(\vec{\mathbf{B}}_j, t) = \frac{P_r(\vec{\mathbf{B}}_j, t)}{P_r^*(\vec{\mathbf{B}}_j, t)}$$

is called a weight. In the above process, the analog probabilities are replaced by the biased probabilities. To maintain the estimator unbiased the analog estimator $\eta(\vec{\mathbf{B}}_j)$ is multiplied by a weight $W(\vec{\mathbf{B}}_j, t)$ yielding a new estimator $\eta^*(\vec{\mathbf{B}}_j)$. While maintaining the first moment unchanged, it is desired to change the second moment so that it is smaller then the analog second moment, leading to a reduction of the variance.

The second moment takes the form

$$\begin{aligned} S^{*2}(\eta) &= \sum_{k=0}^{n} \sum_{\vec{\mathbf{B}} \in \Lambda_k} \left(\eta(\vec{\mathbf{B}}_j) \frac{P_r(\vec{\mathbf{B}}_j, t)}{P_r^*(\vec{\mathbf{B}}_j, t)} \right)^2 P_r^*(\vec{\mathbf{B}}_j, t) \\ &= \sum_{k=0}^{n} \sum_{\vec{\mathbf{B}} \in \Lambda_k} \eta(\vec{\mathbf{B}}_j) \left(\frac{P_r(\vec{\mathbf{B}}_j, t)}{P_r^*(\vec{\mathbf{B}}_j, t)} \right) P_r(\vec{\mathbf{B}}_j, t). \end{aligned} \tag{4.1.7}$$

Optimum values for the biased probability set can be obtained using Lagrange multipliers similar to the method used for integration with piecewise constant distributions (equations (3.3.17)–(3.3.20)). Taking the derivative of the second moment with the constraint of the normalization of the biased probabilities yields

$$\frac{dS^{*2}(\eta)}{dP_r^*(\vec{\mathbf{B}}_j, t)} = \frac{d}{dP_r^*(\vec{\mathbf{B}}_j, t)} \left\{ \sum_{j=1}^{2^n} \eta(\vec{\mathbf{B}}_j) \left(\frac{P_r^2(\vec{\mathbf{B}}_j, t)}{P_r^*(\vec{\mathbf{B}}_j, t)} \right) + \lambda \left(\sum_{j=1}^{2^n} P_r^*(\vec{\mathbf{B}}_j, t) - 1 \right) \right\} = 0, \tag{4.1.8}$$

yielding

$$-\eta(\vec{\mathbf{B}}_j) \left(\frac{P_r^2(\vec{\mathbf{B}}_j, t)}{P_r^{*2}(\vec{\mathbf{B}}_j, t)} \right) + \lambda = 0 \quad \text{or} \quad P_r^*(\vec{\mathbf{B}}_j, t) = \sqrt{\frac{\eta(\vec{\mathbf{B}}_j) P_r^2(\vec{\mathbf{B}}_j, t)}{\lambda}}.$$

Upon substitution to the normalization condition, and using the fact that $\eta(\vec{\mathbf{B}}_j)$ is a binomial estimator, one obtains the optimal probabilities in the form

$$P_r^*(\vec{\mathbf{B}}_k)_{\text{opt}} = \frac{\eta(\vec{\mathbf{B}}_k) P_r(\vec{\mathbf{B}}_k, t)}{\sum_{j=1}^{2^n} \eta(\vec{\mathbf{B}}_j) P_r(\vec{\mathbf{B}}_j, t)}. \tag{4.1.9}$$

The product of the analog estimator and the probability represents the true importance of the state vector. Optimal biasing is obtained when the biased probabilities are proportional to the true importance for each point in the space. Note that all the state vectors for which the system is not failed have zero importance. The recipe in expression (4.1.9) is simple to interpret but difficult to apply. One has no direct control over the sampling of the state vector.

Substituting the optimal probabilities back in (4.1.7) reveals the same effect as in the case of optimal biasing of integrals (equation (3.3.11a)). A zero variance is obtained, but the denominator of (4.1.9) is simply the unavailability. Thus, a zero variance MC is again obtained if the final answer is known, this time on a different space.

This is thus a general observation. Zero variance is obtained if the biased probabilities (or distributions) are proportional to the true importance of every point in space with respect to the target quantity. To obtain such proportionality and maintain the necessary normalization of the biased probabilities (distributions), the average of the target quantity must be known. This average is, on the other hand, the target of the calculation.

To apply biasing and improve the efficiency (FOM) of the calculation, some approximating assumptions must be used, as shown in the following.

Single parameter biasing

Assuming that the components have identical availabilities, namely $p_i(t) = p \ \forall i = 1, \ldots, n$, the probability of any state $\vec{\mathbf{B}} \in \Lambda_k$ is $P_k = (1-p)^k p^{(n-k)}$. The probability depends only upon the number of failed components in the state. Expression (4.1.6) then takes the form

$$E(\eta) = \sum_{k=0}^{n} (1-p)^k p^{(n-k)} \left[\sum_{\vec{\mathbf{B}} \in \Lambda_k} \eta(\vec{\mathbf{B}}_j) \right] = \sum_{k=0}^{n} (1-p)^k p^{(n-k)} A_k, \tag{4.1.10}$$

where

$$A_k = \sum_{\vec{\mathbf{B}} \in \Lambda_k} \eta(\vec{\mathbf{B}}_j)$$

is the number of states in Λ_k which results in a failed system. Let $p^* = \alpha p$ with $\alpha \leq 1$. Reducing the availability of each component will increase the probability of subsets Λ_k of higher k to appear, thereby increasing the probability of sampling a failed state vector. With the above formulation, $P_r^*(\vec{\mathbf{B}}) = (1-p^*)^k p^{*(n-k)}$. Substituting in (4.1.6a) and (4.1.7) yields

$$E(\eta) = \sum_{k=0}^{n} (1-p^*)^k p^{*(n-k)} \left[\sum_{\vec{\mathbf{B}} \in \Lambda_k} \frac{(1-p)^k p^{(n-k)}}{(1-p^*)^k p^{*(n-k)}} \eta(\vec{\mathbf{B}}_j) \right], \tag{4.1.11a}$$

and

$$\begin{aligned} S^{*2}(\eta) &= \sum_{k=0}^{n} \sum_{\vec{\mathbf{B}} \in \Lambda_k} \eta(\vec{\mathbf{B}}_j) \left(\frac{(1-p)^k p^{(n-k)}}{(1-p^*)^k p^{*(n-k)}} \right) (1-p)^k p^{(n-k)} \\ &= \sum_{k=0}^{n} \frac{(1-p)^{2k} p^{(n-k)}}{(1-\alpha p)^k \alpha^{(n-k)}} A_k. \end{aligned} \tag{4.1.11b}$$

To realize this biasing technique, sampling of components states is done with a probability $p^* = \alpha p$. The estimator $\eta(\vec{\mathbf{B}}_j)$ is multiplied, for each component, by $(1-p)/(1-p^*)$ if the state is failed, or by p/p^* if the state is not failed. The accumulated weight will become

$$w(\vec{\mathbf{B}}_j) = \frac{(1-p)^k p^{(n-k)}}{(1-p^*)^k p^{*(n-k)}}$$

in accordance with equation (4.1.11a). By estimating A_k the second moment can be calculated as function of α, and an optimal value of α can be estimated.

For the serial case discussed above, $A_0 = 0$ and for $k > 0$ $A_k = \binom{n}{k}$. Since all vector states with at least one failed component are failed states, the second moment takes the form

$$S^{*2}(\eta) = \sum_{k=1}^{n} \binom{n}{k} \frac{(1-p)^{2k} p^{(n-k)}}{(1-\alpha p)^k \alpha^{(n-k)}} = \left(\frac{\alpha - 2\alpha p + p}{(1-\alpha p)\alpha}\right)^n - \left(\frac{p}{\alpha}\right)^n. \tag{4.1.12}$$

For $\alpha = 1$ the second moment yields the unbiased result of $1-p^n$, which equals the unavailability. As α approaches zero, the second moment diverges indicating that no information can be obtained. An optimum value can be found by plotting $S^{*2}(\eta)$ as function of α. This plot is shown in Figure 4.1.4 for a number of values of p ranging from 0.99 to 0.99999.

Two effects can be noted from the figure. The 'optimum zone' is wide. Values of α in the range (0.5, 0.95) will produce near-optimum results. A sharp increase of the second moment occurs outside this zone (note the logarithmic scale). Also, the improvement brought about by biasing is increasing with the 'rarity' of the calculated event. These two effects are typical of biasing methods. The optimal results, for the five components

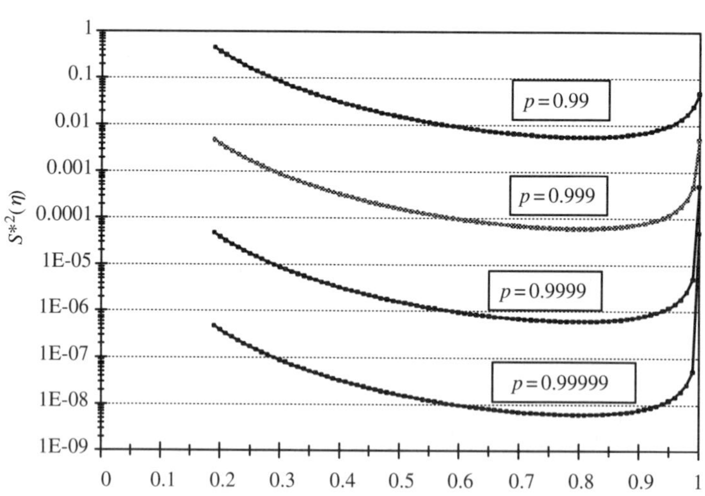

Figure 4.1.4 The biased second moment as function of the biasing parameter α

Table 4.1.3 Expected benefit at optimal single parameter biasing

P	Unavailability U	α Optimal value	$S^{*2}(\eta)$ biased second moment	σ_0 unbiased $\sqrt{U(1-U)}$	$\sigma_0^* = \sqrt{S^{*2}(\eta) - U^2}$	$B_e = \sigma_0^2\tau/\sigma_0^{*2}\tau$
0.9	0.40951	0.87	0.2884	0.4917	0.3475	1.99
0.99	0.04901	0.81	0.005637	0.2158	0.05687	13.7
0.999	0.00499	0.8	6.055E-05	0.0704	0.00597	139
0.9999	0.0004999	0.8	6.0986E-07	0.02235	0.0005999	1387
0.99999	4.9999E-05	0.8	6.103E-09	0.0070	6.00258E-05	13683

Table 4.1.4 Monte Carlo results with single parameter biasing

Component availability	Estimated unavailability $\langle\eta\rangle$	Biasing parameter α	PRSD	T calculation time (sec)	FOM	Benefit: $\frac{FOM_{Biased}}{FOM_{analog}}$
0.9	0.4099	1.0	0.3793	3.41	2.038	–
0.9	0.4084	0.87	0.2689	4.34	3.186	1.56
0.99	0.04915	1.0	1.39	3.02	0.171	–
0.99	0.04892	0.81	0.367	4.29	1.73	10.12
0.999	0.00483	1.0	4.54	3.19	0.0152	–
0.999	0.00497	0.8	0.379	4.12	1.689	111.1
0.9999	0.00043	1.0	15.24	3.02	0.001425	–
0.9999	0.0004959	0.8	0.3805	4.23	1.632	1145
0.99999	0.00005	1.0	44.72	2.97	0.000168	–
0.99999	0.000049595	0.8	0.381	4.23	1.628	9693

serial case with $p=0.9$, 0.99, 0.999, 0.9999 and 0.99999, are summarized in Table 4.1.3. The 'benefit ratio B_e' is defined as the ratio of the FOMs of the biased and unbiased (analog) cases. It indicates the time ratio of the two methods in achieving any given PRSD. It is assumed here that the processing time of a history does not change due to biasing.

In Table 4.1.4 detailed results are shown for the above five cases. For each case, two calculations are presented. The first with $\alpha = 1$ (i.e. the analog case), and the second with the optimal α value taken from Table 4.1.3.

The results in the table are shown in pairs: first the analog, and then the optimally biased case. Calculations were done with 100 000 histories in each case. Note that the benefit ratios of Table 4.1.4, derived from the actual calculations, are in good agreement with those predicted in Table 4.1.3. This confirms the assumption that the processing times are not increased by the biasing. Also, for the fifth case the unavailability is small and biasing provides a dramatic benefit of about 10 000. Every second of calculation using near optimum biasing is worth about 10 000 seconds of analog calculation. Biasing can be a very useful tool for rare events.

Although the above example provides some general features of biasing and demonstrates its efficiency, the above case can still be easily calculated analytically.

The above serial case with equal probabilities is indeed a trivial case, but it is always useful to study such simple cases since the results obtained may provide useful guidelines and insight into more elaborate cases. In more general cases, the system is not serial and the component's availabilities are not identical. The single parameter biasing can still be used as above, except that it is not that simple to obtain an optimal value of α or an exact evaluation of the expected benefit. These can only be approximated.

Consider a system of 28 components in a 12 out of 28 structures, i.e. if at least 12 components are operating the system is operating. The availabilities of the individual components are not identical. The first five components have an availability of 0.95, the next five 0.98, the next six 0.995, the next nine components 0.975, and the last three 0.921667. The average of these availabilities is 0.97. The unavailability of the system as a whole would be quite difficult to calculate analytically, because all 268 435 456 possible states of the system must be accounted for. This would require over 11 computer hours with the analytic algorithm. The analog Monte Carlo algorithm will also be extremely inefficient, since the unavailability is of the order of 10^{-20} and a failed state is extremely rare. 10^{22} histories will be required to obtain a 10% PRSD (equation (3.2.25)), and this will require about 1.13×10^{18} seconds.

Single parameter biasing will be useful in this case. To obtain an approximate prediction of the second moment equation (4.1.11b) is used with an average availability value of $p = 0.97$ for each component. This is done only as an approximate method for obtaining the optimal range of biasing. Also, $A_k = 0$ for $k < 12$ and equals $\binom{n}{k}$ for $k \geq 12$. Since the optimal region is usually wide, it is expected that the approximate expression will yield a reasonable range of the optimal parameter and an approximate value of the expected benefit. Figure 4.1.5 shows the estimated second moment (left scale). A set of calculations was done for α in the range 0.1–0.8 and the square of the PRSD and the reciprocal of the FOM are shown on the right scale. It can be seen that both the optimal range and a good estimate of the expected benefit are obtained. The experimental point at $\alpha = 1$ cannot be calculated, and was estimated using the binomial nature of the estimator. It should be noted that in the general case, when each component has a different availability, the weight accumulation is such that for each

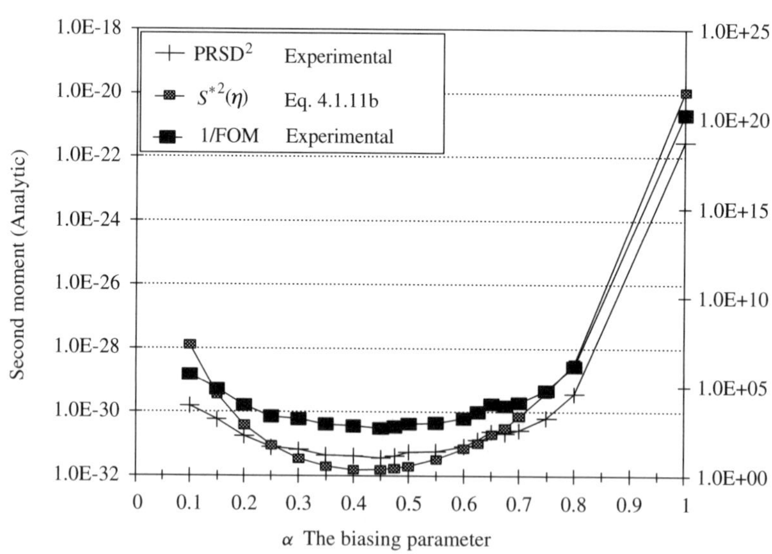

Figure 4.1.5 Estimated and calculated (by the Monte Carlo method) second moments and efficiency

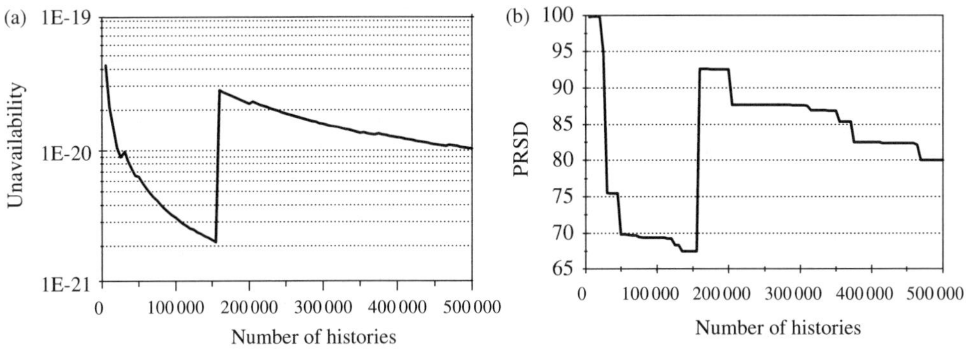

Figure 4.1.6 (a) The behavior of the estimate and (b) the PRSD when overbiasing is applied

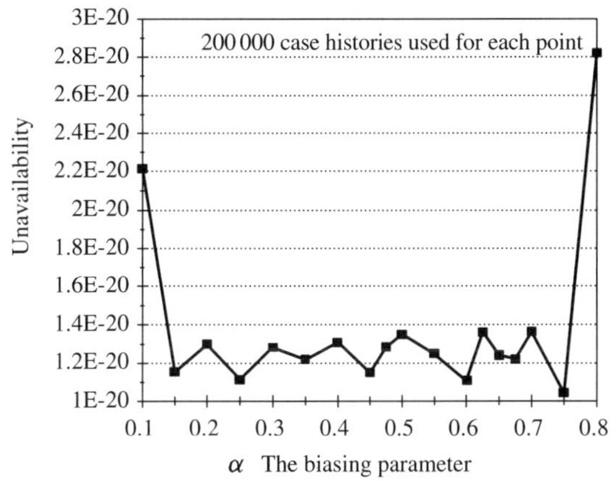

Figure 4.1.7 Estimates of system unavailability obtained with various values of the biasing parameter

failed component the estimator $\eta(\vec{\mathbf{B}}_j)$ is multiplied by $(1 - p_i)/(1 - p_i^*)$ and by p_i/p_i^* if $b_i = 1$.

All the calculations are done with 200 000 histories. The optimal range is 0.25–0.8, and the benefit is of the order of 10^{19}. At the optimum, $\alpha = 0.45$, the result is 1.149×10^{-20} with a statistical error of 3.5%. This calculation required 42.5 seconds. Biasing, in this case, enables the calculation; without it, the calculation becomes practically impossible.

Outside the optimal range, overbiasing may result, as seen in the case of $\alpha = 0.1$. In this case, the availability of each component is strongly reduced such that the probability of the appearance of low importance, rare state vectors is enhanced far above their true importance. The result is the chainsaw pattern similar to that discussed in Section 3.3 (Figure 3.3.2). Figures 4.1.6(a,b) display the behavior of the estimate and the PRSD as function of the number of histories. In Figure 4.1.7 the results for the first moments obtained in the above calculations are shown.

Note the fluctuations resulting for estimates outside the optimal range. In the above case an exact calculation of the 'biased second moment' is not possible because different

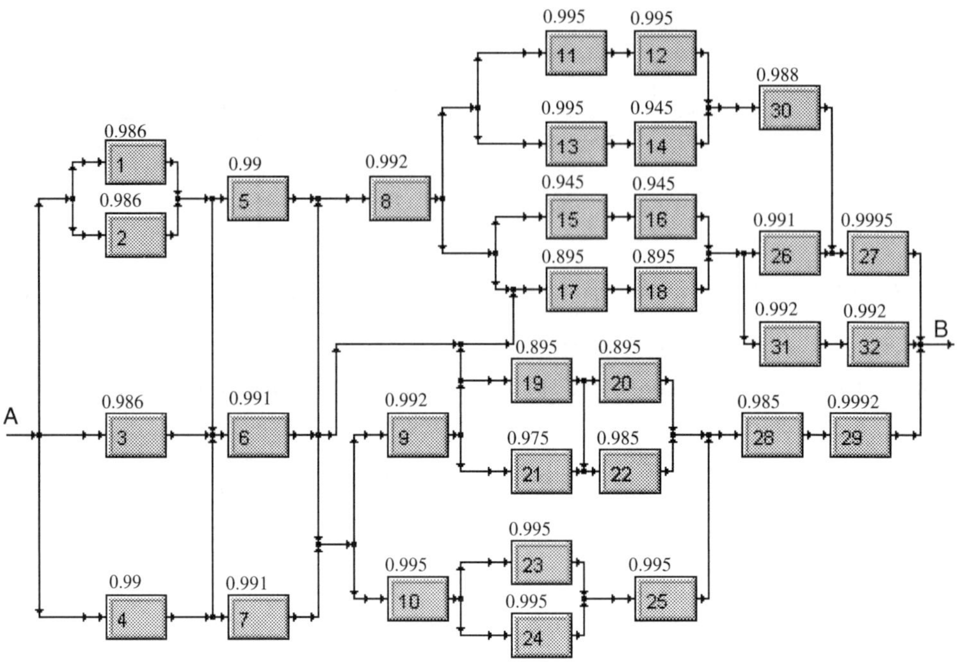

Figure 4.1.8 A redundant communication system (reproduced by permission from Malchi Science Ltd.)

Table 4.1.5 Biased calculations for a general system

α – Biasing parameter	Unavailability	PRSD (%)
0.6	2.91×10^{-5}	66.5
0.65	5.137×10^{-5}	40.79
0.7	8.911×10^{-5}	41.67
0.75	7.59×10^{-5}	17.6
0.8	7.00×10^{-5}	12.73
0.85	7.14×10^{-5}	7.84
0.9	7.104×10^{-5}	5.52
0.95	7.36×10^{-5}	5.94
1.0 (Analog)	7.011×10^{-5}	26.7

components have different availabilities. Still, a good estimate is obtained both for the optimal range of the biasing parameter and the expected benefit. This case is still simplified by the fact that the A_ks are simple and known. In the next example a more realistic situation is considered.

Consider the system block diagram shown in Figure 4.1.8. The component's availabilities appear on each component. The result of calculations with values of α from 0.6 to 1 are shown in Table 4.1.5. A benefit of 23 is realized with the biasing parameter in the range 0.9–0.96.

An approximate prediction of the biased second moment and optimal range may be obtained using equation (4.1.11a). Note that if a state vector with k failed components

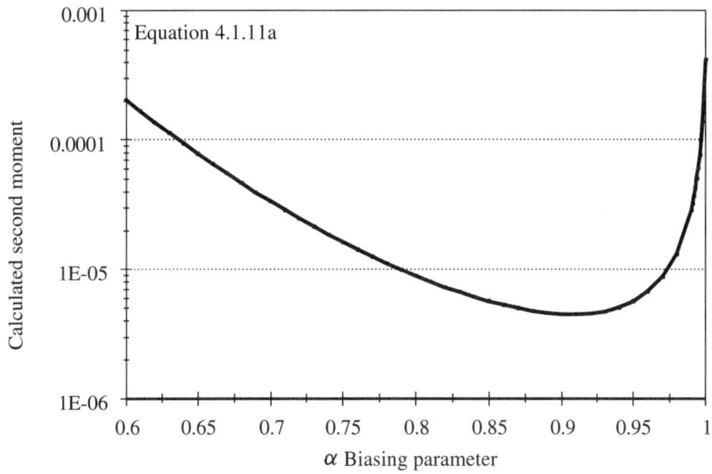

Figure 4.1.9 Estimation of the optimal range and second moment by an approximation based on equation (4.1.11a)

is sampled randomly, then the probability $P_f(\Lambda_k)$ that the system is failed for this state vector is given by the ratio of A_k to the total number of states in Λ_k, namely

$$A_k = \binom{n}{k} \times P_f(\Lambda_k), \tag{4.1.13}$$

where n is the number of components and $k \leq n$.

The probability $P_f(\Lambda_k)$ can be calculated by MC as follows:

1. Sample k component numbers at random. Set the states of these components and obtain a random uniformly sampled state vector $\vec{\mathbf{B}}(k)$.
2. Set the binomial estimator $\eta(k) = 1$ if $S(\vec{\mathbf{B}}(k)) = 0$ and zero otherwise. $\eta(k)$ is an unbiased estimator of $P_f(\Lambda_k)$.

For monotonous systems (i.e. systems for which adding a component to a cut set keeps the set a cut-set) the estimation process can be expedited. Starting with $k = 1$, add failed components at random until a failed system state is reached at k_0. For each $k < k_0$ set $\eta(k) = 0$ and for $k \geq k_0$ set $\eta(k) = 1$. Thus, a single 'history' is used to obtain the estimator for every value of k.

Once the A_ks are estimated they can be used in equation (4.1.11a) to estimate the second moment. For the value of the availability, the average availability of the components can be used. This will yield a reasonable estimate of the optimal region, as shown in Figure 4.1.9.

4.2 Considerations in Optimizing Spare Parts Allocation

In this section a simplified model of spare parts required for a set of identical systems is considered. A hybrid concept of analytic calculation, based on a Markovian model and the Monte Carlo method, will be considered.

Consider a set of m identical systems, referred to as a 'field'. Each system contains k components of the same type, such as three identical radios or two identical engines, etc. Thus, the total number of components of the same type in the field is $q = m \times k$. Upon failure, the failed component must be replaced. Such a component is sometimes called an LRU (Line Replaceable Unit). The failed component is shipped to a remote location, repaired there, and returned to the storage of spare parts at the field site. The field storage is sometimes referred to as the level A storage, and the remote repair site as level B. The storage initially contains n spare units. The recycling time is the time elapsed since shipment to level B until the LRU returns to the level A storage, and it is denoted by T_r. This scenario is identical in its logic to that of Example 2.2.3. To apply a Markov solution to the problem, one must assume a constant failure rate λ and a constant recycling rate $\alpha = 1/T_r$. To simplify the problem, replacement is assumed to be immediate (so that the mean time to repair of the component in the system is zero). This assumption may be justified if the replacement time is much smaller then the mean time to failure and recycling, and by arguing that one is interested only in the impact of spare parts on the system's availability.

A state of the field will be comprised of three integer indicators $(i_1; i_2; i_3)$, the first indicating the number of operational units in systems in the field, the second indicating the number of available spares in the level A storage, and the third indicating the number of units in recycling. The sum $i_1 + i_2 + i_3$ is fixed at $mk + n$, since no components are added or discarded. The states are numbered by $j = i_3$ and there are $mk + n + 1$ states.

The initial state (state 0) is $(q; n; 0)$. This state transfers only into state number 1, denoted by $(q; n-1; 1)$ at a rate $q\lambda$. From state number 1 a return to state zero is possible at rate α or transfer into state number 2 $(q; n-2; 2)$, at rate $q\lambda$. Thus, the first $n+1$ states are of the form $(q; n-j; j)$, with $j = 0, 1, \ldots, n$. The rate of flow from state j into state $j+1$ is $q\lambda$, and the rate back is $j \times \alpha$. In all the first $n+1$ states no units are missing in the field. The last state in this group is $(q; 0; n)$. From this state with a rate $q\lambda$, a transfer occurs into the state $(q-1; 0; n+1)$, in which one spare is missing in the field. Following this state there are $q = mk$ states of the form $(q-i; 0; n+i)$, with $i = 1, \ldots, mk$. Each such state transfers to the next state at a rate $(q-i)\lambda$, and to the previous state at a rate $(n+i) \times \alpha$. The resulting Markov diagram is shown in Figure 4.2.1.

The steady state solution of this model is readily obtained with the following consideration. From Figure 4.2.1, it can be seen that $\alpha P(1) = q\lambda P(0)$, hence $P(1) = uP(0)$, where $u = q\lambda/\alpha$.

The equation for the second state reads $(q\lambda + \alpha)P(1) = q\lambda P(0) + 2\alpha P(2)$. Substituting $P(1)$ from the above relation yields

$$P(2) = \frac{u^2}{2} P(0).$$

It is straightforward to show by induction that for the first $n+1$ states the state probability is given by

$$P(k) = \frac{u^k}{k!} P(0) \quad \text{for } k = 1, \ldots, n+1 \quad \text{and } P(n+1) = \frac{u}{n+1} P(n). \tag{4.2.1}$$

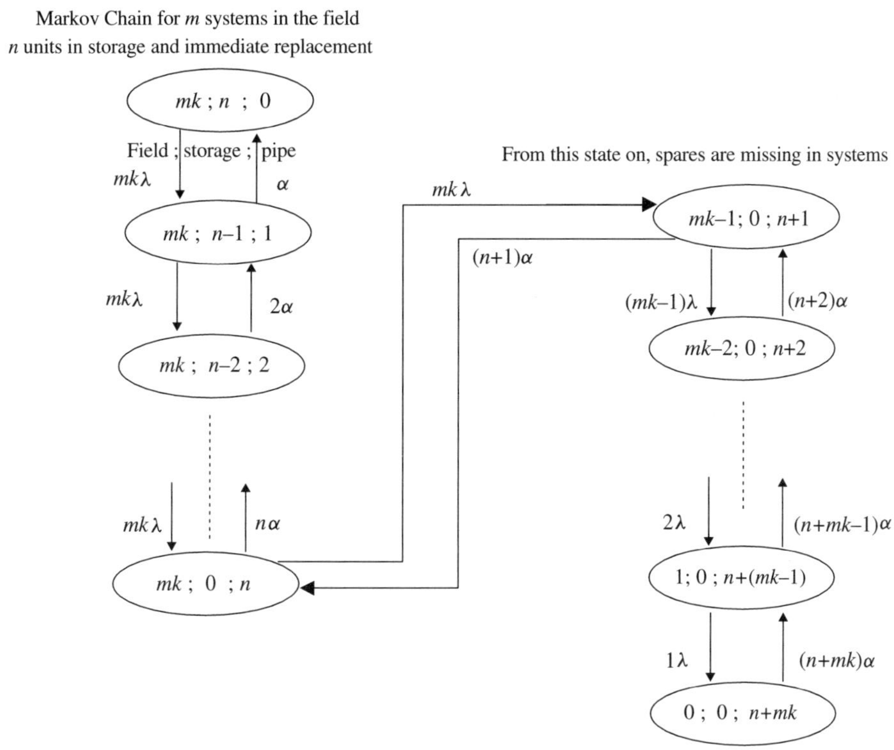

Figure 4.2.1 Markov diagram of the field with recycling

The equation for state $n+1$ is $[(q-1)\lambda+(n+1)\alpha]P(n+1)=q\lambda P(n)+(n+2)\times \alpha P(n+2)$, leading to

$$P(n+2)=\frac{u-y}{n+2}P(n+1)$$

where $y=\lambda/\alpha$ and, in general,

$$P(n+j+1)=\frac{u-j\times y}{n+j+1}P(n+j)=\frac{(mk-j)\times y}{n+j+1}P(n+j)=C_jP_{n+j}\quad j=0,\dots,(mk-1).$$

Denoting now

$$d_k=\prod_{j=0}^{k}\left(\frac{u-j\times y}{n+j+1}\right)$$

one may write

$$P(n+k+1)=d_kP(n)=\frac{u^n}{n!}d_kP(0)\quad k=0,\dots,(mk-1). \tag{4.2.1a}$$

$P(0)$ may now be obtained from the normalization relation $\sum_{i=0}^{mk+n} P(i) = 1$. Substituting (4.2.1) and (4.2.1a) into this relation and extracting $P(0)$ yields

$$P(0) = \left[1 + \sum_{i=1}^{n} \frac{u^i}{i!} + \frac{u^n}{n!} x \sum_{k=0}^{mk-1} d_k\right]^{-1} \quad \text{with } u = \frac{mk\lambda}{\alpha} \quad \text{and } y = \frac{\lambda}{\alpha}. \tag{4.2.1b}$$

Equations (4.2.1), (4.2.1a) and (4.2.1b) allow determination of the state probabilities.

Example 4.2.1

A field contains a total of 10 transmitters. The failure rate of a transmitter is $\lambda = 1.075 \times 10^{-4}$. The recycling time is 1400 hours, thus $\alpha = 7.1429 \times 10^{-4}$. What is the probability that no spare parts will be missing as function of the number of spares kept in the storage?

In this scenario there are $10 + n + 1$ possible states, where n is the number of spares. The probability that (at any given point in time) no spares are missing is the sum of the first $n + 1$ states. This probability is calculated using equations (4.2.1a,b). [`program marksp.for`] The time-dependent state probabilities are also calculated for the diagram of Figure 4.2.1 using the procedure of equation (2.2.34). The behavior of the first five states for the case of zero spares is shown in Figure 4.2.2, and the steady state probability of no spare part missing is shown in Table 4.2.1.

Setting four spare parts guaranties that, in steady state, the probability that no spare parts will be missing is 0.9818. Still, this result does not provide any information as to the performance of the system resulting from any spare parts strategy, because a system is not defined. The situation here is similar to that of Examples 2.2.1 and 2.2.2. It is thus necessary to consider the connection of the Markov state probabilities and system availability.

Having defined the probability for a given number of missing spares, it is now necessary to connect these states to the availability of systems in the field, thereby creating a connection between the number of spares and the unavailability. Given that L spares are missing in the field, the probability, $P_s(r|L)$, that exactly r units are missing in a given system is given by

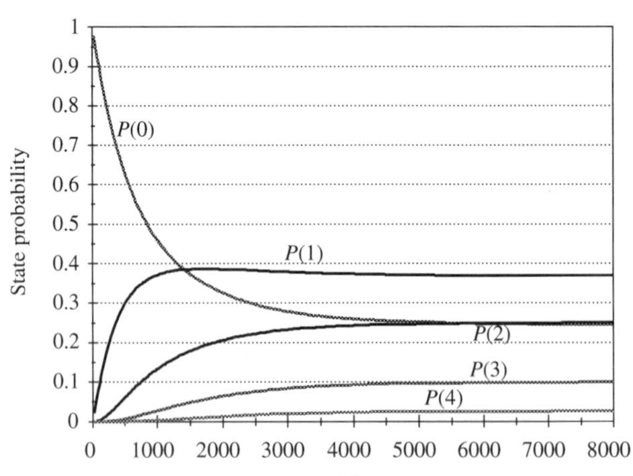

Figure 4.2.2 Probability of being in the first four states

Table 4.2.1 Probability of not missing a spare

Spares	Probability that spare is available
0	0.2451
1	0.5771
2	0.8167
3	0.9364
4	0.9818
5	0.9956
6	0.99908

(see Example 1.1.2)

$$P_s(r|L) = \frac{\binom{q-k}{L-r} \times \binom{k}{r}}{\binom{q}{L}} \qquad r = 0, \ldots, \min(L, k). \tag{4.2.2}$$

Let $U(L)$ denote the unavailability of the system with L spares missing in the field. The average unavailability can then be expressed by

$$\begin{aligned} U &= \sum_{j=0}^{n} P(j) \times U(0) + \sum_{L=1}^{q} P(n+L) \times U(L) \\ &= \sum_{j=0}^{n} P(j) \times U(0) + \sum_{L=1}^{q} P(n+L) \left[\sum_{r=0}^{\min(k;L)} P_s(r|L) U(r) \right], \end{aligned} \tag{4.2.3}$$

where $U(L)$ is expressed as an average over all the events of r units missing in a given system, and $U(r)$ is the unavailability given that r units are missing in a system. Noting that L cannot take values smaller then r, the summation limits can be changed to

$$\sum_{L=0}^{q} \sum_{r=1}^{\min(k;L)} = \sum_{r=1}^{k} \sum_{L=r}^{q},$$

yielding the expression

$$U = \sum_{r=1}^{k} \left[\sum_{L=r}^{q} P(n+L) P_s(r|L) \right] U(r) = \sum_{r=1}^{k} P_a(r) U(r). \tag{4.2.4}$$

The probability, averaged over the possible values of L, that r units are missing in a given system is

$$P_a(r) = \sum_{L=r}^{q} P(n+L) P_s(r|L).$$

In expression (4.2.4), immediate repair is assumed so that $U(0) = 0$ and the first term of (4.2.3) is zero. If no immediate repair is assumed, then $U(0)$ can be obtained by the methods described in Section 4.1.

Table 4.2.2 Spare parts availability and system availability

Spares	Spare parts availability	System availability
0	0.2451	0.869
1	0.5771	0.936
2	0.8167	0.974
3	0.9364	0.991
4	0.9818	0.997
5	0.9956	0.9994
6	0.99908	0.9998

The simplest possible case is that of a system composed of a single component. This implies $k = 1$, in which case $r = 1$ is the only term in (4.2.4). In this case, $U(1) = 1$, and thus

$$U = P_a(1) = \sum_{L=1}^{q} P(n+L)\frac{L}{q}. \tag{4.2.5}$$

Applying equation (4.2.5) to Example 4.2.1 yields the results shown in Table 4.2.2, in which, for comparison, the spare parts availability of Table 4.2.1 is also displayed.

Noting that the reason for having spares is performance, Table 4.2.2 highlights the fact that it is almost meaningless to set the number of purchased spares according to a 'required' spare parts availability, although it is easier to calculate. The demand for availability should determine the number of spare parts needed, nothing else.

Example 4.2.2: Buffer tanks in a plant (an m out of k units systems)

In a plastic products plant there are three identical production lines. In each line six buffer tanks are arranged such that if at least four of them are operational the production is not interrupted. The cost of each spare tank is \$65 K. The MTTF of a tank is 18 200 h and the recycling time is 4200 h. The value of one hour of production in each line is \$520. How many spare parts should be placed in the plant considering one production year?

In this example, $k = 6$, $q = 18$ and $U(r) = 0$ for $r \leq 2$, hence expression (4.2.4) takes the form

$$U = \sum_{r=3}^{6}\left[\sum_{L=r}^{18} P(n+L)P_s(r|L)\right], \tag{4.2.6}$$

where for any given r and L if $L - r > q - k = 12$, then $P_s(r|L) = 0.0$, and for

$$L - r \leq q - k = 12 \quad P_s(r|L) = \frac{\binom{12}{L-r}\times\binom{k}{r}}{\binom{18}{L}} \quad r \leq \min(L, k).$$

The steady states probabilities for the case of zero spares are shown in Table 4.2.3.

The optimization process is described in Table 4.2.4. The steady state unavailability obtained from equation (4.2.6) is multiplied by 8760 h to obtain the average yearly down time, and then by

Table 4.2.3 Markov steady state probabilities with zero spares

State number	State probability	State number	State probability
0	0.0238134	10	0.0004463
1	0.098917	11	7.491E−05
2	0.1940295	12	1.008E−05
3	0.2388056	13	1.074E−06
4	0.2066587	14	8.852E−08
5	0.1335333	15	5.447E−09
6	0.0667667	16	2.357E−10
7	0.0264132	17	6.399E−12
8	0.0083811	18	8.204E−14
9	0.002149	–	–

Table 4.2.4 Optimization of total cost with a number of spares

Number of spares	Unavailability	Cost of production loss	Cost of spares	Total cost
0	0.08412	1 159 550	0	1 159 000
1	0.04893	668 657	65 000	733 000
2	0.02657	343 094	130 000	493 000
3	0.01343	183 529	195 000	378 000
4	0.006309	86 216	260 000	346 000
5	0.002742	37 471	325 000	362 000
6	0.001102	15 059	390 000	405 000
7	0.0004096	5597	455 000	460 000

the cost of each production hour for all three production line. The total cost is the sum of the production loss cost and the cost of spare parts purchased.

So far, the discussion has involved systems with a single type of component. The treatment will now be extended to include multiple types and the general system's structure.

Multiple types of components Assuming a system with m different components types, if the different parts of the system do not interact with each other (i.e. they are statistically independent), then each type in the system can be treated independently, and the availability of spares for each type may be obtained. However, the availability considerations become somewhat elaborate. Consider the system shown in Figure 4.2.3. It is a model of a redundant communication system. In the first line (LRUs 1,4,7) there are power supply units; the second line is transmitters; and the third line is a set of modems. The right part of the system is composed of a general LRU, called a "line", which represents the wiring of the system. The MTTFs and the recycling time of each type are printed below the type name. A failure in the line unit does not produce a recyclable unit. The failed part is discarded. Also, at least two of the series on the left part of the system must operate for the system to function properly.

The application of expression (4.2.4) to such a system has the following elements to consider. The unavailability may become a complex function of the missing spares. For example, if one unit of each of the types PS, Transmit and Modem is missing, the

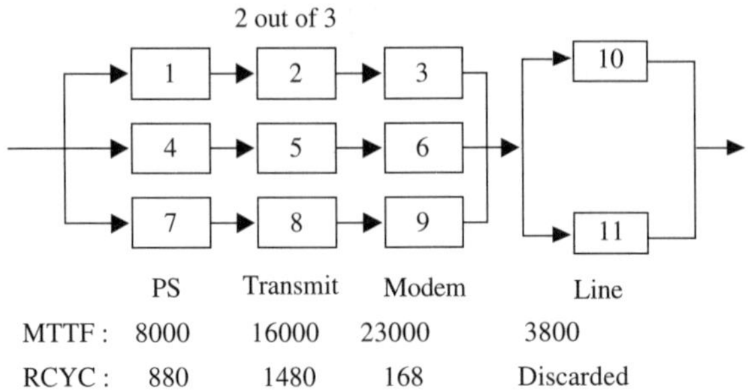

Figure 4.2.3 A communication system with redundancy, recycling and discarded units

system may be operational or not, depending on the exact location of the missing unit. If the missing units are 1, 2 and 3, the system is up, as is the case for 4, 5 and 6. However, if the missing locations are, say, 1, 5 and 6, then the system is down. Let $\langle U(r_1, r_2, \ldots, r_m)\rangle$ denote the average unavailability of a system, where r_i is the number of missing spares of type i. This unavailability can be evaluated using the Monte Carlo method with the following process. Let $S(\vec{\mathbf{B}})$ denote the system function. For the current example, it takes the form

$$S(\vec{\mathbf{B}}) = K(S_1, S_2, S_3) \times (b_{10} + b_{11} - b_{10}b_{11})$$

with $S_1 = b_1 \times b_2 \times b_3$, $S_2 = b_4 \times b_5 \times b_6$ and $S_3 = b_7 \times b_8 \times b_9$. Also,

$$K(S_1, S_2, S_3) = \begin{cases} 1 & \text{if } S_1 + S_2 + S_3 \geq 2 \\ 0 & \text{otherwise.} \end{cases}$$

Let the array $K_j = (n_{1_j}, n_{2_j}, \ldots, n_{k_j})$ be the array of the component numbers that belong to type j, and k_j is the number of components of type j in the system. In the above system, for example, $K_1 = (1, 4, 7)$, $K_2 = (2, 5, 8)$, $K_3 = (3, 6, 9)$ and $K_4 = (10, 11)$. For any given array of missing spares, $r_1, r_2, \ldots, r_m$, one may uniformly sample r_1 elements from K_1, r_2 elements from K_2, and so on, until a set of failed components waiting for a missing spare is created in the system. Repeating this sampling process and calculating $S(\vec{\mathbf{B}}_i)$ for the ith sample, the average unavailability can be calculated as

$$U(r_1, r_2, \ldots, r_m) = \frac{1}{N}\sum_{i=1}^{N}(1 - S(\vec{\mathbf{B}}_i)). \tag{4.2.7}$$

The statistical error is binomial, since $(1 - S(\vec{\mathbf{B}}_i))$ is a binomial estimator.

Discarded units Equation (4.2.4) is based on the Markovian process of Figure 4.2.1, and the assumption that the steady state probabilities of the states can serve as an adequate measure of the true state of the system. In the case of a discarded unit, the time period T_m for which spares are purchased must be given, and the time dependent states probabilities $P(j, t)$ must be averaged over the time period. Thus, the steady state probability will be replaced by

$$P(j) = \frac{1}{T_m} \int_0^{T_m} P(j, t)dt. \tag{4.2.8}$$

$P(j)$ is the average probability to find the field in state j and $T_m \times P(j)$ is the average amount of time the field is in this state.

The average unavailability can be generalized from expression (4.2.4) in the form

$$U(n_1, n_2, \ldots, n_m) = \sum_{r_1=0}^{k_1} \cdots \sum_{r_m=0}^{k_m} \left[\sum_{L_1=r_1}^{q_1} \cdots \sum_{L_m=r_m}^{q_m} \left(\prod_{i=1}^{m} P(n_i + L_i) P_s(r_i|L_i) \right) \right] U(r_1, r_2, \ldots r_m), \tag{4.2.9}$$

with m being the number of different types in a system, and with the understanding that for $L_i = 0$, $P(n_i + L_i) = \sum_{k=0}^{n_i} P(k)$, which is the total probability that no spares are missing. Unlike the situation in the single type case, the case $r_i = 0$ must be included in the averaging process. Thus, the sum includes all the possible states of the field and the systems. Therefore, the normalization equation (4.2.10)

$$\sum_{r_1=0}^{k_1} \cdots \sum_{r_m=0}^{k_m} \left[\sum_{L_1=r_1}^{q_1} \cdots \sum_{L_m=r_m}^{q_m} \left(\prod_{i=1}^{m} P(n_i + L_i) P_s(r_i|L_i) \right) \right] = 1 \tag{4.2.10}$$

must hold. It is interesting to note that the number of terms to be considered in the sum of (4.2.9) is of the order of

$$\left(\prod_{i=1}^{m} (k_i + 1) \right) \times \left(\prod_{j=1}^{m} q_i \right).$$

(It is, in fact, slightly smaller and would be exact if the summation for each L_i were performed from zero to q_i, rather then from r_i.) In the present example of Figure 4.2.3, this means 839 000 terms, where the exact number is 707 472 terms. Note that, for a single type, the number of terms involved is q_i, thus the dimensional problem is again presented in this expression. The calculation time of expression (4.2.9) for the current example takes about 10 seconds, thus a field of 20 systems, each with 30 different types of units and two units of each type in each system, would mean about 1.27×10^{45} terms, and a calculation time of 4.8×10^{32} *years*. Increasing the number of components in the system and taking a rigorous account of the features of the system implies a multi-dimensional problem. The state vector of the field has $2^{(q_1+q_2+\cdots+q_m)}$ different values, and any realistic calculation 'explodes' in terms of the number of terms needed to be considered in the solution. Indeed, a 'linearization' of the problem seems necessary and tempting in such cases. Linearization, in this context, means that the solution can be presented as a simple function of one-dimensional solutions. For example, in

Table 4.2.5 Average availability for the system of Figure 4.2.3 for various spare parts mixes

Storage	Total number of Spares	1 year – ⟨Unavailability⟩
0, 0, 0, 0	0	0.485
0, 0, 0, 10	10	0.130
1, 1, 0, 14	16	0.0286
0, 0, 0, 20	20	0.0867
0, 0, 0, 30	30	0.0864
0, 0, 10, 30	40	0.0806
1, 1, 0, 20	22	0.0197
2, 2, 0, 20	24	0.00410
2, 2, 1, 20	25	0.00324
3, 3, 1, 21	28	0.000625

the current problem one may assume that $U(n_1, n_2, \ldots, n_m) = U_1(n_1)U_2(n_2)\cdots U_m(n_m)$, i.e. that the average availability of the m-dimensional problem can be presented as a product of m one-dimensional solutions. This assumption may hold if the system has a serial structure and the components are statistically independent. In that case, equation (4.2.4) can be applied to each type, and the overall number of terms will be of the order of $\sum_{i=1}^{m} k_m q_m$, which is manageable for any realistic number of components. This kind of approximation allows an increase in the number of components in the system while still being able to perform analytic calculations. In most realistic cases, the results obtained are simply wrong and the approximation is made not because it is justified, but because it is the only 'sensible' way to address the problem – as was stated before, the Monte Carlo method, being independent of the dimension of the problem, allows an exact solution with no need for uncontrolled approximations. Table 4.2.5 displays results of the availability for a number of spare parts strategies. The availability is calculated using equations (4.2.7)–(4.2.9). The computer program used is listed as `multyp.for`.

A 'pipe line' approach to the spare parts problem This methodology is brought as a simple example of an approximate, single unit approach. Also, the Monte Carlo availability calculation described in Section 4.1 is utilized.

The starting point of this simple approach suggests a single item (LRU) reaches a storage facility at a rate λ. This rate is assumed to be constant, representing the flow of units from the field. Upon arrival, the failed unit is repaired and returned to storage. The time required for this process is the recycling time T_r. If there are initially n spares in the storage and the recycling time is denoted as T_r, then a lack of spares will be caused only if upon the arrival of the $(n+1)$th unit to the storage, the first one has not yet returned from the recycling process (see Figure 4.2.4). This condition can be expressed as

$$T = \sum_{i=2}^{n+1} t_i < T_r, \quad n > 0, \tag{4.2.11}$$

where t_i is the time interval between the arrival of the $(i-1)$th and ith units.

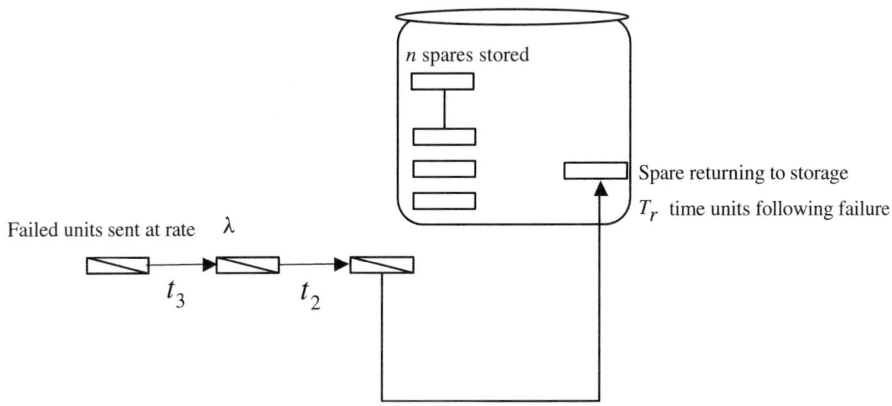

Figure 4.2.4 The recycling process

Since t_i is exponentially distributed, T has a Gamma distribution with n degrees of freedom:

$$f(T) = G_n(T) = \frac{\lambda(\lambda T)^{n-1} e^{-\lambda T}}{\Gamma(n)}; \quad \Gamma(n) = (n-1)! \quad n \geq 1.$$

The probability $P_s(n)$ that a spare part will be available is the probability that $T > T_r$, hence

$$P_s = 1 - \int_0^{T_r} G_n(T) dT = 1 - CG_n(T_r) \quad \text{for } n = 0, \; P_s = 0, \tag{4.2.12}$$

where $CG_n(x)$ stands for CDF of the Gamma distribution. Equation (4.2.12) also applies to discarded units, where T_r is replaced by T_m, which is the time period for which spares are purchased.

Given that a spare part is not available, the average waiting time for a spare is given by

$$T_w = \int_0^{T_r} (T_r - T) \frac{\lambda(\lambda T)^{n-1} e^{-\lambda T}}{\Gamma(n)} dT = T_r CG_n(T_r) - \frac{n}{\lambda} CG_{(n+1)}(T_r) \tag{4.2.13}$$

for $n = 0$, $T_w = T_r$.

Thus, if a spare part is available the repair time of a unit is just its given MTTR. If a spare is not available, the waiting time of equation (4.2.13) is added to the nominal repair time. The above discussion suggests the following calculation steps:

1. Calculate, for each type in the system, the spare availability $P_s(n_i)$, where n_i is the number of spares stored for type i and the conditional waiting time $T_w(n_i)$ from expressions (4.2.12) and (4.2.13). The value of λ should be the total rate of failures of components of type i in the field.

2. Use the method of Section 4.1 for a Monte Carlo calculation of the system availability. In each history the steady state availability of each component is calculated as MTTF/(MTTF+TTR), where TTR equals the nominal Mean Time To Repair (MTTR) of that type (for the current example, the nominal MTTR is zero, since immediate repair is assumed) with probability $P_s(n_i)$ and with probability $1 - P_s(n_i)$, $TTR = MTTR + T_w(n_i)$.

In Table 4.2.6, results obtained with this method (with 10 000 histories for the system availability calculation) are compared to those of the multidimensional method of

Table 4.2.6 System availability as a function of a spare parts strategy

Storage	Pipeline method	Equation (4.2.10) ⟨Unavailability⟩	Monte Carlo SPAR
0, 0, 0, 0	0.526	0.485	0.477
0, 0, 0, 10	0.2245	0.130	0.123
1, 1, 0, 14	0.0400	0.0286	0.0297
0, 0, 0, 20	0.091	0.0867	0.0792
0, 0, 0, 30	0.0877	0.0864	0.0783
0, 0, 10, 30	0.0867	0.0806	0.0733
1, 1, 0, 20	0.0155	0.0197	0.0171
2, 2, 0, 20	0.0013	0.00410	0.00398
2, 2, 1, 20	0.00091	0.00324	0.00309
3, 3, 1, 21	0.000122	0.000625	0.000653

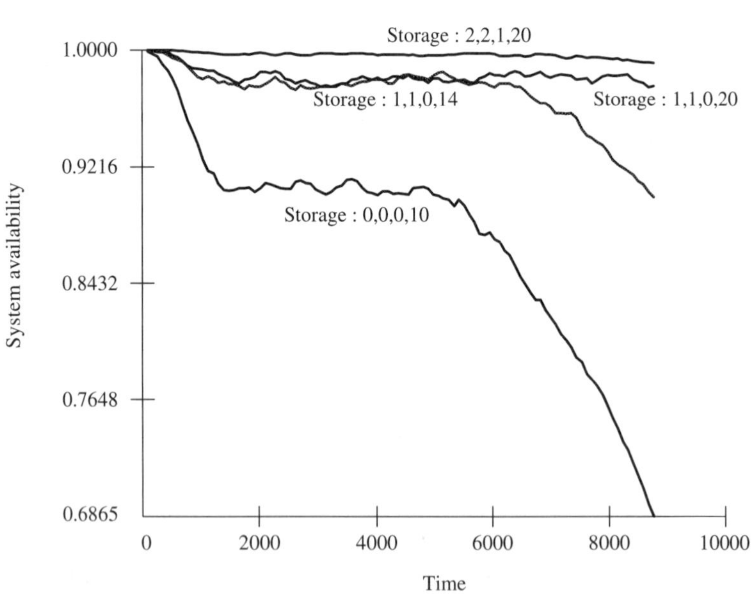

Figure 4.2.5 System time-dependent availability with a number of storage allocations

equation (4.2.10) and those obtained with an exact Monte Carlo calculation (Dubi, A. and Goldfeld, A. (1996). *SPAR – A Multipurpose MC Computer Code for General System Engineering Analysis – Manual*, Malchi Science Publications), with respect to the problem described in Figure 4.2.3.

The availability as function of time for a number of storage allocations, as obtained with the multi-purpose MC code SPAR, are shown in Figure 4.2.5.

5
Elements of Uncertainty and Uncertainty Analysis

5.1 General Discussion

From previous chapters it may be concluded that knowledge of the various distributions involved in the system model, such as the failure and repair distributions, is an essential requirement for any solution to be obtained. Not much can be done if this data is not available. In most realistic cases, no exact knowledge on the distributions is available: at best, one has a listing of events, failures, repairs, maintenance operations, inspections, etc. from which one would like to obtain an idea about the distributions. A large body of well established statistical inference and estimation methods have been developed for this purpose (see Lawless, J. F. (1982). *Statistical Models and Methods for Lifetime*, John Wiley & Sons, Chichester; Blanks, H. S. (1993). *Reliability in Procurement and Use*, John Wiley & Sons; Dudewicz, E. J. (1976). *Introduction to Probability*, Holt, Rhinehart & Winston, New York; Asher, H. and Feingold, M. (1984). *Repairable Systems Reliability – Modeling, Inference, Misconceptions and their Causes*, Marcel Dekker, New York; Kumamoto, H. and Henely, E. J. (1996). *Probabilistic Risk Assessment and Management for Engineers and Scientists*, IEEE Press, New York; Chen, Z. (1997). Statistical inference about the shape parameter of the Weibull distribution, *Statistical & Probability Letters*, **36**, 85–90; and the references therein).

In most of the literature, methods are presented from the statistical standpoint, i.e. given a set of independent failure time points along with some censoring conditions (meaning, in general, that an experiment was stopped before all its participant failed). This standpoint is not always suitable for the engineer in the plant, who rarely has the opportunity to conduct a well organized experiment. At best, historical data pertaining to the events occurring in various components is available. Here, not all the relevant methods will be considered, but the standpoint will be that of the engineer in the plant. Thus, rather than looking at a sequence of failure time points, we look at a scenario, one which may involve one or more identical components and their known life history. It will become obvious that when a general scenario is considered, not only are the failure time points of importance, but so is the nature of the repair that was applied in each case. Minimal repair, which keeps the component as bad as old or, in contrast, full repair, which yields a component as good as new with zero age, produce completely different results. It is attempted here to bring a comprehensive methodology in which the MC method and sampling play an essential role.

Consider a situation in which it is known that four identical components had the following failure times: 760 h, 1300 h, 1760 h and 2330 h. What can one say about the failure distribution?

Knowing the exact distribution requires an infinite amount of information. This can be deduced simply from the way in which an average converges to the exact value. The $1/\sqrt{N}$ convergence implies that an infinite number of samples is required for the statistical error to be reduced to zero. Indeed, the distribution must be some function $f(t, \theta)$ with one or more parameters. The knowledge one wishes to extract from the above four numbers is: What is the distribution, and what is the value of the parameter?

This is a lot to ask in view of the fact that even when the distribution is known, just to learn the exact value of θ requires an infinite sample. Inevitably, some degree of *arbitrariness* must be assumed. For example, we will assume that the failure distribution is a Weibull (which includes the exponential distribution), just because this distribution has been successfully used in the past for similar types of components. Such considerations must be invoked if one wishes to obtain any form of failure distributions. Basic to all estimation procedures is also the assumption that the quantities calculated are not overly dependent on the parameters, so that even if the estimates are not exact, or even if they are considerably off, the results produced in the final system analysis are still useful (in the sense that our predictions about the system behavior will be close to its true future reality). Such assumptions cannot be proven, but make sense – after all, any system that is to be extremely sensitive to the value of the parameters will suffer from extreme variations in its behavior upon small natural changes of the parameters, and will probably not stand the test of time.

Suppose now that one 'knows' that the failure distribution of the above four points is exponential. Thus, one is left with the task of estimating λ, the failure rate, or $\theta = 1/\lambda$, the MTTF of the distribution.

How about assuming that θ equals the sample average of the failure times, namely 1537.5 h? The justification for this step is the fact that the sample average is an unbiased estimate of the MTTF. Not only that we cannot say that the value estimated is the true value, but we know with certainty that it is an incorrect value, because the *a priori* probability of any continuous random variable to assume a given value is zero. All we can really say about 1537.5 is that it makes some sense! The sample second moment is 2.6985×10^6 and the estimated standard deviation will therefore yield (using equation (3.3.6) and dividing the variance by $(n-1)$ rather than by n, for small samples) 333.98 hours. We cannot even say that a 65% confidence interval is of a size of 334 hours, because the sample size is not large enough to assume that the central limit theorem can be applied. Still, the results obtained seem reasonable. It would be unreasonable to assume that the MTTF is 0.2, or 23, or 230 000 000, because the probability of obtaining the above four realizations with these three estimates is imaginatively small. An interesting concept emerges from this argument. Intuition leads us to think that 0.2 is not plausible, because the probability to have a failure at 2330 h is very, very small. Can this statement be reversed? Instead of a 'given 0.2' and a 'probability for 2330', can we have a given 2330 and a probability for 0.2, or 23, or 1200, etc.? This is a key to the understanding of how inference is performed. We will return to this concept later.

One can try another approach for the treatment of the above four numbers. Assume that the median of the probability is such that 50% of the realizations are below it. A sensible choice of the median would be the mid-point between the second and the third realizations, namely 1530. This yields the equation $1 - e^{-(1530/\theta)} = 0.5$, with the solution 2207.3 hours. If the median point is chosen at just above the second event 1300 h, the solution is 1875.5 – somewhat closer to the sample average estimate.

A more elaborate consideration can be made. For each given time point the CDF is unknown and given by $P_i = F(t_i, \theta)$. P_i is a random variable since, for each experiment, t_i may take a different value. The PDF of P_i, $g(P_i)$, can be easily established. For a sample of n time points, the probability that $(i - 1)$ time points will be smaller than t_i and $(n - i)$ points larger than it is

$$\binom{n}{i-1} P_i^{i-1} \times (n - i + 1) \times (1 - P_i)^{n-i},$$

since the preceding $(i - 1)$ points can be chosen in $\binom{n}{i-1}$ ways, while the ith point can be chosen in $(n - i + 1)$ different ways from the remaining points. Hence:

$$g(P_i) = \frac{n!}{(n-i)!(i-1)!} P_i^{i-1} (1 - P_i)^{n-i}. \tag{5.1.1}$$

One may chose, for each P_i, the value for which $g(P_i)$ is maximized. Taking the derivative of $g(P_i)$ and comparing to zero yields

$$P_{i,0} = \frac{i-1}{n-1} \quad (\text{for } i > 1 \text{ and } i < n).$$

This yields $1 - e^{-(1300/\theta)} = 0.333$, and an estimate of 3206 h for the second point and 1604 h for the third point, the average being 2405. Looking for θ, that yields a minimum for the square of the differences, namely

$$\left(1 - e^{-(t_2/\theta)} - P_{2,0}\right)^2 + \left(1 - e^{-(t_3/\theta)} - P_{3,0}\right)^2,$$

yields a general estimate θ, which is the solution of

$$\sum_{i=2}^{n-1} \left(Q_{i,0} - e^{-(t_i/\theta)}\right) t_i e^{-(t_i/\theta)} = 0, \tag{5.1.2}$$

where $Q_i = 1 - P_i$. A numerical solution of (5.1.2) yields $\theta = 2155$.

In this estimation two data points were explicitly used. A different choice of estimating the P_is will allow use of all the data points. Assume that P_i is taken to be the median of the distribution $g(P_i)$. The median $\hat{P}_i$ is obtained from the equation

$$\int_0^{\hat{P}_i} g(P_i)\,dP_i = \frac{n!}{(n-i)!(i-1)!}\int_0^{\hat{P}_i} P_i^{i-1}(1-P_i)^{n-i}\,dP_i = 0.5. \tag{5.1.3}$$

Values for this integral may be obtained from the incomplete beta function (see, for example, Pearson, E. S. and Hartley, H. O. (eds.) (1954). *Biometrica Tables for Statisticians, Vol. 2*, Cambridge University Press, p. 142):

$$B(i,k,x) = \int_0^x y^{i-1}(1-y)^{k-1}\,dy \quad \text{with } k = n-i+1. \tag{5.1.4}$$

In the present case, the values of $\hat{P}_i$ are 0.1591, 0.38573, 0.61427 and 0.8409 for $i = 1, 2, 3, 4$, respectively. A simple well known approximation can be used for equation (5.1.3) in the form

$$\hat{P}_i = \frac{i-0.3}{n+0.4},$$

yielding the respective values 0.159, 0.3863, 0.6136 and 0.8409. Equation (5.1.2) can now be used with all four values, and the estimate minimizing the sum of squares is $\theta = 2025$. The last method is called the *median rank method.* With all the estimation ideas outlined so far, the numbers obtained are 1537, 2207, 1875, 2405, 2155 and 2025. All we can really say about these estimates is that

1. They are, with certainty, not equal to the 'true value'.
2. They all make sense in terms of the values obtained.
3. They all make sense in terms of the method used.

Such estimation methods are called *point estimates*, as they attempt to provide a single value for the parameter estimated. Attempting to obtain a single value leads to certainty of its being wrong! This is a plausible result of the fact that an infinite amount of information is required to set the true value. Having only a finite amount of information means that the information ratio is zero, and so is the probability that our estimate is 'true'. Is there an alternative?

Is it possible that, if instead of obtaining a single value one were to settle for an interval, some certainty (or rather, a finite uncertainty) will be attributed to the possibility that the true value is in the interval? It is plausible to assume that the probability that the true MTTF is in the interval [0, 10] is very small. There seems to be a higher probability for the interval [1000, 1600], and an even higher probability that it is in the interval [1100, 2800] – all one needs is a way to translate this plausibility into rigorous statements about confidence intervals and confidence levels. Methods that yield confidence intervals are called *interval estimation methods* (what else?).

In the next two sections, a more elaborate treatment is introduced: first for point estimation suitable for analyzing results of scenarios; and then for interval estimation.

5.1.1 The maximum likelihood method (MLE)

It has already been seen in the preceding discussion that any point estimation method includes an arbitrary, yet plausible, assumption which underlies the estimate. The Maximum Likelihood Method (MLE) is based on the plausible idea that the likely outcome of an experiment is the result that has the highest probability of appearance. This idea is, of course, wrong in the strict mathematical sense. In a continuous space an outcome of an experiment will be any result within an interval that has a non-zero probability, but it is plausible since results with higher probability will have a higher tendency to appear in experiments. A plausible, wrong, yet plausible assumption is hidden behind any point estimate.

The simplest case is that of independent results. Assume that the failure distribution is given by $f(t)$ with a single parameter λ. Let $t_1, t_2, \ldots, t_n$ be n independent failure points for n independent components. The probability to obtain this result in the experiment is called the *likelihood function*, and is given, for this case, as

$$L(t_1, \ldots, t_n|\lambda) = n!f(t_1)f(t_2)\cdots f(t_n) = n!\prod_{i=1}^{n} f(t_i). \tag{5.1.5}$$

For the exponential distribution, the likelihood function takes the form

$$L(t_1, \ldots, t_n|\lambda) = n!\,\lambda^n \mathrm{e}^{-\lambda \sum_{i=1}^{n} t_i} = n!\,\lambda^n \mathrm{e}^{-\lambda T},$$

where T is the sum of the failure times. Taking the derivative of the likelihood function with respect to λ and equating to zero yields $n\lambda^{n-1} - \lambda^n T = 0$, resulting in the MLE estimate of λ as $\lambda = n/T$, and the MTTF estimate

$$\theta_{\mathrm{MLE}} = \frac{1}{n}\sum_{i=1}^{n} t_i. \tag{5.1.6}$$

The sample average discussed in the preliminary example is also the MLE estimate. θ_{MLE} of equation (5.1.6) is also an unbiased estimator of θ, which means that θ_{MLE} as a random variable fluctuates around its average, the true value. It is not at all obvious that the MLE estimate is in every case an unbiased estimator.

In many realistic situations, the data is not obtained as a result of a planned and well designed experiment, but rather as a record of actual events that happened over a period of time. This is referred to as a *scenario*. The analysis must then fit the specific scenario that produced the data.

5.1.2 Scenario MLE

Consider four independent identical components operating in a plant. The following information is recorded for each of them. The first component had a first failure at time $t_{1,1}$; minimal repair was applied so that the age of the component upon its repair remained the same as it was just before the failure. A second failure occurred at $t_{1,2}$, to which minimal repair was also applied, and since then the component has been operating uninterruptedly up to time $t_{1,3}$. The time points are measured from the beginning of the operation of each component. The probability density for such a history to occur is

$$f(t_{1,1}) \times \frac{f(t_{1,2})}{1-F(t_{1,1})} \times \frac{1-F(t_{1,3})}{1-F(t_{1,2})}.$$

The second component had no failure up to $t_{2,1}$, and the probability for this is $1-F(t_{2,1})$. The third component had a first failure at $t_{3,1}$, in which it was completely repaired, thus restarting with age zero. It then continued to operate without failure up to time $t_{3,2}$. The probability density for such a history is $f(t_{3,1}) \times [1-F(t_{3,2}-t_{3,1})]$.

The fourth component had a single failure at time $t_{4,1}$. It was then removed and never resumed operation. The PDF for that failure is $f(t_{4,1})$. The likelihood function of this scenario is the product of the probability densities for the scenarios of the independent components, namely:

$$L(t_{1,1}, \ldots, t_{4,1}) = f(t_{1,1}) \times \frac{f(t_{1,2})}{1-F(t_{1,1})} \times \frac{1-F(t_{1,3})}{1-F(t_{1,2})} \times [1-F(t_{2,1})] \times f(t_{3,1}) \times [1-F(t_{3,2}-t_{3,1})] \times f(t_{4,1}).$$

Assuming a Weibull failure distribution yields

$$L(t_{1,1}, \ldots, t_{4,1}|\lambda, \beta) = \lambda\beta\left(t_{1,1}^{(\beta-1)}\right) \times \lambda\beta\left(t_{1,2}^{(\beta-1)}\right) \times e^{-\lambda t_{1,3}^{\beta}} \times e^{-\lambda t_{2,1}^{\beta}} \lambda\beta\left(t_{3,1}^{(\beta-1)}\right) e^{-\lambda t_{3,1}^{\beta}} \times e^{-\lambda(t_{3,2}-t_{3,1})^{\beta}} \lambda\beta\left(t_{4,1}^{(\beta-1)}\right) e^{-\lambda t_{4,1}^{\beta}},$$

or

$$L(t_{1,1}, \ldots, t_{4,1}|\lambda, \beta) = (\lambda\beta)^4 (t_{1,1} \times t_{1,2} \times t_{3,1} \times t_{4,1})^{(\beta-1)} \times e^{-\lambda\left[t_{1,3}^{\beta}+t_{2,1}^{\beta}+t_{3,1}^{\beta}+(t_{3,2}-t_{3,1})^{\beta}+t_{4,1}^{\beta}\right]}. \tag{5.1.7}$$

This is a rather general form. It contains the product $(t_{1,1} \times t_{1,2} \times t_{3,1} \times t_{4,1})^{(\beta-1)}$, which is a product of all the failure points in the scenario. The failure points are measured from the last point at which the age was set to zero, thus if a failure occurred at time t_f and the last time at which the age was set to zero is t_0, then the value $(t_f - t_0)$ must appear in the product. In the above specific scenario, all the former zero age points are at the beginnings of the observations.

The sum $[t_{1,3}^{\beta} + t_{2,1}^{\beta} + t_{3,1}^{\beta} + (t_{3,2}-t_{3,1})^{\beta} + t_{4,1}^{\beta}]$ appears in the exponent. Each time point in this sum is a point at which the age is set to zero. It can be either a failure point at which complete repair (or replacement by a new component) was applied, or a maintenance point at which the age was reduced to zero or an endpoint of the observation. Again, each time point is measured from its preceding nearest zero age point.

Thus, a general scenario in this context involves the following types of time points: failure points with minimal repair; failure points with complete repair; maintenance points with complete age rejuvenation; and end of observations which are also considered as zero age points. Every scenario should be described in terms of these points. The time points in the scenario are then divided into two sets: a set of m failure time points $(t_1, \ldots, t_m)$, and a set of k time points at which the age was set to zero $(\tau_1, \ldots, \tau_k)$. It should be noted that each time is measured from the last time at which the age was set to zero. The likelihood function will then have the general form:

$$L(t_{1,1}, \ldots, t_{4,1}|\lambda, \beta) = (\lambda\beta)^m a(t_1, \ldots, t_m)^{(\beta-1)} \times \mathrm{e}^{-\lambda g(\tau_1,\ldots,\tau_k)}, \tag{5.1.8}$$

where

$$a(t_1, \ldots, t_m) = t_1 \times t_2 \times \cdots \times t_m \quad \text{and} \quad g(\tau_1, \ldots, \tau_k) = \sum_{i=1}^{k} \tau_i^\beta.$$

This scenario's likelihood function does not include situations in which partial age reduction is involved, because in most cases data concerning such changes is scarce. Note that the number of components involved in the scenario does not appear explicitly, because a point at which the age is set to zero and a new component are identical, so that k components tested up to the first failure point and one component with k failures and complete repair at each failure are in fact the same scenario.

Differentiating the likelihood function with respect to λ and β, and equating to zero, yields the equations

$$\frac{m + \beta \ln[a(t_1, \ldots, t_m)]}{\beta(dg(\tau_1, \ldots, \tau_k)/d\beta)} = \frac{m}{g(\tau_1, \ldots, \tau_k)} \quad \text{and} \quad \lambda = \frac{m}{g(\tau_1, \ldots, \tau_k)}, \tag{5.1.9}$$

which takes the explicit form

$$\frac{m + \beta \sum_{i=1}^{m} \ln(t_i)}{\beta \sum_{j=1}^{k} \tau_j^\beta \ln(\tau_j)} = \frac{m}{\sum_{j=1}^{k} \tau_j^\beta} \quad \text{and} \quad \lambda = \frac{m}{\sum_{j=1}^{k} \tau_j^\beta}. \tag{5.1.10}$$

A numerical solution can be easily applied to the first equation to obtain β, which can then be substituted in the second equation to yield λ. The expression for λ holds, however, for any value of β. The advantage of the MLE method is its flexibility. It can be quite easily applied to a large class of scenarios, but it still suffers from the basic deficiencies of every point estimation method as described above. In particular, one should be cautious about possible bias in the estimate. It was shown that the estimate of equation (5.1.6) is unbiased. No such general statement can be made for the results of equation (5.1.10). If the estimate is biased, then one may obtain results that tend to consistently over- or under-estimate the value of β. An over-estimation of the value of β by a factor of two may have severe consequences in terms of making incorrect decisions concerning the future and the system. This point therefore deserves additional attention. Next, a number of special cases will be considered. They may reveal interesting properties of the MLE.

5.1.3 As bad as old

Consider the case of a censored experiment in which a component is allowed to have r failures and the experiment is stopped upon the last failure. Minimal repair is applied at every failure, so that the age is not reduced following repair. Upon repair the component remains as bad as old. The preceding discussion suggests that in this scenario, $a = t_1 t_2 \cdots t_r$ and $g = t_r^\beta$. The resulting equations for λ and β are quite simple:

$$\beta_{\text{MLE}} = \frac{r}{\left[r \ln(t_r) - \ln\left(\prod_{i=1}^{r} t_i\right)\right]} \quad \text{and} \quad \lambda_{\text{MLE}} = \frac{r}{t_r^\beta}. \tag{5.1.11}$$

Every set of observation time points on which the MLE is based is a sample of an experiment or a scenario in which a 'true' β_{tr} exists, *but is unknown.* What can be said about the relation between β_{MLE} and β_{tr}? This relation is not explicitly apparent from the MLE method. The MLE estimates do not have a direct or a simple relation with the true values; they just maximize the likelihood function of the scenario sampled from the true values. This point is further developed in the following example.

Example 5.1.1

A set of four failure times with minimal repair are 1637 h, 3430 h, 4900 h and 6614 h. Using equations (5.1.11), the MLE estimates are $\beta_{\text{MLE}} = 1.7$ and $\lambda_{\text{MLE}} = 1.28 \times 10^{-6}$. It is sometimes convenient to use the parameter $\eta = \lambda^{1/\beta}$, which in this case is 0.000342.

How does one verify that the value of 1.7 is anywhere close to the true value? A feasible way would be to check if the MLE estimate is invariant under the MLE procedure. Assume that the above values are the 'true' values. Then sample a set of four failure points, with minimal repair, according to the relevant scenario, and use these points to obtain a new MLE value of β_{MLE}^1; repeat this operation N times to obtain a set of values $\{\beta_{\text{MLE}}^1, \ldots, \beta_{\text{MLE}}^N\}$. If the MLE estimate is invariant, then the set of values should fluctuate around the true value. The set of values can be used to obtain an unbiased estimate of the average value

$$\langle \beta_{\text{MLE}} \rangle_N = \frac{1}{N} \sum_{i=1}^{N} \beta_{\text{MLE}}^i$$

and a histogram of the PDF of β_{MLE}. If the method is unbiased, then the average estimate, $\langle \beta_{\text{MLE}} \rangle_N$, should converge to the true value. To sample a scenario in which, at each failure, minimal repair is applied, sampling from an aged distribution must be considered.

5.1.4 Sampling from an aged distribution

The above scenario calls for sampling from a Weibull distribution. However, the age of the component at the point from which sampling of the next event is done is not zero. Thus, the sampling equation (3.2.5) must be modified. Let a component have a birth point at t^b (this is the time point at which the age is zero) and the component is now at time t_c. Then (see equation (2.2.37) and the discussion therein), the conditional PDF for an event at time $t > t_c$ is given by

$$f^*(t|t^b, t_c) = \frac{f(t - t^b)}{1 - F(t_c - t^b)}$$

which, according to the age definition (expression (2.2.36)) can be written as

$$f^*(t|t^b, t_c) = \frac{f(t - t^b)}{1 - \text{Age}(t_c - t^b)}.$$

The CDF is obtained by integrating $f^*(t|t^b, t_c)$ from t_c to t, hence the CDF sampling equation takes the form

$$F^*(t|t^b, t_c) = \frac{F(t - t^b) - \text{Age}(t_c - t^b)}{1 - \text{Age}(t_c - t^b)} = \xi \tag{5.1.12}$$

or

$$\begin{aligned} F(t - t^b) &= \xi \times [1 - \text{Age}(t_c - t^b)] + \text{Age}(t_c - t^b) \\ &= \xi + \text{Age}(t_c - t^b) \times (1 - \xi) = \xi^*. \end{aligned} \tag{5.1.13}$$

This expression is convenient since existing sampling routines from $F(t)$ can be used by merely modifying the sampled random number from ξ to ξ^*. Equation (5.1.13) is general and applies to any distribution. Note that as the age approaches zero (or t_c approaches t^b), the regular sampling equation (3.2.2) is obtained.

To sample r consecutive minimal repaired failure points, let ξ_1 be the first random number. The first failure point is obtained from the equation $F(t_1) = \xi_1$. The second failure point is obtained from the equation $F(t_2) = \xi_2(1 - \xi_1) + \xi_1 = \xi_2^*$, since the age at the first event point is ξ_1 and, in general,

$$F(t_i) = \xi_i(1 - \xi_{i-1}) + \xi_{i-1} = \xi_i^* \tag{5.1.14}$$

(with the definition $\xi_0 \equiv 0$ to include the case $i = 1$). Applying equation (5.1.14) to the Weibull distribution yields

$$t_i = -\frac{[\ln(1 - \xi_i^*)]^{1/\beta}}{\lambda^{1/\beta}}. \tag{5.1.15a}$$

Denoting now

$$t_{i(1)} = -\frac{[\ln(1 - \xi_i^*)]}{\lambda},$$

so that $t_i = [t_{i(1)}]^{1/\beta}$, it becomes evident that the ith failure point can be obtained by taking the corresponding failure point with $\beta = 1$ to the power of $1/\beta$. Yet the ith failure point with $\beta = 1$ is simply the sum of i exponential failure times, thus (5.1.15a) can be replaced by the equivalent expression:

$$t_i = \left[\frac{\sum_{j=1}^{i}(-\ln(\xi_j))}{\lambda}\right]^{1/\beta}. \tag{5.1.15b}$$

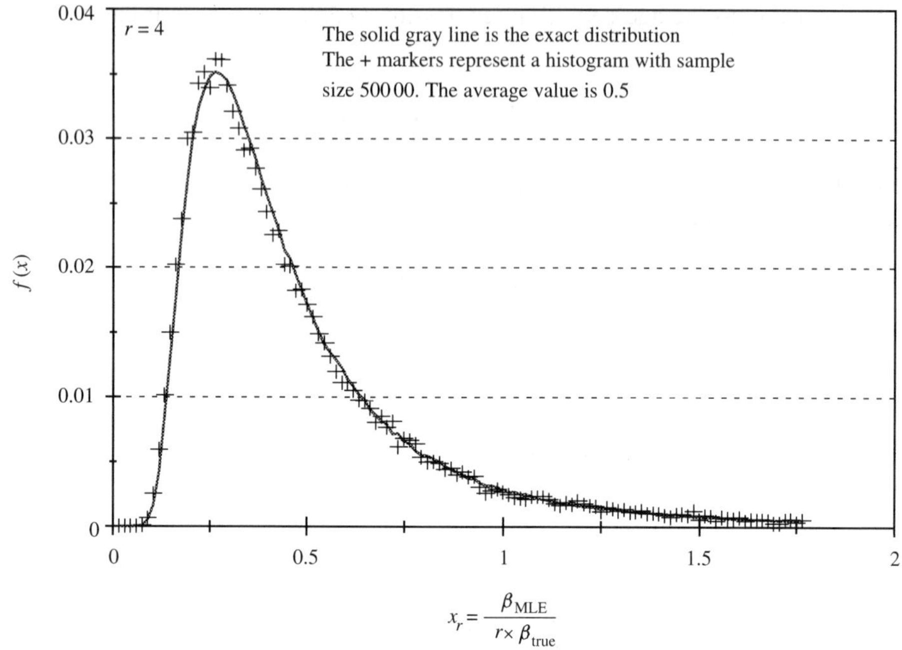

Figure 5.1.1 PDF of the MLE estimate of β for the case of $r = 4$ of Example 5.1.1

Expressions (5.1.15a) and (5.1.15b) are not identical; they represent random variables that have the same PDF. With regard to example (5.1.1), equation (5.1.15b) is used with $i = 1, \ldots, 4$ to generate four minimal repair failure points. Equation (5.1.11) is then used to obtain the MLE estimate. This process is repeated N times, and in Figure 5.1.1 the resulting histogram for $N = 50\,000$ is displayed along with the exact distribution. Substituting equation (5.1.15b) into (5.1.11) yields

$$\beta_{\text{MLE}} = \frac{r \times \beta_{\text{true}}}{\left[r \times \ln[y_r] - \ln\left(\prod_{i=1}^{r} y_i\right)\right]} = \frac{r \times \beta_{\text{true}}}{\ln[y_r^{r-1}/(y_1 y_2 \cdots y_{r-1})]} \tag{5.1.16}$$

where $y_i = \sum_{j=1}^{i} -\ln(\xi_j)$ is a sum of i standard exponential variables.

The average value of x_r is exactly 0.5. This means that the average value of the MLE estimate is twice the true value, i.e. $\beta_{\text{true}} = \frac{1}{2}\langle\beta_{\text{MLE}}\rangle$, an obvious bias exists (in fact, a very large bias); enough to make most decisions based on the MLE estimate risky and irrelevant to the future reality. Equation (5.1.16) indicates that the ratio of the MLE estimate to the true value is independent of the true value, thus the bias does not depend on β_{true} but only on r. This means that in the above example, the MLE estimate of β is, in fact, over-estimated by a factor of two, and the true value is around 0.85. This means, for example, that the hazard is decreasing as a function of time, rather than increasing. The bias dependence on r is defined in the following theorem.

Theorem 5.1.1: The bias of the minimal repair MLE estimator *Let*

$$x_r = \frac{\beta_{\text{MLE}}}{r\beta_{\text{true}}}$$

where the MLE estimate is based on r consecutive failures with minimal repair derived from equation (5.1.11), *then the PDF of* x_r *is given by*

$$f(x_r) = \frac{\mathrm{e}^{-(1/x_r)}}{(r-2)!x_r^r} \quad \textit{and} \quad \langle x_r \rangle = \frac{1}{r-2}.$$

(The reader may skip the details of the proof of this theorem without disturbance to the flow of the material. The proof is brought mainly because of its statistical beauty, and because it involves some useful statistical transformation principles.)

Proof From equation (5.1.16) it is clear that

$$x_r = \frac{1}{\ln[y_r^{r-1}/(y_1 y_2 \cdots y_{r-1})]},$$

where y_j is the sum of j standard, independent exponential variables, i.e. $y_j = \sum_{i=1}^{j} v_i$, where $v_i = -\ln \xi_i$. Also, $y_j = y_{j-1} + v_j$. Denoting

$$s_r = \frac{y_r^{r-1}}{y_1 y_2 \cdots y_{r-1}}$$

leads to $x_r = 1/(\ln(s_r))$. The distribution of x_r will be derived by investigating the distribution of s_r.

Consider first the case of $r = 3$. Define the variable $s_3 = y_3^2/(y_1 y_2)$. Consider the PDF of s_3, $g(s_3)$, by obtaining the probability $g(s_3)\,ds_3$ that s_3 is in an infinitesimal interval ds_3 around a given value. From the definition of s_3, it is clear that $y_3 = \sqrt{s_3 y_1 y_2}$. Note also that from the definition of y_i one may write $y_{i+1} = y_i + v_1$, where $v_1 = -\ln(\xi)$ is a single independent standard exponential variable. Thus, $y_{i+1} \geq y_i$, hence $\sqrt{s_3 y_1 y_2} \geq y_2$, yielding $s_3 y_1 \geq y_2 \geq y_1$. The domain of y_1, y_2, y_3 for which s_3 maintains a given value is when $\infty \geq y_1 \geq 0$, $\sqrt{s_3 y_1 y_2} \geq y_2$ and $y_3 = \sqrt{s_3 y_1 y_2}$; also,

$$dy_3 = \frac{\sqrt{y_1 y_2}}{2\sqrt{s_3}}\,ds_3,$$

hence one may write

$$g(s_3)ds_3 = \int_0^\infty dy_1 \int_{y_1}^{s_3 y_1} dy_2\, f(y_1, y_2, \sqrt{s_3 y_1 y_2})\, dy_3$$
$$= \int_0^\infty dy_1 \int_{y_1}^{s_3 y_1} dy_2\, f(y_1, y_2, \sqrt{s_3 y_1 y_2}) \frac{\sqrt{y_1 y_2}}{2\sqrt{s_3}}\, ds_3,$$

where $f(y_1, y_2, y_3)$ is the joint PDF of (y_1, y_2, y_3). It is also clear that the conditional PDF $f(y_{i+1} | y_i) = e^{-(y_{i+1} - y_i)} = e^{-v_1}$, hence

$$f(y_1, y_2, \ldots, y_i) = f(y_i | y_1, y_2, \ldots, y_{i-1}) f(y_{i-1} | y_1, y_2, \ldots, y_{i-2}) \cdots f(y_2 | y_1) f(y_1)$$
$$= e^{-(y_i - y_{i-1})} e^{-(y_{i-1} - y_{i-2})} \cdots e^{-(y_2 - y_1)} e^{-y_1} = e^{-y_i}. \tag{5.1.17}$$

To obtain the PDF of y_i, integration must be carried over all the other variables, recalling that $y_{j-1} \le y_j \le y_{j+1}$ for $j = 2, \ldots, i$, the integration takes the form

$$f(y_i) = \int_0^{y_i} dy_1 \int_{y_1}^{y_i} dy_2 \ldots \int_{y_{i-1}}^{y_i} dy_{i-1} f(y_1, y_2, \ldots, y_i) = \frac{y_i^{i-1} e^{-y_i}}{(i-1)!}, \tag{5.1.18}$$

which could have been deduced without the integration, since y_i is a sum of i exponential variables (see equation (2.2.8)). Substituting $f(y_1, y_2, y_3)$ from (5.1.17) into the expression for $g(s_3)$ yields

$$g(s_3) = \frac{1}{2\sqrt{s_3}} \int_0^\infty dy_1 \int_{y_1}^{s_3 y_1} \sqrt{y_1 y_2}\, dy_2\, e^{-\sqrt{s_3 y_1 y_2}}.$$

Using the transformation $\sqrt{y_2} = w$, the integration over y_2 takes the form

$$\int_{y_1}^{s_3 y_1} \sqrt{y_1 y_2}\, e^{-\sqrt{s_3 y_1 y_2}}\, dy_2 = 2\sqrt{y_1} \left\{ \frac{e^{-\sqrt{s_3 y_1} w}}{(s_3 y_1)^{3/2}} [s_3 y_1 w^2 + 2\sqrt{s_3 y_1}\, w + 2] \Bigg|_{\sqrt{s_3 y_1}}^{\sqrt{y_1}} \right\}.$$

Elaborating the above expression, the distribution takes the form:

$$g(s_3) = \frac{2}{s_3^2} \left[\int_0^\infty \frac{e^{-\sqrt{s_3} y_1} - e^{-s_3 y_1}}{y_1}\, dy_1 \right].$$

The integral in the brackets can be viewed as a function of s_3, $h(s_3)$. Taking its derivative with respect to s_3, one obtains the differential equation

$$\frac{dh(s_3)}{ds_3} = \frac{1}{2s_3}$$

with the boundary condition $h(1) = 0$, for which a unique solution is $h(s_3) = \frac{1}{2}\ln(s_3)$, leading to

$$g(s_3) = \frac{\ln(s_3)}{s_3^2}. \tag{5.1.19}$$

Noting that

$$x_3 = \frac{1}{\ln(s_3)} \quad \text{and} \quad \left|\frac{ds_3}{dx_3}\right| = \frac{1}{x_3^2}\mathrm{e}^{1/x_3}$$

yields the distribution of x_3:

$$f(x_3) = g(s_3)\left|\frac{ds_3}{dx_3}\right| = \frac{\mathrm{e}^{-1/x_3}}{x_3^3} \tag{5.1.20}$$

which proves the case for $r = 3$.

In general, the considerations used above may be extended. Namely, denote

$$s_r = \frac{y_r}{y_1}\frac{y_r}{y_2}\cdots\frac{y_r}{y_{r-1}}$$

then for any value of $1 \leq s_r \leq \infty$, $y_r = [y_1 y_2 \cdots y_{r-1} s_r]^{1/(r-1)}$. Considering the fact that $y_r \geq y_{r-1} \geq y_{r-2} \geq \cdots \geq y_2 \geq y_1$ and applying this relation to the above expression successively yields the range of each of the variables:

$$\begin{aligned}
[y_1 y_2 \cdots y_{r-1} s_r]^{1/(r-1)} \geq y_{r-1} &\Rightarrow [y_1 y_2 \cdots y_{r-2} s_r]^{1/(r-2)} \geq y_{r-1} \\
[y_1 y_2 \cdots y_{r-2} s_r]^{1/(r-2)} \geq y_{r-2} &\Rightarrow [y_1 y_2 \cdots y_{r-3} s_r]^{1/(r-3)} \geq y_{r-2} \\
&\vdots \\
[y_1 y_2 y_3 s_r]^{1/3} \geq y_3 &\Rightarrow [y_1 y_2 s_r]^{1/2} \geq y_3 \\
[y_1 y_2 s_r]^{1/2} \geq y_2 &\Rightarrow [y_1 s_r] \geq y_2.
\end{aligned}$$

Also,

$$dy_r = [y_1 y_2 \cdots y_{r-1}]^{1/(r-1)} \frac{s_r^{(2-r)/(r-1)}}{(r-1)} ds_r$$

and $f(y_1, y_2, \ldots, y_r) = \mathrm{e}^{-y_r}$, so that

$$g(s_r) = \frac{1}{r-1} s_r^{(2-r)/(r-1)} \int_0^\infty dy_1 \int_{y_1}^{y_1 s_3} dy_2 \int_{y_2}^{\sqrt{y_1 y_2 s_r}} \cdots$$
$$\times \int_{y_{r-2}}^{(y_1 \cdots y_{r-2} s_r)^{1/(r-2)}} (y_1 \cdots y_{r-1})^{1/(r-1)} \mathrm{e}^{-(y_1 \cdots y_{r-1} s_r)^{1/(r-1)}} dy_{r-1}. \qquad (5.1.21)$$

The explicit integration of (5.1.21) is quite lengthy; mathematical induction may be used instead. For this purpose, the following lemma will be useful:

Lemma 5.1.1 *Define*

$$z_{r,i} = \ln\left(\frac{y_r}{y_i}\right)$$

with $i < r$, then the PDF of $z_{r,i}$ is given by

$$f(z) = \frac{(r-1)!\,\mathrm{e}^{-(r-1)z}\,[\mathrm{e}^z - 1]^{(r-i-1)}}{(i-1)!(r-i-1)!}. \qquad (5.1.22)$$

Proof Recall that y_i is a sum of i standard exponential variables. Also, $y_r = y_i + V_{r-i}$, where V_{r-i} is again a sum of $(r-i)$ exponential variables which are independent of the sum of the first i variables. To indicate this point, we will also use the notation $y_i = V_i$. So $z_{r,i}$ may be written in the form

$$z_{r,i} = \ln\left(\frac{y_r}{y_i}\right) = \ln\left(1 + \frac{V_{r-i}}{V_i}\right).$$

Consider the distribution of

$$x = \left(\frac{V_{r-i}}{V_i}\right).$$

Keeping x fixed, as V_i moves in the range $[0, \infty)$, V_{r-i} takes the value $V_{r-i} = V_i x$ and $dV_{r-1} = V_i\,dx$. Also, any V_j has a standard Gamma distribution

$$G_j(V_j) = \frac{V_j^{(j-1)}\,\mathrm{e}^{-V_j}}{(j-1)!},$$

hence the PDF of x is given by

$$f(x) = \int_0^\infty V_i G_i(V_i) G_{r-i}(V_i x)\,dV_i = \frac{x^{r-i-1}}{(i-1)!(r-i-1)!} \int_0^\infty V_i^{r-1}\,\mathrm{e}^{-V_i}\,dV_i.$$

The integral on the right-hand side is the Gamma function $\Gamma(r) = (r-1)!$, hence

$$f(x) = \frac{(r-1)!\, x^{r-i-1}}{(i-1)!\,(r-i-1)!}. \tag{5.1.23}$$

The distribution of $w = 1 + x$ is then

$$f(w) = \frac{(r-1)!(w-1)^{r-i-1}}{(i-1)!(r-i-1)!}.$$

Using the relation $z_{r,i} = \ln(w)$ with the transformation relation

$$f(z_{r,i}) = f(w)\left|\frac{dw}{dz_{r,i}}\right|,$$

(note that the notation $f(\cdot)$ is used here as an indication of the PDF of the argument variable and the two functions are not the same) yields equation (5.1.22), which completes the proof of the lemma.

It was shown that for $r = 3$,

$$f(x_3) = \frac{e^{-1/x_3}}{x_3^3}.$$

Denoting

$$s_r = \frac{y_r}{y_1}\frac{y_r}{y_2}\cdots\frac{y_r}{y_{r-1}}$$

and $a_r = \ln(s_r)$, it is clear that

$$x_3 = \frac{1}{a_3}$$

and the relation

$$f(x_3) = f(a_3)\left|\frac{da_3}{dx_3}\right|$$

yields $f(a_3) = a_3\, e^{-a_3}$.

In the induction process, one assumes that

$$f(a_r) = \frac{a_r^{r-2}\, e^{-a_r}}{(r-2)!},$$

and shows that this leads to the relation

$$f(a_{r+1}) = \frac{a_r^{r-1}\,\mathrm{e}^{-a_{r+1}}}{(r-1)!}.$$

From its definition, a_{r+1} can be expressed as

$$\begin{aligned} a_{r+1} &= \ln(s_{r+1}) = \ln\left(\frac{y_{r+1}}{y_1}\frac{y_{r+1}}{y_2}\cdots\frac{y_{r+1}}{y_r}\right) \\ &= \ln\left(\frac{y_r}{y_1}\frac{y_r}{y_2}\cdots\frac{y_r}{y_{r-1}} \times \left(\frac{y_{r+1}}{y_r}\right)^r\right) = a_r + r\ln\left(\frac{y_{r+1}}{y_r}\right). \end{aligned}$$

The distribution of a_r is known from the induction assumption, and the distribution of

$$\ln\left(\frac{y_{r+1}}{y_r}\right) = z_{r+1,r}$$

is given by Lemma 5.1.1 in equation (5.1.22). To obtain the PDF of a_{r+1}, note that the relation $a_{r+1} = a_r + r \times z_{r+1,r}$ implies that

$$\frac{da_{r+1}}{r} = dz_{r+1,r}$$

and the domain of a_r is in the range $[0, a_{r+1}]$, thus the distribution of a_{r+1} is given as

$$f(a_{r+1}) = \int_0^{a_{r+1}} \frac{1}{r} f(a_r) f_z\left(\frac{a_{r+1} - a_r}{r}\right) da_r.$$

Substituting the induction assumption for $f(a_r)$ and expression (5.1.22) for

$$f_z\left(\frac{a_{r+1} - a_r}{r}\right)$$

with $i = r$ yields

$$\begin{aligned} f(a_{r+1}) &= \int_0^{a_{r+1}} \frac{1}{r} f(a_r) f_z\left(\frac{a_{r+1} - a_r}{r}\right) da_r = \int_0^{a_{r+1}} \frac{1}{r}\frac{a_r^{r-2}\,\mathrm{e}^{-a_r}}{(r-2)!}\frac{(r)!\,\mathrm{e}^{-(a_{r+1}-a_r)}}{(r-1)!}\, da_r \\ &= \int_0^{a_{r+1}} \frac{a_r^{r-2}}{(r-2)!}\mathrm{e}^{-a_{r+1}}\, da_r = \frac{a_{r+1}^{(r-1)}\,\mathrm{e}^{-a_{r+1}}}{(r-1)!} \end{aligned}$$

which completes the induction.

It is thus proven that the variable

$$a_r = \ln(s_r) = \ln\left(\frac{y_r}{y_1}\frac{y_r}{y_2}\cdots\frac{y_r}{y_{r-1}}\right)$$

follows a Gamma distribution with $r-1$ degrees of freedom. To complete the theorem, note that $x_r = 1/a_r$, hence

$$f(x_r) = f(a_r)\left|\frac{da_r}{dx_r}\right| = \frac{a_r^{r-2}\,\mathrm{e}^{-a_r}}{(r-2)}\frac{1}{x_r^2} = \frac{\mathrm{e}^{-1/x_r}}{(r-2)!\,x_r^r},$$

as stated in the theorem. Also,

$$x_r = \frac{\beta_{\mathrm{MLE}}}{r\beta_{\mathrm{true}}}$$

with the mean value

$$\langle x_r\rangle = \left\langle\frac{\beta_{\mathrm{MLE}}}{r\beta_{\mathrm{true}}}\right\rangle = \int_0^\infty \frac{x_r}{(r-2)!}\frac{\mathrm{e}^{-1/x_r}}{x_r^r} = \int_0^\infty \frac{(r-3)!}{(r-2)!}\frac{\mathrm{e}^{-1/x_r}}{(r-3)!x_r^r} = \frac{1}{r-2},$$

which completes the proof.

The PDF of

$$x_r = \frac{\beta_{\mathrm{MLE}}}{r\beta_{\mathrm{true}}}$$

is plotted in Figure 5.1.1, along with the sampled histogram. The sampling technique is shown to produce a good agreement with the analytic function. In this case, an analytic expression could be produced for distribution of the MLE estimate. In general scenario cases, an analytic expression is either unknown or prohibitively difficult to calculate. One would not seriously consider a numerical calculation of expression (5.1.21). This, however, is exactly what was done with the Monte Carlo method with extreme ease. It is again the high dimensions of the integral that makes it prohibitively difficult for analytic methods and very easy for the Monte Carlo method. The above relation suggests that

$$\langle\beta_{\mathrm{MLE}}\rangle = \frac{r\beta_{\mathrm{true}}}{r-2}. \tag{5.1.24}$$

Thus, in this case the bias is known. This suggests that a correction should be applied to the MLE estimate in the form

$$\beta_{\mathrm{ub}} = \frac{(r-2)\beta_{\mathrm{MLE}}}{r}. \tag{5.1.25}$$

β_{ub} is an unbiased estimate of the true beta based on the MLE estimate. Returning now to the results of Example 5.1.1 yields $\beta_{ub} = 0.5\beta_{MLE} = 0.85$ and, accordingly, also $\lambda_{ub} = 2.262 \times 10^{-3}$ and $\eta = 7.724 \times 10^{-4}$.

Recall that λ_{ub} is the value of λ that maximizes the likelihood function with the relevant β. To further demonstrate the biasing correction, a set of four time points is again generated using the same procedure that was used for Figure 5.1.1 (using equation (5.1.15b)) with $\beta = 0.85$ and $\lambda = 2.262 \times 10^{-3}$. The points obtained are 1233, 2278, 2414 and 4336, with the estimates $\beta_{ub} = 0.804$, and $\lambda_{ub} = 4.744 \times 10^{-3}$. Looking at the process of sampling and MLE corrected estimates as a statistical transformation, one may say that the true parameters are invariant under the transformation.

5.1.5 As good as new

The case discussed above of minimal repair turns to be the simplest, because the bias can be evaluated analytically and the distribution of β_{MLE}/β_{true} is explicitly known. It should be noted that the bias factor is independent of β_{true}, and is a function of the sample size only. The case of 'as good as new' following repair is somewhat more complicated.

Consider an experiment in which a component had r such failures at times $t_1, t_2, \ldots, t_r$ and the experiment is terminated at the last failure. The failure times are each measured from the preceding failure, since the age is set to zero at each failure. Also, the zero age time points coincide, in this case, with the failure points, so that the MLE equation takes the form

$$\frac{r + \beta \sum_{i=1}^{r} \ln(t_i)}{\beta \sum_{j=1}^{r} t_j^{\beta} \ln(t_j)} = \frac{r}{\sum_{j=1}^{r} t_j^{\beta}} \quad \text{and} \quad \lambda = \frac{r}{\sum_{j=1}^{k} t_j^{\beta}}. \tag{5.1.26}$$

This equation cannot be solved analytically, and requires a numerical solution [`program genmle.for`]. When solving equation (5.1.10), an overflow can occur due to the form of the denominators. In such cases, one may use a different scale for the time points to reduce their numerical value. If a scale factor C is used, hence $t' = Ct$, then the shape parameter β is not affected, but the scale parameter must be adjusted according to

$$\lambda = \lambda' \times C^{\beta}, \tag{5.1.27}$$

where λ' is the scale parameter obtained with the transformed time points.

Each time point in this scenario is obtained as a sample of the form

$$t_i = \left[\frac{(-\ln(\xi_i))}{\lambda_{tr}}\right]^{1/\beta_{tr}},$$

where λ_{tr} and β_{tr} are the true parameters of the scenario. Substituting this expression into equation (5.1.26) yields

$$\frac{r + x\sum_{i=1}^{r} \ln y_i}{\sum_{i=1}^{r} x y_i^x \ln y_i} = \frac{r}{\sum_{i=1}^{r} y_i^x}, \tag{5.1.28}$$

where y_i is a scaled independent exponential variable, i.e. $y_i = -\ln \xi_i / \lambda_{tr}$, and $x = \beta_{MLE}/\beta_{tr}$. Thus, the distribution of the MLE estimate is independent of the true shape parameter, and so is the bias factor.

Referring now to the time points of Example 5.1.1 and assuming that they represent four points of independent failures, one obtains $\beta_{MLE} = 2.48$, $\lambda_{MLE} = 7.848 \times 10^{-10}$, and hence $\eta_{MLE} = 2.1307 \times 10^{-4}$. Are these values biased?

To answer this question, the MLE values are now assumed to be the true values. N sets of new r event points are sampled, and from each sample j the β^j_{MLE} is calculated. This allows a Monte Carlo estimation of the average $\langle \beta_{MLE} \rangle$ and a histogram estimation of the distribution of $x = \beta_{MLE}/\beta_{tr}$. The value of the bias factor $B(r)$ for this case is 1.62, hence

$$\beta_{ube} = \frac{\beta_{MLE}}{B(4)} = \frac{\beta_{MLE}}{1.62}$$

is an unbiased estimator of β_{tr}. This implies that, for the present case, the above results must be modified to $\beta_{ube} = 1.53$, $\lambda_{ube} = 2.698 \times 10^{-6}$ and $\eta_{ube} = 2.2922 \times 10^{-4}$. The bias factor as function of r is shown in Figure 5.1.2.

The solid thin line represents a useful approximation to the curve of the form

$$B(r) = \frac{r - 0.75}{r - 2}. \tag{5.1.29}$$

The PDF of $x = \beta_{MLE}/\beta_{tr}$, as obtained with a sample of 200 000 histories (each set of event points is referred to as a history) is shown in Figure 5.1.3. In Figure 5.1.4, the behavior of the distribution of β_{MLE} for a number of values of r with a given $\beta_{tr} = 2.48$ is demonstrated.

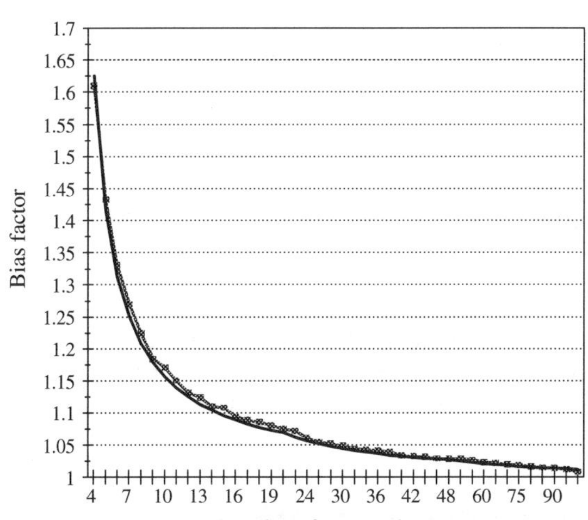

Figure 5.1.2 The bias factor as a function of r, the number of event points (based on 10 000 histories for each value of r)

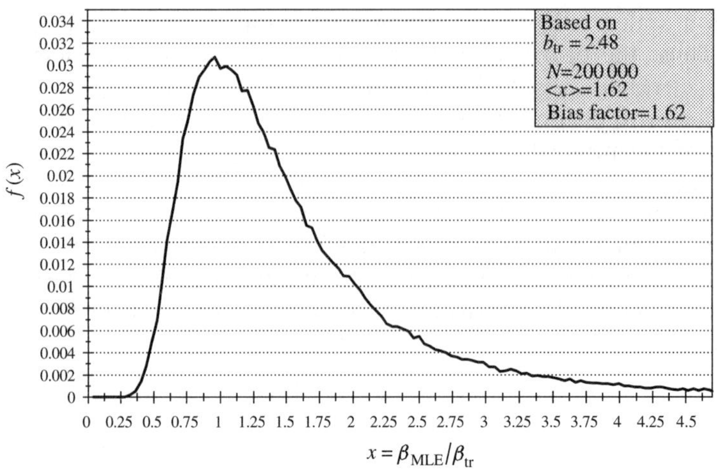

Figure 5.1.3 The PDF of the shape parameter ratio for $r = 4$

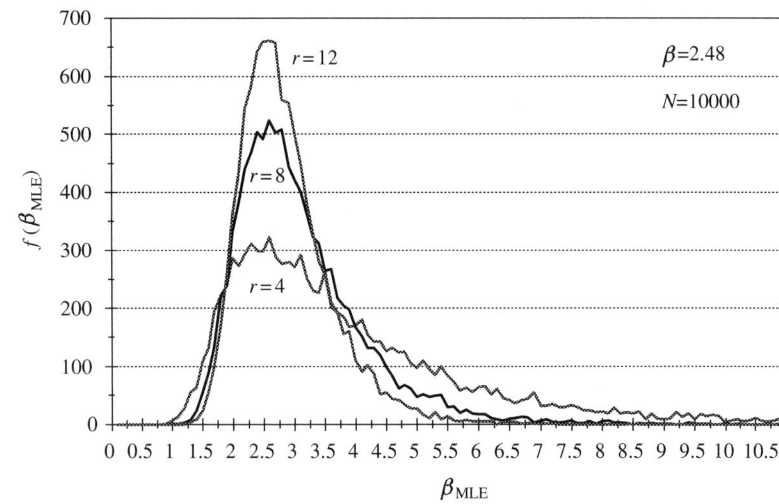

Figure 5.1.4 The distribution of β_{MLE} for three values of r

It should be noted that as r increases, the distributions become pointed, its variance decreases and it becomes symmetric. This is how the bias factor converges to unity. Thus β_{MLE} is a *consistent estimator*. This is another example of a consistent estimator which is not a UBE.

5.1.6 Time truncated scenario

The above two cases are important, but are far from being general. In both cases, experiment is terminated by the number of failures. In reality, the scenario is usually truncated by the time at which the data is being observed and used. Consider an extension of the 'as good as new' case, in which the experiment is not terminated at the rth failure, but rather continues up to a given time T_o. Since the event points represent the time difference between events, it follows that an additional zero age time point is added, namely

$$t_{r+1} = T_o - \sum_{i=1}^{r} t_i.$$

Substituting this into the MLE equation yields

$$\frac{m + \beta \sum_{i=1}^{r} \ln(t_i)}{\beta \sum_{j=1}^{r} t_j^{\beta} \ln(t_j) + \beta\left[T_o - \sum_{i=1}^{r} t_i\right]^{\beta} \ln\left[T_o - \sum_{i=1}^{r} t_i\right]} = \frac{r}{\sum_{j=1}^{r} t_j^{\beta} + \left[T_o - \sum_{i=1}^{r} t_i\right]^{\beta}}, \tag{5.1.30a}$$

and

$$\lambda_{\text{MLE}} = \frac{r}{\sum_{j=1}^{r} t_j^{\beta} + \left[T_o - \sum_{i=1}^{r} t_i\right]^{\beta}}. \tag{5.1.30b}$$

The problem in this expression is that, upon substitution of the sampling expression of

$$t_i = \left[\frac{(-\ln(\xi_i))}{\lambda_{\text{tr}}}\right]^{1/\beta_{\text{tr}}}$$

it becomes obvious that β_{MLE} depends both upon β_{tr} and λ_{tr}. It is thus a two-dimensional function of the true parameters. This is a serious practical difficulty. The difficulty can be relaxed using the approximation

$$\lambda_{\text{tr}} = \lambda_{\text{MLE}}(\beta_{\text{tr}}), \tag{5.1.31}$$

which means that for each value of β_{tr}, λ_{tr} is obtained from equation (5.1.30b) with the given time points, and where β_{tr} is assumed for β. This introduces some arbitrariness into the estimation, and is based on the fact that for any value of β the λ obtained from the MLE relation maximizes the likelihood function. Still, the bias factor will now depend upon β_{tr}, and no single distribution can be found for the shape parameter ratio. Moreover, whereas in the former cases the number of events was fixed, it now becomes a random variable. In sampling the scenario, a set of failure time point is sampled until the next event point is larger than T_o, at which point the history is terminated. With the above remarks, the MLE problem may now be restated as follows: *In any estimation process there are no data points. There is a* ***scenario*** *which includes data points.*

The scenario involves a number of identical components with some given rules of operation. The rules apply to failures, repairs, maintenance and the endpoints of each scenario. It is now clear that the nature of the repair is of essence in considering the data. This is not the most general possible scenario: partial repair, in which the age is reduced to an intermediate level upon repair, and other forms of partial maintenance and age dependent maintenance, are not considered. It is also assumed that the component has a Weibull failure distribution with unknown parameters β_{tr} and λ_{tr}. These are the quantities to be estimated. The data emerging from the scenario is grouped into two sets of time points. A set of failure points $(t_{o1}, \ldots, t_{or})$ (the index 'o' stands to indicate that these are the actually observed points, differentiating it from time points that will be sampled in subsequent steps) and a set of zero age points $(\tau_{o1}, \ldots, \tau_{ok})$. β_{MLE} and λ_{MLE} are calculated by equations (5.1.10) using the sets $(t_{o1}, \ldots, t_{or})$ and $(\tau_{o1}, \ldots, \tau_{ok})$.

$\beta_{\text{MLE}}(t_1, \ldots, t_r; \tau_1, \ldots, \tau_k)$ can be viewed as an estimator of β_{tr}. As such, it is also a random variable, and it makes sense to ask if it is an unbiased estimator of β_{tr}, i.e. if

$$\langle \beta_{\text{MLE}}(t_1, \ldots, t_r; \tau_1, \ldots, \tau_k) \rangle = \beta_{\text{tr}}?$$

In most cases this is not the case, β_{MLE} is not an unbiased estimator of β_{tr}. It is a consistent estimator, so that if r is large enough the bias may be neglected.

Now let $M = \{(t_1, \ldots, t_m; \tau_1, \ldots, \tau_k); m \geq 4; k \geq 1\}$ be a subset of all the histories that can be sampled for the given scenario with the given β_{tr} and λ_{tr} where λ_{tr} is obtained from (5.1.10) as

$$\lambda_{\text{tr}} = \frac{r}{\sum_{j=1}^{k} \tau_{oj}^{\beta_{\text{tr}}}}, \tag{5.1.32}$$

and with the constraint that there are at least four event points. This subspace will be referred to as the *MLE sample space*. It is on this space that the average of β_{MLE} is performed. To realize this process the following steps are taken.

The MLE scenario sampling procedure

1. Given β_{tr}, calculate λ_{tr} using equation (5.1.32).
2. Given β_{tr}, λ_{tr} and the conditions of the scenario, sample N sets of time points $(t_1^i, \ldots, t_m^i; \tau_1^i, \ldots, \tau_k^i)$, $i = 1, \ldots, N$. Note that histories that have less than four events are rejected and are not counted.
3. Estimate the Bias factor:

$$\tilde{B}(\beta_{\text{tr}}) = \frac{1}{\text{N}\beta_{\text{tr}}} \sum_{i=1}^{N} \beta_{\text{MLE}}^i(t_1^i, \ldots, t_m^i; \tau_1^i, \ldots, \tau_k^i).$$

   In those cases where the bias factor is independent of β_{tr} (such as the first two cases discussed above), apply the correction $\beta_{\text{ub}} = \beta_{\text{MLE}}/B$, where β_{MLE} is the MLE estimate obtained from the set of points observed. If the bias factor is a function of β_{tr}, obtain the unbiased estimate by solving the equation $\beta_{\text{ub}} B(\beta_{\text{ub}}) = \beta_{\text{MLE}}$.
4. For large enough N, it is also recommended to generate the approximate PDF of $x = \beta_{\text{MLE}}/\beta_{\text{tr}}$ in the form of a histogram.

Example 5.1.2

Consider the case of as good as new repair with time truncation discussed above. The time points are those of Example 5.1.1, namely 1637 h, 3430 h, 4900 h and 6614 h. These points are viewed as consecutive failures (with full repair) happening at times 1637, 5067, 9967 and 16 581. With the truncation of the experiment at 20 000 hours, which adds a time point of 3419 hours to the zero age points. With these values, one obtains $\beta_{\text{MLE}} = 2.686$, $\lambda_{\text{MLE}} = 1.996 \times 10^{-10}$ and $\eta_{\text{MLE}} = 2.205 \times 10^{-4}$.

The MLE sampling scenario procedure is used for values of β in the range 0.5–6. The bias as function of β is shown in Figure 5.1.5(a) and the product $\beta B(\beta)$ in Figure 5.1.5(b). All the results are based on samples of 20 000 case histories.

The bias factor has a small variability around a value of 1.43. Using the results of Figure 5.1.5(b) for the solution of $\beta_{\text{ub}} B(\beta_{\text{ub}}) = \beta_{\text{MLE}}$, one obtains $\beta_{\text{ube}} = 1.905$ and $\lambda_{\text{ube}} = 9.562 \times 10^{-8}$ resulting $\eta = 2.066 \times 10^{-4}$. In Table 5.1.1, the results of the above three cases are summarized. They are interesting since, in all three cases, the same event points are considered. They differ only in the scenario leading to the events. The histograms for a range of values of the shape parameter are shown in Figure 5.1.6.

It may be noted that the distributions are similar in the range [1, 6] and do not vary much as β_{tr} changes. This property will be of significance when the question of interval estimation is considered. The detailed steps for calculating the average MLE and the histogram for this example are shown below:

Scenario analysis steps

Scan over a range of values of β.

- For a given value of β, set sample size $N = 0$.
  1. Set the values of the histogram boundaries: $a_1 = 0.0$; $\Delta a = 0.1$; $a_i = a_1 + i \times \Delta a$ for $i = 1, \ldots, 100$.
  2. Set the value of λ to

$$\lambda = \frac{4}{1637^{\beta} + 3430^{\beta} + 4900^{\beta} + 6614^{\beta} + 3419^{\beta}}.$$

  3. Sample a scenario:
     3.1. Set $t_c = 0.0$ and number of events $r = 0$.

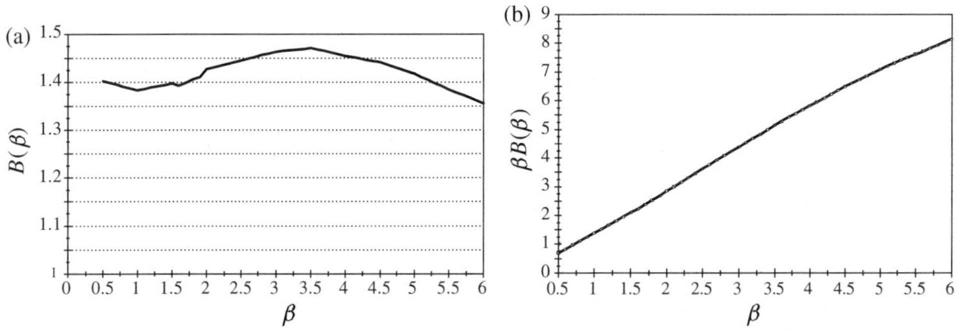

Figure 5.1.5 (a) The bias factor and (b) the bias product, $\beta B(\beta)$, as a function of the shape parameter

Table 5.1.1 Minimum likelihood and unbiased estimates of three scenarios with identical event points

Case	β_{MLE}	λ_{MLE}	η_{MLE}	Bias factor	β_{ube}	λ_{ube}	η_{ube}
Minimal repair, End at last failure	1.7	1.28×10^{-6}	3.42×10^{-4}	2	0.85	2.262×10^{-3}	7.72×10^{-4}
Full repair, End at last failure	2.48	7.848×10^{-10}	2.130×10^{-4}	1.62	1.53	2.698×10^{-6}	2.292×10^{-4}
Full repair, End at time point	2.686	1.99×10^{-10}	2.205×10^{-4}	1.41	1.905	9.562×10^{-8}	2.066×10^{-4}

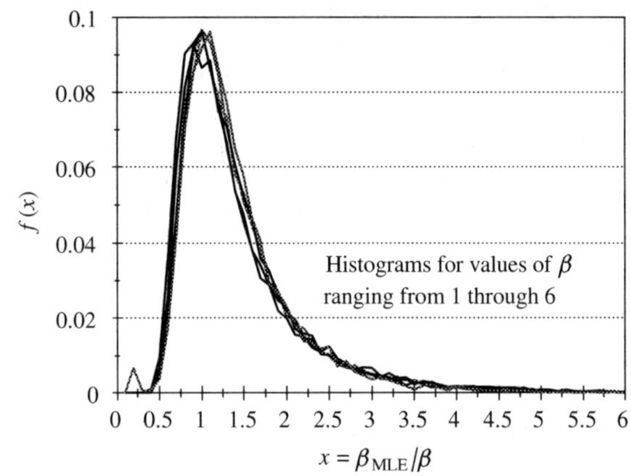

Figure 5.1.6 Histograms obtained with samples of 20 000 case histories for a range of values of β_{tr}

3.2. Sample next event:

$$\Delta t = -\left[\frac{\ln(\xi)}{\lambda}\right]^{1/\beta};$$

set $t_c = t_c + \Delta t$.

3.3. If $t_c \geq T_{\max}$ (in the present case, $T_{\max} = 20\,000$). Go to step 3.5, otherwise proceed to 3.4.

3.4. Set $r = r + 1$; set $t_r = \Delta t$, set $t_r = \Delta t$ and $\tau_r = \Delta t$. Go to 3.2.

3.5. Set $\tau_{r+1} = T_{\max} - \tau_0$, where τ_0 is the last zero age point. Go to step 3.6.

3.6. Terminate scenario sampling, go to step 4.

4. If $r < 4$ reject the scenario sample, go to step 3, otherwise go to step 5.
5. Increase the sample size: set $N = N + 1$.
6. Calculate $\beta_{\text{MLE}}(t_1, \ldots, t_r; \tau_1, \ldots, \tau_{r+1})$ by solving equation (5.1.30a).
7. Update the histogram: set $x = \beta_{\text{MLE}}/\beta$. Find a_k such that $a_k \leq x < a_{k+1}$; increase $n_k = n_k + 1$.
8. Update sum of estimator $S_N = S_N + \beta_{\text{MLE}}$.

9. If N reached a predetermined limit $N_{\max}$, go to step 10, otherwise go to 3.
10. Evaluate average $\tilde{\beta}_{\mathrm{AV}} = S_N/N$; bias factor $\tilde{\mathrm{B}} = \tilde{\beta}_{\mathrm{AV}}/\beta$. The set $\{n_k\}$ contains the histogram.

This procedure may be easily extended to any general scenario.

So far, the examples have involved a single component with an identical type of repair. The method discussed above is now summarized in a more general and realistic example.

Example 5.1.3 Mixed scenario of two components

Consider a plant in which two identical pressure pump units are operating. The first unit failed twice at points 270 days and 432 days, in which minimal repair was applied. At time 520 days, maintenance was applied and the component was completely rejuvenated. No other failures were observed up to time 710 days. The second component was operating for 815 days. It failed at 315 days, at which point minimal repair was applied. The next failure occurred at 471 days, with full repair applied. The last failure occurred at 690 days, and minimal repair was applied. The scenario is illustrated in Figure 5.1.7.

The likelihood function takes the form

$$L(t_{1,1},\ldots,t_{2,4}|\lambda,\beta) = f(t_{1,1}) \times \frac{f(t_{1,2})}{1-F(t_{1,1})} \times \frac{1-F(t_{1,3})}{1-F(t_{1,2})} \times [1-F(t_{1,4}-t_{1,3})]$$

$$\times f(t_{2,1}) \times \frac{f(t_{2,2})}{1-F(t_{2,1})} \times f(t_{2,3}-t_{2,2}) \times \frac{1-F(t_{2,4}-t_{2,2})}{1-F(t_{2,3}-t_{2,2})}$$

$$= (\lambda\beta)^5 \left[t_{1,1}t_{1,2}t_{2,1}t_{2,2}(t_{2,3}-t_{2,2})\right]^{\beta-1} \times \mathrm{e}^{-\lambda\left\{t_{1,3}^{\beta}+[t_{1,4}-t_{1,3}]^{\beta}+t_{2,2}^{\beta}+(t_{2,4}-t_{2,2})^{\beta}\right\}}.$$

The event points are (270, 432, 315, 471, 219), and the zero age points are (520, 190, 471, 344). This yields $\beta_{\mathrm{MLE}} = 2.925$, $\lambda_{\mathrm{MLE}} = 2.68 \times 10^{-8}$ and $\eta_{\mathrm{MLE}} = 2.578 \times 10^{-3}$.

To sample the scenario one must specify the rules. It should be noted that the data from such scenarios is often obtained without much reference to the 'rules of the game'. For example, in the above scenario in some cases minimal repair is applied, and in others full repair – was there an *a priori* rule that defined when each type of repair should be applied? Indeed, a non-chaotic organization would have set such rules that

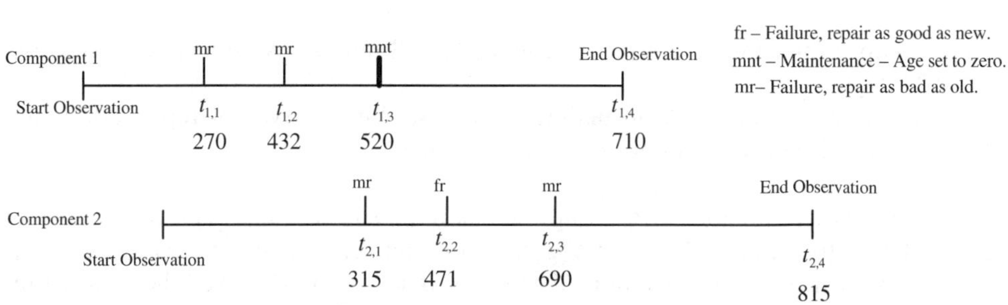

Figure 5.1.7 A two component scenario

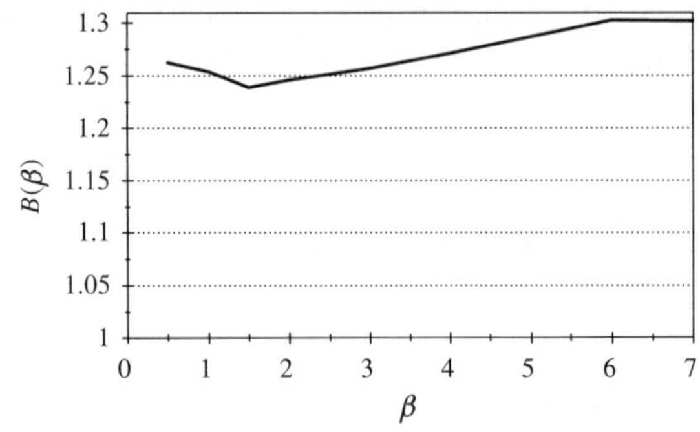

Figure 5.1.8 The bias factor as a function of the true beta

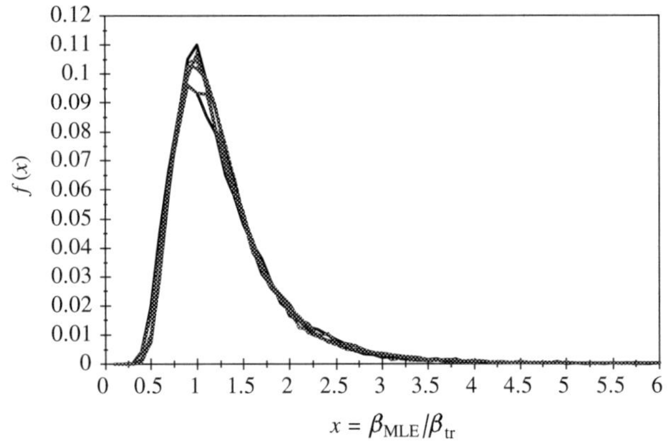

Figure 5.1.9 Histograms of the beta ratio for a range of values of β_{true}, for six values of the true beta: 0.5, 1, 2, 3, 4, 6. Samples larger than 2000 are used

will, hopefully, optimize its operation. Yet reality is not always ideal; one may find that different types of repair are applied due to local temporary conditions and constraints. Such a realistic situation may force the introduction of fuzzy elements into the scenario by which types of repairs are determined with some probabilities. For the current scenario the rules are:

1. For component 1 minimal repair is applied for all failures. Maintenance was preset at 520 days.
2. For component number 2, no maintenance is set but the type of repair alternates. Every second failure a complete repair is applied.

The bias factor as function of β_{true} and the histograms of $x = \beta_{\text{MLE}}/\beta_{\text{tr}}$ for a range of values of β_{true} (0.5–6) are shown in Figures 5.1.8 and 5.1.9 [`Program allmle.for`].

It can be seen that the variation of the bias is less than 4% over the range, thus one may take it as a constant of 1.275. Dividing the MLE estimate by this factor yields $\beta_{\text{ube}} = 2.294$, $\lambda_{\text{ube}} = 1.287 \times 10^{-6}$ and $\eta_{\text{ube}} = 2.706 \times 10^{-3}$. The variation of the histograms over the range of the shape parameter is minimal.

5.2 Interval Estimation

In the previous section point estimation was discussed. Recall that the material discussed in Chapters 3 and 4 also relates to point estimates, yet these estimates are considered for a situation in which the number of 'observations' can be increased at will, paying for it only with computer time. Thus, if the estimator is unbiased its convergence is guaranteed. In the case of data uncertainty, no such convergence can be sought. The finite, usually small, number of observations forces the introduction of additional, rather arbitrary considerations, such as the maximum likelihood principle. Some degree of the arbitrariness may be removed by setting the estimator as unbiased. Still, the main virtues of the method are that it makes sense, it is objective and it is about all that can be done. It is when point estimation is combined with interval estimation that a meaningful treatment of uncertainty is obtained. Unlike point estimation, an interval estimate is founded on solid theoretical grounds. The basic premise of interval estimation is that, based on the data observed, one may identify a confidence interval *which includes the estimated quantity* with a known confidence level. Thus, looking again at the distribution $f(x, \theta)$ with an unknown parameter θ, it is desirable to identify an interval $[\theta_L, \theta_U]$ and a confidence level $1 - \alpha$ such that

$$P_r\{\theta_L \le \theta \le \theta_U\} = 1 - \alpha.$$

This statement, however, raises an interesting question. It is obvious that with regard to any random variable one may seek an interval and the probability with which the random variable will have a realization in that interval. However, there is nothing random about θ. It is not a random variable, nor is its value set by a PDF; it is a fixed valued parameter; it is unknown but it is not random. In the above statement the location and width of the interval $[\theta_L, \theta_U]$ are the random elements. Rather than asking about the probability that θ is in the interval, one may (and should) ask about the probability that the random interval $[\theta_L, \theta_U]$ covers, or contains, the unknown fixed parameter θ.

Assume now that θ is the average value of the random variable x; it is intuitively plausible that (assuming a finite variance) realizations of x which are very remote from θ are unlikely to appear. If θ is 200 then one would not expect to see x in the vicinity of 20 000 000 (unless it is a very special distribution such as the wave function of a free electron with precisely known momentum, which is uniform over the universe). The intuitive understanding works both ways. It is also unlikely that θ is far from a measured realization of x. A quantification of this intuitive understanding may lead to a definition of the confidence interval.

Let $\vec{x} = (x_1, \ldots, x_n)$ be a set of n random variables. Let $g(\vec{x})$ be any function of $\vec{x}$ with some general properties that will become apparent shortly. $g(\vec{x})$, being a function of a set of random variables, is also a random variable. Also, since the distribution of each x_i depends upon θ, so is the distribution of $g(\vec{x})$, which will be denoted as $f(g, \theta)$. Further, define a percentile p, denoted by g_p, or $g(p)$ as the value of the variable such that

$$\int_{-\infty}^{g(p)} f(g, \theta)\, dg = p,$$

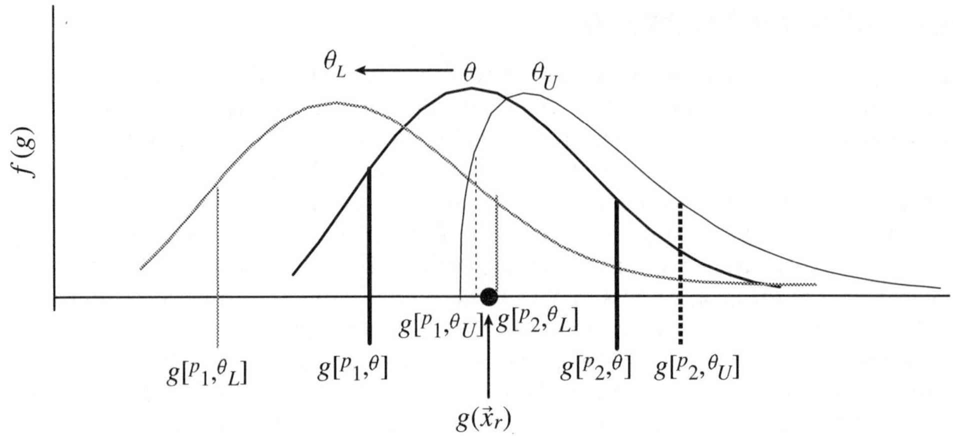

Figure 5.2.1 The relation between a realization $g(\vec{x}_r)$, the confidence interval and θ

i.e. the probability that the r.v. is smaller or equal to the p percentile is exactly p. Clearly, given two percentiles $g(p_1)$ and $g(p_2)$, where $p_1 < p_2$ the interval $[g(p_1), g(p_2)]$ is a $(p_2 - p_1)$ confidence interval of g. Since the distribution $f(g, \theta)$ depends upon the parameter θ, so do the percentiles, and henceforth they will be denoted as $g(p, \theta)$, i.e. the p percentile of the distribution with a given value of θ.

Consider now a realization of the observations $\vec{x}_r = (x_{1r}, \ldots, x_{nr})$ leading to a realization $g_r(\vec{x})$. Consider one value of θ for which $g_r(\vec{x})$ lies well within the confidence interval $[g(p_1, \theta), g(p_2, \theta)]$. This situation is shown in Figure 5.2.1. $f(g, \theta)$ is shown in the thick black line and the percentiles by the vertical thick lines. Suppose now that θ decreases; the distribution is shifted to the left. This means that the percentiles are increasing functions of θ. As a result of the left shift, the upper boundary of the interval, $g(p_2, \theta)$, will decrease until it reaches a point at which it equals $g_r(\vec{x})$. The value of θ for which this happens is denoted by θ_L, thus at this point $g_r(\vec{x}) = g(p_2, \theta_L)$. In Figure 5.2.1, this is shown with the light gray distribution, and the gray vertical lines indicate the percentiles. Clearly, for any θ smaller than θ_L, $g_r(\vec{x})$ will be outside the confidence interval. Namely, for any $\theta < \theta_L$, $g_r(\vec{x}) > g(p_2, \theta)$, and clearly $g_r \notin [g(p_1, \theta), g(p_2, \theta)]$. $g_r(\vec{x})$ is shown in the figure as a large black dot, and $g(p_2, \theta_L)$ is drawn just to the right of the point, for clarity.

Similarly, if θ is increased, $g(p, \theta)$ increases and the distribution shifts to the right. Accordingly, $g(p_1, \theta)$ increases until at some point, θ_U, $g(p_1, \theta_U)$ equals the realization $g(\vec{x}_r)$. Thus, at this point $g_r(\vec{x}) = g(p_1, \theta_U)$. In Figure 5.2.1 this is shown with the thin black distribution.

Clearly, for any θ larger than θ_U, $g_r(\vec{x})$ will be outside the confidence interval, i.e. for any $\theta > \theta_u, g_r(\vec{x}) < g(p_1, \theta)$, and thus $g_r \notin [g(p_1, \theta), g(p_2, \theta)]$. In the figure $g(p_1, \theta_U)$ is drawn just to the left of $g(\vec{x}_r)$. This leads to the conclusion that if $\theta_L \leq \theta \leq \theta_U$, then also $g(p_1, \theta) \leq g_r(\vec{x}) \leq g(p_2, \theta)$, and if θ is not in the interval $[\theta_L, \theta_U]$ then $g(\vec{x}_r)$ is not in the interval $[g(p_1, \theta), g(p_2, \theta)]$. A connection was thus created between intervals in the domain of $g(\vec{x})$ and intervals in the domain of θ, i.e. the event $\{g_r$ is in the interval $[g(p_1, \theta), g(p_2, \theta)]\}$ and the event $\{\theta$ is in the interval $[\theta_L, \theta_U]\}$ are identical, hence they have the same probability to appear. Thus, by the definition of the percentiles

$P_r\{\theta_L \le \theta \le \theta_U\} = p_2 - p_1$ the only property required from the distribution $f(g,\theta)$ is that the percentiles are a strictly monotonic increasing (or decreasing) function of θ. The above discussion and proof can be summarized in the following theorem.

Theorem 5.2.1: The basic inference theorem *Let $\vec{x} = (x_1, \ldots, x_n)$ be a set of n random variables (note that it is not required that the variables are independent or even identical) such that each x_i for $i = 1, \ldots, n$ has a possibly different distribution $f(x_i, \theta)$ with an unknown parameter θ common to all the distributions. Let $g(\vec{x}, \theta)$ be any function of $\vec{x}$ with a distribution $f(g,\theta)$ and such that the percentiles $g(p, \theta)$ of the distribution are strict monotonic functions (increasing or decreasing) of θ. Then for any given value of a realization g_r of $g(\vec{x},\theta)$ and for any two percentiles $p_1 < p_2$ the interval $[\theta_L, \theta_U]$ where θ_L and θ_U are the solutions of the equation*

$$g(p_2, \theta_L) = g_r \quad \textit{and} \quad g(p_1, \theta_U) = g_r. \tag{5.2.1a}$$

$[\theta_L, \theta_U]$ is a $(p_2 - p_1)$ confidence interval of θ, i.e.

$$P_r\{\theta_L \le \theta \le \theta_U\} = p_2 - p_1. \tag{5.2.1b}$$

The proof of this theorem was given above.

It is common to define a parameter $0 < \alpha < 1$ and choose $p_1 = \alpha/2$ and $p_2 = 1 - \alpha/2$, so that the solution of the equations

$$g\left(1 - \frac{\alpha}{2}, \theta_L\right) = g_r \quad \text{and} \quad g\left(\frac{\alpha}{2}, \theta_U\right) = g_r \tag{5.2.1c}$$

yields a $(1 - \alpha)$ symmetric confidence interval.

To apply the theorem, it is necessary to define the function $g(\vec{x},\theta)$ so that it is a monotonic function of θ, and obtain either the percentile $g(p,\theta)$ of its distribution or the percentiles for a given value of α. The solution of equation (5.2.1a) immediately yields the desired confidence interval of θ. We will refer to $g(\vec{x},\theta)$ as the 'observable' of the estimation process.

The exponential case For the exponential distribution $f(t) = (1/\theta)\,\mathrm{e}^{-(t/\theta)}$, the simplest option would be to take $g = t$. Having an inference based on a single observation is not very useful, but it does demonstrate a simple application of the theorem. Recall that t, as an exponential variable, follows the sampling expression $t = -\theta \times \ln(\xi)$, hence g is an increasing (linear) function of θ as required by the conditions of the theorem. The percentile $g(p,\theta)$ is easily derived from the equation $1 - \mathrm{e}^{g(p)/\theta} = p$, yielding

$$g(p,\theta) = -\theta \ln(1 - p); \tag{5.2.2}$$

substituting this relation in the left-hand side of equation (5.2.1c) yields $g_r = -\theta_L \,\mathrm{Ln}(\alpha/2)$ and $g_r = -\theta_U \,\mathrm{Ln}(1 - \alpha/2)$, where g_r is a single realization, or a single event point, of t. A $(1 - \alpha)$ confidence interval is seen to be given by the simple form

$$\theta_L = \frac{g_r}{\mathrm{Ln}(\alpha/2)} \quad \text{and} \quad \theta_U = \frac{g_r}{\mathrm{Ln}(1 - \alpha/2)}. \tag{5.2.3}$$

If, for example a single event is given at $g_r = t_r = 870_{\text{hours}}$ then a 95% confidence interval for θ is obtained by taking $\alpha = 0.05$, resulting in $\theta_L = 235.8_{\text{h}}$ and $\theta_U = 34363.1_{\text{h}}$. A smaller confidence interval of 90% confidence is obtained with $\alpha = 0.1$, namely [290.4, 16961]. This also means that the probability of θ being in the range [16961, 34363] is 0.025, and the same probability applies for the range [235.8, 290.4]. In fact, by calculating a number of confidence intervals with increasing confidence levels, one may obtain an approximate piecewise constant PDF for θ. The choice $p_2 \to 1$ yields, by equation (5.2.2), $\theta_L \to 0$ and, for any $p_1 < 1$,

$$\theta_U = \frac{g_r}{\text{Ln}(1 - p_1)}.$$

Thus, θ_U is a $q = 1 - p_1$ percentile of θ.

In this form, θ is presented as a simple random variable with a known CDF, since its percentiles are known. Indeed, θ is not a random variable, yet the simple basic theorem creates a relation between the random variable domain and the θ domain, enabling the transformation of our uncertainty in the value of θ into a statistical uncertainty identical to that of a random variable. It is thus possible, in general, to treat any *fixed unknown* quantity as if it is a random variable, if it can be presented as a parameter in a distribution of a true measurable random variable. This is a rather profound notion that justifies the term 'basic' in the name of Theorem 5.2.1.

The pivot variable

Another point of view of the same issue emerges as one accepts the above notion that, from the calculation aspect, θ can be treated as a random variable. From the above discussion, one has $g = -\theta \times \ln(\xi)$, and denoting $x = g/\theta$ also yields $x = -\ln(\xi)$, hence x is a standard exponential variable. x is a function of θ and its distribution is independent of θ. Such a variable is called a *pivot variable*. The PDF of x is $f(x) = e^{-x}$, and that of $y = 1/x$ is

$$f(y) = f(x)\left|\frac{dx}{dy}\right| = \frac{e^{-(1/y)}}{y^2} \tag{5.2.4}$$

with a CDF $F(y) = e^{-(1/y)}$. For any observation g_r, $\theta = g_r y$, resulting in a PDF and a CDF for θ

$$f(\theta) = \frac{g_r\, e^{-(g_r/\theta)}}{\theta^2}; \quad F(\theta) = e^{-(g_r/\theta)}. \tag{5.2.5}$$

In this case, an exact distribution function can be explicitly obtained for the unknown parameter. The p percentile of θ is obtained from $F(\theta(p)) = e^{-(g_r/\theta(p))} = p$, hence

$$\theta(p) = -\frac{g_r}{\ln(p)}.$$

Taking now the $(1-\alpha)$ symmetric confidence interval with $p_1 = \alpha/2$ and $p_2 = 1-\alpha/2$ again yields the confidence interval of equation (5.2.3). This discussion implies that if a pivot variable $y(g, \theta)$ can be found with a distribution $F(y)$ independent of θ, then it would be easy to obtain the percentiles of y, namely, $y(p_1)$ and $y(p_2)$, such that $P_r\{y(p_1) \leq y \leq y(p_2)\} = p_2 - p_1$. Then for a given observation g_r, substituting $y(g, \theta)$ in terms of θ and g_r, the expression can be presented as a confidence interval for θ. This confidence interval is identical to expression (5.2.1b).

The above example has no real practical value. A confidence interval of 235 to 34 000 hours for the MTTF is meaningless for any realistic purpose, but the steps of the procedure are of general validity. To obtain meaningful results more than one event must be observed and considered.

Exponential case $n > 1$ events Let $t_1, t_2, \ldots, t_n$ be a set of n independent failure points observed from an exponential distribution with an unknown parameter θ. It is convenient, in this case, to define $g(\vec{t}) = \sum_{i=1}^{n} t_i$. As in the case of the single observation, it is easy to show that $g(\vec{t})$ is an increasing function of θ. Furthermore, it is also possible in this case to find a pivotal with the following considerations. Since each of the failure times is exponentially distributed, the PDF of $g(\vec{t})$ is the Gamma distribution (see equation (2.2.8))

$$G_n(g) = \frac{\lambda(\lambda g)^{n-1} e^{-\lambda g}}{\Gamma(n)}.$$

Consider the transformation $y = 2\lambda g = 2g/\theta$. The PDF of y is called a Chi-square (χ^2) distribution with $2n$ degrees of freedom. The χ^2 distribution with r degrees of freedom has the form

$$\chi_r^2(y) = \frac{y^{(r/2-1)} e^{-(y/2)}}{2^{r/2}\Gamma(r/2)} \tag{5.2.6}$$

(with first moment r and variance $2r$).

y is a pivot variable since it is independent of θ. Denoting the p percentile of χ^2 by $\chi^2_{r,p}$, a general confidence interval takes the form $P_r[\chi^2_{2n,p_1} \leq y \leq \chi^2_{2n,p_2}] = p_2 - p_1$. Now, using the procedure described for the pivot variable and substituting $y = (2g_r)/\theta$ yields

$$P_r\left[\frac{2g_r}{\chi^2_{2n,p_2}} \leq \theta \leq \frac{2g_r}{\chi^2_{2n,p_1}}\right] = p_2 - p_1 \quad \text{or} \quad P_r\left[\frac{\chi^2_{2n,p_1}}{2g_r} \leq \lambda \leq \frac{\chi^2_{2n,p_2}}{2g_r}\right] = p_2 - p_1. \tag{5.2.7}$$

A double-sided symmetric $(1-\alpha)$ interval is obtained by taking $p_1 = \alpha/2$ and $p_2 = 1 - \alpha/2$, yielding the commonly known confidence interval

$$P_r\left[\frac{2g_r}{\chi^2_{2n,1-\alpha/2}} \leq \theta \leq \frac{2g_r}{\chi^2_{2n,\alpha/2}}\right] = 1 - \alpha. \tag{5.2.7a}$$

The percentiles of the χ^2 distribution are listed in statistical tables.

Example 5.2.1: MTTF specification for new pumps

A stainless steel factory purchased three new pumps. The manufacturer claims that the MTTF of a pump is 1100 hours. The observed failure times of the pumps were 720 h, 1230 h and 660 h. Are these results in agreement with the specifications of the manufacturer?

The value of g_3 is the sum of the failure times, namely 2610 h. The 0.05 and 0.95 percentiles of the χ^2 distribution with 6 degrees of freedom are 1.635 and 12.59, respectively, yielding a 90% confidence interval for the MTTF as [414.6 h, 3192 h]. The quoted MTTF is well within the confidence interval, and the results do not supply enough information to either justify the manufacturer's claim or discard it. Additional follow up of failure times are required. A second approach to the same problem would be to look for the one-sided interval starting at 1100 hours. Setting $2g_3/\chi^2_{6,p} = 1100$ yields $\chi^2_{6,p} = 4.7455$. A linear interpolation of the χ^2 percentiles indicates that $p = 0.42$. Choosing in 5.2.7 $p_1 = 0$ and $p_2 = 0.42$ yields a one-sided confidence interval $P_r[1100 \leq \theta \leq \infty] = 0.42$. This means that with the probability of 0.42, the actual MTTF is larger than specified, while with a probability of 0.58 it is smaller than promised. This analysis emphasizes the importance of the conditions set in the purchase contract. Such conditions must include the type of testing and the details of the confidence interval on which a verification will be made. For example, if the condition is that a symmetric two-sided 90% confidence interval should include the specified MTTF, then obviously the supplier met the terms of the specification. Yet if it is agreed *a priori* that a one-sided interval $[1100 \leq \theta \leq \infty]$ should have a confidence level over 50%, after three observations, the customer is entitled to claim that the system does not meet the promised specification.

The results of Theorem 5.2.1 also have a graphical interpretation worth noticing. The percentiles $g(p_1, \theta)$ and $g(p_2, \theta)$ can be drawn as functions of θ. The observation g_r can be drawn as a horizontal line on that graph, and the solution for the equations $g(p_2, \theta_L) = g_r$ and $g(p_1, \theta_U) = g_r$ which determine the confidence interval is obtained by the intersection of that horizontal line and the percentile curves. This is shown in Figure 5.2.2 for Example 5.2.1.

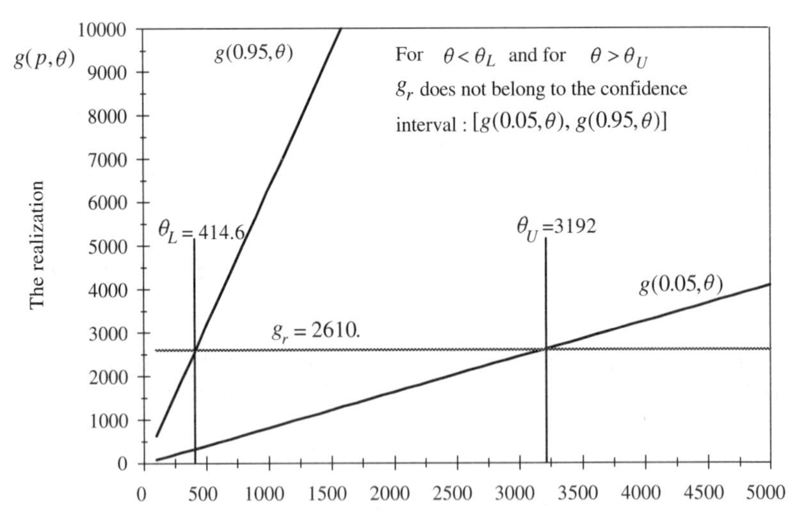

Figure 5.2.2 Graphical solution for the confidence interval according to Theorem 5.2.1

5.2.1 Inference for the shape parameter of the Weibull distribution

The consequences of the basic inference theorem are far reaching. They provide a very simple way to obtain confidence intervals for unknown parameters. It is only required to construct a function $g(\vec{x})$ of the realization observed, make sure that it is a monotonic function of the unknown parameter, find the percentiles of its distribution and construct the confidence interval. Henceforth, this function will be referred to as the *observable function*, or the *inference function*. It is advantageous to choose a realization function that can be easily turned into a pivot variable, since then the CDF of the unknown parameter is easily obtained and a considerable saving in computation effort is achieved. The Weibull distribution presents an additional difficulty, since it involves two independent unknowns, λ and β. In discussing the MLE, this problem was handled either by having a β_{MLE} which is independent of λ, or by taking λ to be the MLE estimate for any set of realizations and given β. In the following, the use of β_{MLE} as an inference function will be discussed, and an additional function will be introduced.

The mean to variance ratio function Consider n failure time observations $(t_1, \ldots, t_n)$. These are not necessarily independent failure points – they may result from a general scenario as discussed in the previous section. Consider the observable function

$$\eta_n(t_1, \ldots, t_n) = \frac{\left[1/n \sum_{i=1}^{n} t_i\right]^2}{\left[1/n \sum_{i=1}^{n} t_i^2\right] - \left[1/n \sum_{i=1}^{n} t_i\right]^2}. \tag{5.2.8}$$

This is simply the ratio of the sample average squared and the sample variance.

Consider the general scenario discussed in the preceding section. The sampling of the next event point is done (see equation (5.1.15a) and its preceding discussion) using the expression

$$t_i = -\frac{[\ln(1-\xi_i^*)]^{1/\beta}}{\lambda^{1/\beta}}, \tag{5.2.9}$$

where ξ_i^* is either a uniformly distributed variable in $[0, 1]$, for the case of complete renewal, or it is given in the form (see equation (5.1.13)) $\xi^* = \xi x[1 - A] + A$, where ξ is uniform in $[0, 1]$ and A is the age at the point from which the next event is sampled. In any case, the argument $v_i = (1 - \xi_i^*)$ is such that $0 \le v_i \le 1$. Also, with unit probability every observation in the set $(v_1, \ldots, v_n)$ is strictly different to any other, i.e. $\forall i \neq j \Rightarrow v_i \neq v_j$. With these conditions the following properties can be proven.

Theorem 5.2.1 *The observable function $\eta(t_1, \ldots, t_n)$ is independent of λ.*

Proof Substituting t_i from equation (5.2.9) into the definition of the observable function with the notation $y_i = \ln(v_i)$ yields

$$\eta_n(t_1, \ldots, t_n) = \frac{\left[1/n \sum_{i=1}^{n} y_i^{\alpha}\right]^2}{\left[1/n \sum_{i=1}^{n} y_i^{2\alpha}\right] - \left[1/n \sum_{i=1}^{n} y_i^{\alpha}\right]^2}, \tag{5.2.10}$$

where $\alpha = 1/\beta$. λ is thus canceled out in the ratio.

It is also necessary to show that the observable function is a strictly monotonic function of β. This is done in the following theorem.

Theorem 5.2.2 *The ratio observable function is a strictly increasing function of β.*

Proof Considering expression (5.2.10) it is easy to see that it is enough to show that the ratio

$$\frac{\left[1/n\sum_{i=1}^{n} y_i^{\alpha}\right]^2}{\left[1/n\sum_{i=1}^{n} y_i^{2\alpha}\right]}$$

is a decreasing function of α. The ratio can be written in the form

$$\frac{1}{n}\left[1+\frac{\sum_{i=1}^{n}\sum_{j=1\, j\neq i} y_i^{\alpha} y_j^{\alpha}}{\left[\sum_{i=1}^{n} y_i^{2\alpha}\right]}\right],$$

hence it is sufficient to show that

$$g(\alpha)=\left[\frac{\sum_{i=1}^{n}\sum_{j=1\, j\neq i} y_i^{\alpha} y_j^{\alpha}}{\left[\sum_{i=1}^{n} y_i^{2\alpha}\right]}\right]$$

is a decreasing function of α. Taking the derivative of $g(\alpha)$ with respect to α yields

$$g'(\alpha)=\frac{1}{\left[\sum_{i=1}^{n} y_i^{2\alpha}\right]^2}\left[\left[\sum_{i=1}^{n}\sum_{\substack{j=1\\ j\neq i}}^{n}(y_i^{\alpha}y_j^{\alpha})\ln(y_i y_j)\right]\times\left[\sum_{i=1}^{n} y_i^{2\alpha}\right]\right.$$
$$\left.-2\left[\sum_{i=1}^{n}\sum_{\substack{j=1\\ j\neq i}}^{n}(y_i^{\alpha}y_j^{\alpha})\right]\times\left[\sum_{i=1}^{n} y_i^{2\alpha}\ln(y_i)\right]\right].$$

Since the denominator is positive, it is sufficient to show that the numerator is negative, namely

$$\left[\sum_{i=1}^{n}\sum_{\substack{j=1\\ j\neq i}}^{n}(y_i^{\alpha}y_j^{\alpha})\ln(y_i y_j)\right]\times\left[\sum_{i=1}^{n} y_i^{2\alpha}\right]-2\left[\sum_{i=1}^{n}\sum_{\substack{j=1\\ j\neq i}}^{n}(y_i^{\alpha}y_j^{\alpha})\right]\times\left[\sum_{i=1}^{n} y_i^{2\alpha}\ln(y_i)\right]<0$$

or

$$\frac{\left[\sum_{i=1}^{n}\sum_{j=1\, j\neq i}^{n}(y_i^{\alpha}y_j^{\alpha})\ln(y_i y_j)\right]}{\left[\sum_{i=1}^{n}\sum_{j=1\, j\neq i}^{n}(y_i^{\alpha}y_j^{\alpha})\right]}<2\frac{\left[\sum_{i=1}^{n} y_i^{2\alpha}\ln(y_i)\right]}{\left[\sum_{i=1}^{n} y_i^{2\alpha}\right]}.$$

The numerator of the left-hand side of the inequality is symmetric in i and j therefore

$$\left[\sum_{i=1}^{n}\sum_{\substack{j=1\\ j\neq i}}^{n}(y_i^{\alpha}y_j^{\alpha})\ln(y_i y_j)\right]=2\left[\sum_{i=1}^{n}\sum_{\substack{j=1\\ j\neq i}}^{n}\ln(y_i)(y_i^{\alpha}y_j^{\alpha})\right]$$

yielding the inequality

$$\frac{\left[\sum_{i=1}^{n}\sum_{j=1\, j\neq i}^{n}(y_i^{\alpha}y_j^{\alpha})\ln(y_i)\right]}{\left[\sum_{i=1}^{n}\sum_{j=1\, j\neq i}^{n}(y_i^{\alpha}y_j^{\alpha})\right]} < \frac{\left[\sum_{i=1}^{n} y_i^{2\alpha}\ln(y_i)\right]}{\left[\sum_{i=1}^{n} y_i^{2\alpha}\right]}.$$

With the notation $z_i = y_i^{\alpha}$ and upon multiplication of both sides by α, the inequality takes the form

$$\frac{\left[\sum_{i=1}^{n}\sum_{j=1\, j\neq i}^{n}(z_i z_j)\ln(z_i)\right]}{\left[\sum_{i=1}^{n}\sum_{j=1\, j\neq i}^{n}(z_i z_j)\right]} < \frac{\left[(n-1)\sum_{i=1}^{n} z_i^2\ln(z_i)\right]}{\left[(n-1)\sum_{i=1}^{n} z_i^2\right]}.$$

$z_1, \ldots, z_n$ is a set of strictly positive non-identical real numbers. One may thus assume, with no loss of generality, that they are arranged in increasing order, thus, $z_1 < z_2 < \cdots < z_{n-1} < z_n$.

Consider next the two sets of n discrete probabilities defined as

$$p_i = \frac{z_i \sum_{j=1\, j\neq i}^{n}(z_j)}{\left[\sum_{i=1}^{n}\sum_{j=1\, j\neq i}^{n}(z_i z_j)\right]} \quad \text{and} \quad q_i = \frac{\left[(n-1)z_i^2\right]}{\left[(n-1)\sum_{i=1}^{n} z_i^2\right]}. \tag{5.2.11}$$

Also, denote $r_i = \ln(z_i)$ so that the inequality now compares two averages of the form

$$\langle r\rangle_p = \sum_{i=1}^{n} r_i p_i < \langle r\rangle_q = \sum_{i=1}^{n} r_i q_i. \tag{5.2.12}$$

The following lemmas will now become useful:

Lemma 5.2.1 *Denote*

$$A = \left[\sum_{i=1}^{n}\sum_{\substack{j=1\\ j\neq i}}^{n}(z_i z_j)\right]$$

and $B = \left[(n-1)\sum_{i=1}^{n} z_i^2\right]$ *then* $A < B$.

Proof For any two positive non-identical real numbers $\frac{1}{2}(z_i^2 + z_j^2) > z_i z_j$, hence

$$\frac{1}{2}\sum_{i=1}^{n}\sum_{\substack{j=1\\ j\neq i}}^{n}(z_i^2 + z_j^2) > \sum_{i=1}^{n}\sum_{\substack{j=1\\ j\neq i}}^{n}(z_i z_j).$$

Also

$$\frac{1}{2}\sum_{i=1}^{n}\sum_{\substack{j=1\\ j\neq i}}^{n}(z_i^2 + z_j^2) = \sum_{i=1}^{n}\sum_{\substack{j=1\\ j\neq i}}^{n} z_i^2 = (n-1)\sum_{i=1}^{n} z_i^2$$

which proves the lemma.

Lemma 5.2.2 *The set of probabilities q_i and p_i, $i = 1, \dots, n$ are increasing functions of i. Also, $p_1 > q_1$ and there exists $1 < k \leq n$ such that for any $i < k$ $p_i > q_i$ and for $i \geq k$ $q_i > p_i$.*

Proof Since $z_1, \dots, z_n$ is strictly increasing and positive, so is

$$q_i = \frac{(n-1)z_i^2}{B}.$$

Also, denote

$$Z_{i,i+1} = \sum_{j=1}^{n} z_j - z_i - z_{i+1}$$

for $i = 1, \dots, n-1$ such that

$$p_i = \frac{z_i Z_{i,i+1} + z_i z_{i+1}}{A} \quad \text{and} \quad p_{i+1} = \frac{z_{i+1} Z_{i,i+1} + z_{i+1} z_i}{A}$$

hence

$$p_{i+1} - p_i = \frac{(z_{i+1} - z_i) Z_{i,i+1}}{A} > 0$$

and p_i is an increasing function of i. Now $z_2 + \cdots + z_n > (n-1)z_1$, which yields $z_1(z_2 + \cdots + z_n) > z_1(n-1)z_1$ and, since by Lemma 5.2.1, $A < B$, obviously

$$\frac{z_1(z_2 + \cdots + z_n)}{A} > \frac{z_1(n-1)z_1}{B},$$

hence $p_1 > q_1$. Assume now that $p_i > q_i$ for each $i = 1, \dots, n$; this results in an immediate contradiction with the normalization condition arising from the definition of the probabilities. Thus, there must be a value $1 < k \leq n$ such that $p_k < q_k$. This means that

$$p_k = \frac{z_k Z_k}{A} < \frac{(n-1)z_k z_k}{B} = q_k \quad \text{with } Z_i = \sum_{j=1}^{n} z_j - z_i.$$

Z_i is a decreasing function of i. The last inequality yields

$$\frac{z_{k+1} Z_k}{A} < \frac{(n-1) z_{k+1} z_k}{B};$$

replacing Z_k on the left-hand side by Z_{k+1} reduced the left-hand side, since $Z_{k+1} < Z_k$, while replacing z_k by z_{k+1} on the right-hand side increases it, hence:

$$\frac{z_{k+1} Z_{k+1}}{A} < \frac{(n-1) z_{k+1} z_{k+1}}{B} \quad \text{or} \quad p_{k+1} < q_{k+1}$$

which completes the proof of the lemma.

Consider the average $\langle r \rangle_p = \sum_{i=1}^{n} r_i p_i$, where $r_i = \ln(z_i)$. r_i is also an increasing function of i. Denote $\Delta_i = p_i - q_i$, then $\Delta_i > 0$ for $i < k$ and $\Delta_i < 0$ for $i \geq k$, where

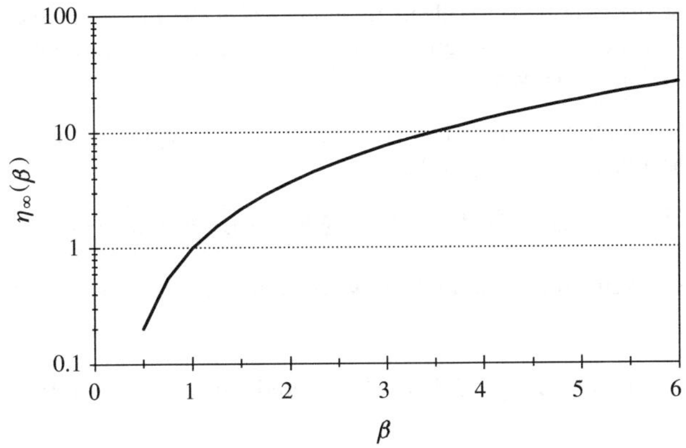

Figure 5.2.3 The mean square to variance ratio, $\eta_\infty(\beta)$, as a function of β

k is the smallest integer for which $p_k < q_k$. Then,

$$\langle r \rangle_p = \sum_{i=1}^{n} r_i p_i = \sum_{i=1}^{n} r_i (q_i + \Delta_i)$$

$$= \langle r \rangle_q + \sum_{i=1}^{k} r_i \Delta_i + \sum_{i=k+1}^{n} r_i \Delta_i < \langle r \rangle_q + r_k \sum_{i=1}^{k} \Delta_i + r_k \sum_{i=k+1}^{n} \Delta_i = \langle r \rangle_q$$

and the theorem is proved.

Thus, the observable function $\eta_n(t_1, \ldots, t_n)$ of equation (5.2.10) is independent of the scale parameter, and is an increasing function of the shape parameter.

Independent failures For independent event points, the ratio observable, $\eta_n(t_1, \ldots, t_n)$ is a consistent estimator of the ratio of the mean square to the variance of the Weibull distribution, namely:

$$\lim_{n \to \infty} \eta_n(t_1, \ldots, t_n) = \eta_\infty(\beta) = \frac{\Gamma^2(1 + 1/\beta)}{\Gamma(1 + 2/\beta) - \Gamma^2(1 + 1/\beta)}. \tag{5.2.13}$$

$\eta_\infty(\beta)$ is shown as function of β in Figure 5.2.3. The ratio equals unity for the exponential case, $\beta = 1$, and thus the value of $\eta(t_1, \ldots, t_n)$ for a given sample may provide an initial clue as to the possibility that the PDF considered is exponential.

The percentiles of $\eta(t_1, \ldots, t_n)$, $\eta(p, \beta)$, can be derived by sampling event points and obtaining a distribution (histogram) for each of a set of values of β. From these distribution, $\eta(p_1, \beta)$ and $\eta(p_2, \beta)$ can be numerically constructed [The program `infers.for`]. This is demonstrated in the following example.

Example 5.2.2

The following five independent failure times have been observed for the internal coating of a pipe used in a chemical process known to have a Weibull failure distribution: 240, 310, 420, 490, 620. It is desired to obtain confidence intervals for both β and λ.

The treatment of the data involves elements discussed in the preceding sections. The MLE is used to obtain both a point estimate and an interval estimate, and the mean to variance ratio observable is used in the same way.

1. *Maximum likelihood point estimate*: the MLE estimate for the above data is (solving equation (5.1.10) or the particular case of (5.1.26)):

$$\beta_{\text{MLE}} = 3.47 \quad \lambda_{\text{MLE}} = 5.508 \times 10^{-10} \quad \text{and} \quad \eta_{\text{MLE}} = \lambda_{\text{MLE}}^{1/\beta} = 2.146 \times 10^{-3}.$$

The bias factor is taken from equation (5.1.29) yielding 1.42 resulting the unbiased MLE point estimates:

$$\beta_{\text{ube}} = 2.45 \quad \lambda_{\text{ube}} = 3.234 \times 10^{-7} \quad \text{and} \quad \eta_{\text{ube}} = 2.243 \times 10^{-3}.$$

2. *Maximum likelihood interval estimate*: the MLE estimate can also serve as an observable function. In this case, the variable $x = \beta_{\text{MLE}}/\beta_{\text{tr}}$ is also a pivot variable, as its PDF is independent of β_{true}. To generate the PDF shown in Figure 5.2.4, 20 000 case histories were sampled using the MLE scenario sampling described in Section 5.1.6.

   From the PDF the 0.05 and 0.95 percentiles are, respectively, 0.668 and 2.917. Thus

$$P_r\left\{0.668 \le \frac{\beta_{\text{MLE}}}{\beta_{\text{tr}}} \le 2.917\right\} = 0.9,$$

yielding

$$P_r\left\{\frac{\beta_{\text{MLE}}}{2.917} \le \beta_{\text{tr}} \le \frac{\beta_{\text{MLE}}}{0.668}\right\} = 0.9 \quad \text{or} \quad P_r\{1.19 \le \beta_{\text{tr}} \le 5.19\} = 0.9.$$

Hence, a 90% confidence interval is [1.19,5.19]. Also $P_r\{1.19 \le \beta_{\text{tr}} \le \infty\} = 0.95$, so that $[1.19, \infty]$ is a one-sided 95% confidence interval. This indicates that with a high probability, the failure PDF is non-exponential. The determination of the confidence intervals can be done graphically (using the results of Theorem 5.1.1 directly), as shown in Figure 5.2.5. This yields, of course, the same result as above.

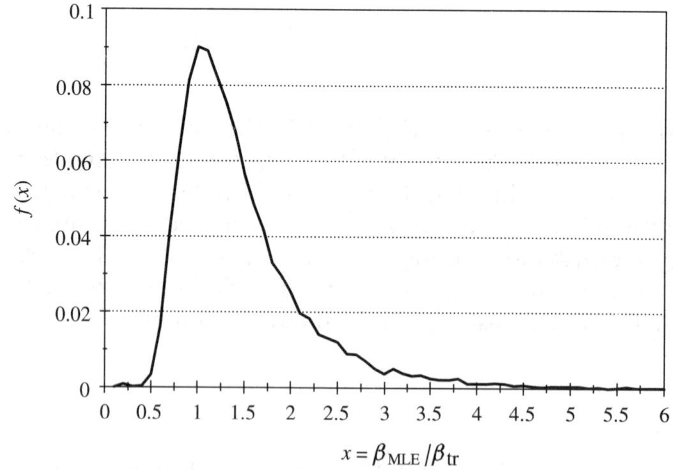

Figure 5.2.4 PDF of $x = \beta_{\text{MLE}}/\beta_{\text{tr}}$ produced by Monte Carlo sampling of event points

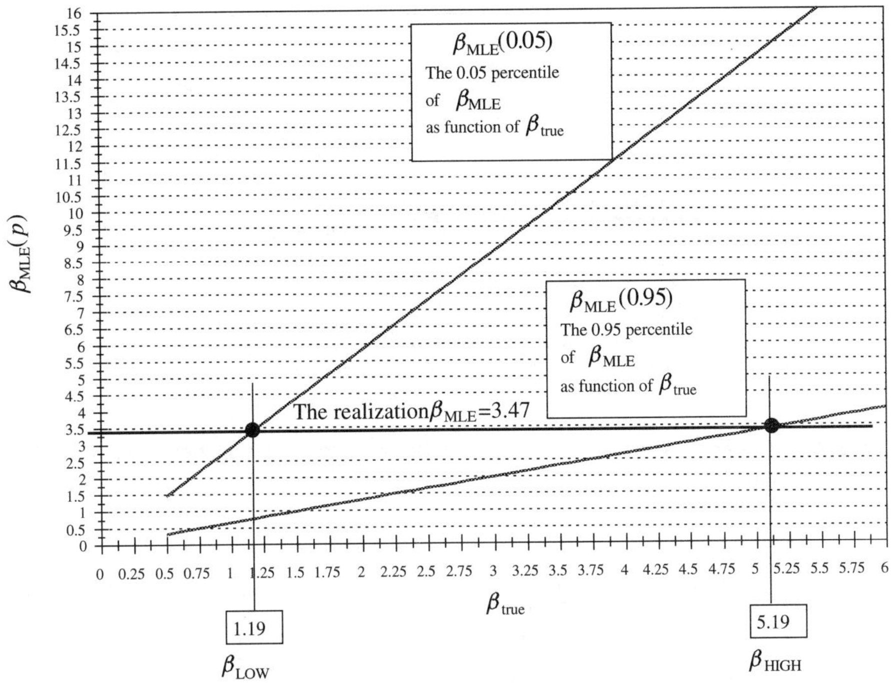

Figure 5.2.5 Graphic determination of the 90% confidence interval for β using the MLE observable

3. *Point estimate with the mean square to variance ratio*: as the scenario is sampled (see the item 'as good as new' in Section 5.1), for each set of five realizations $(t_{r1}^i, t_{r2}^i, t_{r3}^i, t_{r4}^i, t_{r5}^i)$ the mean to variance ratio observable is calculated (for a given β). It is convenient to use the square root in the form

$$\eta_s^i = \sqrt{\eta(t_{r1}^i, t_{r2}^i, t_{r3}^i, t_{r4}^i, t_{r5}^i)} = \left[\frac{\frac{1}{5}\left(\sum_{j=1}^{5} t_{rj}^i\right)^2}{\left(\frac{1}{5}\sum_{j=1}^{5}(t_{rj}^i)^2\right) - \frac{1}{5}\left(\sum_{j=1}^{5} t_{rj}^i\right)^2}\right]^{1/2}. \tag{5.2.14}$$

The index i indicates the ith set of event points sampled. The average value of η_s^i as function of β can be obtained by running N case histories for each value of β. This was done with $N = 20\,000$ for a range of shape parameters from 0.5 through 6. The average value is obtained as

$$\langle \eta_s \rangle = \frac{1}{N}\sum_{i=1}^{N} \eta_s^i.$$

The results are shown in Figure 5.2.6.

The fact that the average value seems to have a linear relation with β indicates that η_s/β is a pivot variable. Calculating now the value of the observable using the event points of the example yields 3.11. The value of β corresponding to 3.11 from Figure 5.2.4 is 2.44, thus $\beta_{evr} = 2.44$, $\lambda_{evr} = 3.315 \times 10^{-7}$ and $\eta_{evr} = 2.243 \times 10^{-3}$, where λ_{evr} is calculated from the maximum likelihood relation with β_{evr} and the example zero age points.

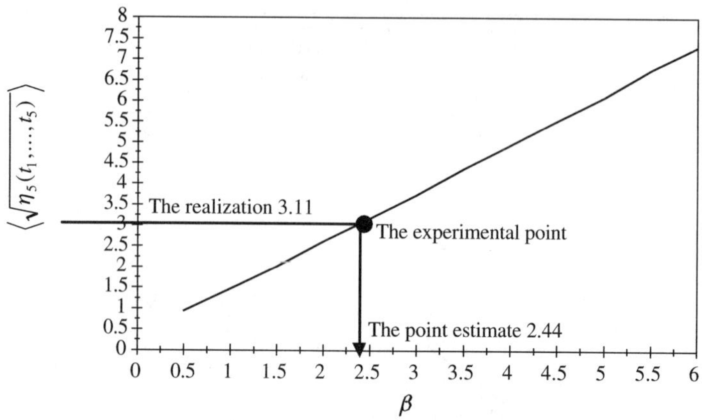

Figure 5.2.6 Average value of the root of the Square Mean to Variance Ratio (SMVR) as a function of β

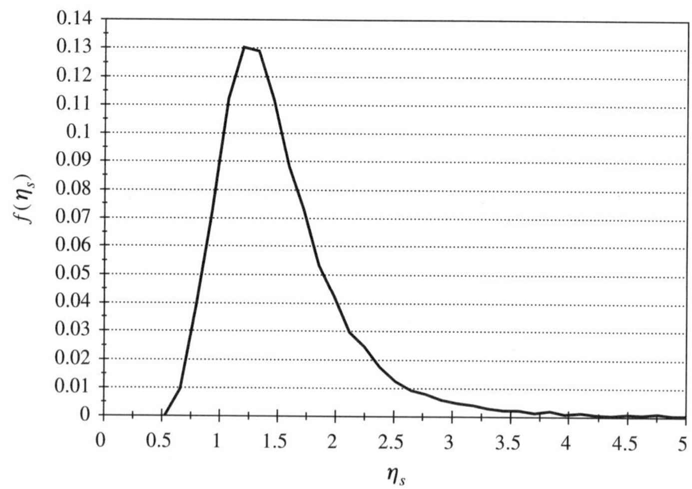

Figure 5.2.7 The PDF of the SMVR for $\beta = 1$

4. *Interval estimate with the SMVR:* using the same sample of 20 000 case histories, an approximate PDF of η_s (the square root of the observable function (5.2.8)) can be constructed. The PDF obtained with $\beta = 1$ is shown in Figure 5.2.7.

From the set of PDFs, the 0.05 and 0.95 percentiles of η_s can be extracted as a function of β, and the procedure of Theorem 5.2.1 used to obtain the 90% confidence interval for β. This is shown in Figure 5.2.8.

Again the percentiles as function of β seem to yield a linear relation, indicating that the SMVR divided by β is a pivot variable. The confidence interval obtained is $P_r\{1.252 \leq \beta_{\text{tr}} \leq 5.49\} = 0.9$.

In summary: two methods are used to analyze the results; one based on the maximum likelihood, and the other on the root of the Square Mean to Variance Ratio (SMVR). In both cases, a point estimate is obtained along with a confidence interval. Although the intervals are relatively large they are consistent for the two methods. Monte Carlo sampling and estimation are extensively used in the process to sample the event points of the scenario for the estimation of the

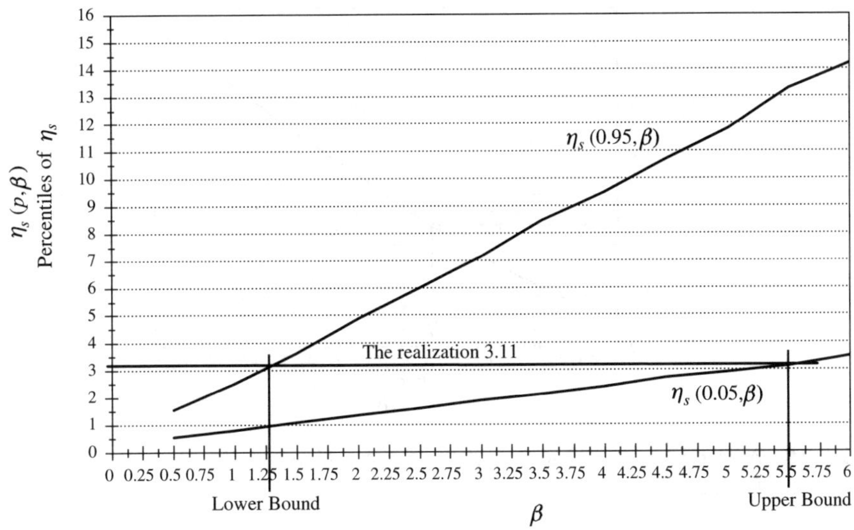

Figure 5.2.8 Construction of the 90% confidence interval for the SMVR observable function

averages, as a function of β, as well as construction of the PDFs and the percentiles from which the confidence intervals are deduced.

Example 5.2.3: Mixed scenario of two components

(Confidence intervals for Example 5.1.3.) Example 5.1.3 is reconsidered. In this example two pressure pumps are considered with an operational scenario involving minimal repair, full repair and maintenance. The event points are (270, 432, 315, 471, 219) and the zero age points are (520, 190, 471, 344). In Example 5.1.3 the MLE estimates were $\beta_{\text{MLE}} = 2.925$, $\lambda_{\text{MLE}} = 2.68 \times 10^{-8}$ and $\eta_{\text{MLE}} = 2.578 \times 10^{-3}$. The bias factor was 1.275, and the unbiased estimates yield $\beta_{\text{ube}} = 2.294$, $\lambda_{\text{ube}} = 1.287 \times 10^{-6}$ and $\eta_{\text{ube}} = 2.706 \times 10^{-3}$. The steps outlined in Example 5.2.2 are followed. The realization of the SMVR observable yields

$$\eta_{rs} = \sqrt{\eta_r} = \sqrt{\frac{\left(\frac{1}{5}(270+432+315+471+219)\right)^2}{\frac{1}{5}\times(270^2+432^2+315^2+471^2+219^2) - \left(\frac{1}{5}(270+432+315+471+219)\right)^2}} = 3.567. \tag{5.2.15}$$

The average value of the SMVR as a function of β is shown in Figure 5.2.9. From this curve, the point EVR estimate is obtained as $\beta_{\text{evr}} = 2.395$ with a corresponding (see equation (5.1.10) for λ)

$$\lambda_{\text{evr}} = \frac{5}{520^{2.395} + 190^{2.395} + 471^{2.395} + 344^{2.395}} = 6.948 \times 10^{-7} \quad \text{and} \quad \eta_{\text{evr}} = 2.683 \times 10^{-3}.$$

In Figure 5.2.10, the 0.05 and 0.95 the percentiles of $\beta_{\text{MLE}}(p,\beta)$ and $\eta_{\text{evr}}^{1/2}(p,\beta)$ are shown with the graphical derivation of the two-sided symmetric 90% confidence intervals. The intervals are $[1.24, 4.72]$ for the MLE and $[1.15, 5.65]$ for the EVR. By successively obtaining the $0.05, 0.1, 0.2$ percentiles of β, a piecewise constant approximation for the PDF of β can be obtained. The two

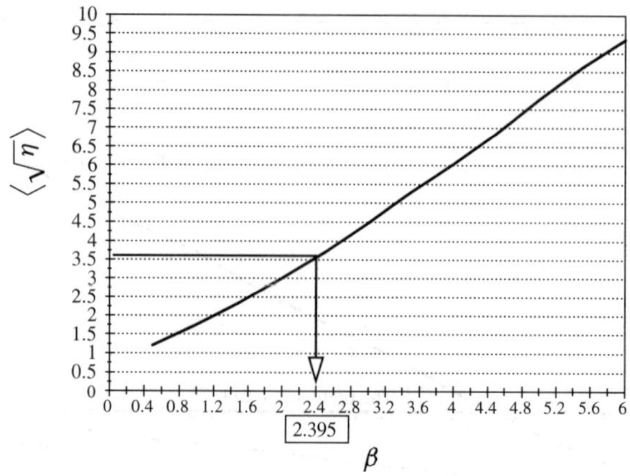

Figure 5.2.9 Average value of the root of the Square Mean to Variance Ratio (SMVR) as a function of β

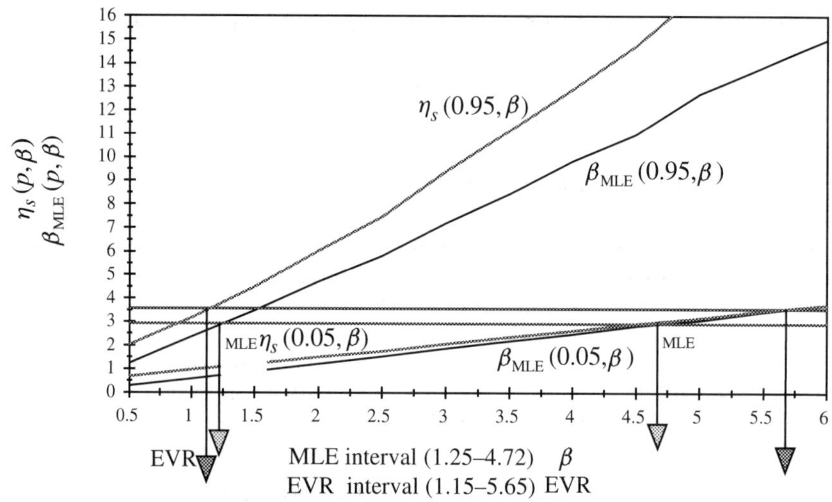

Figure 5.2.10 Construction of the 90% confidence interval for the SMVR and the MLE observables

distributions obtained for the MLE and EVR are shown in Figure 5.2.11. These distributions can be of use if one wishes to sample values of β for use in systems calculations. This will be further discussed in the next section.

Example 5.2.4: Minimal repair and partial maintenance of blade sets

Blade sets are sub-units of gas turbines for electrical power generation, as well as jet engines. A typical blade set contains about 50 rotating blades. A failure of a blade in operation may cause severe damage to the whole system, as fast flying fragments of a broken blade pierce through the system. It is therefore common to apply periodic inspection of such blades in order to identify

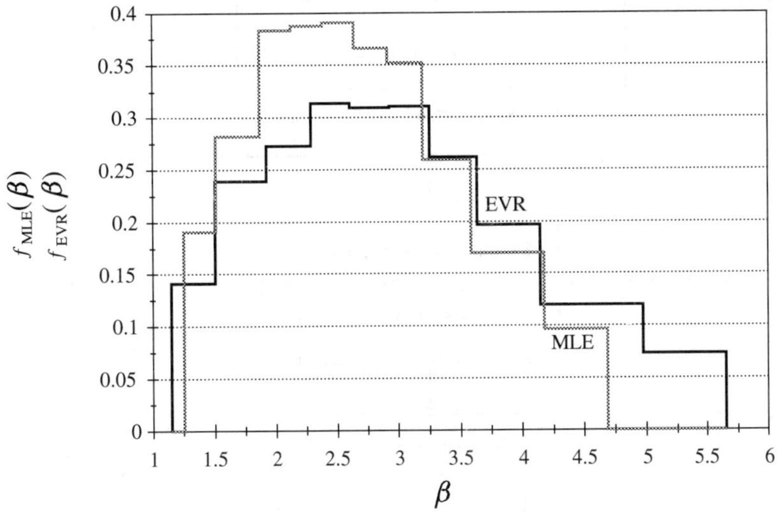

Figure 5.2.11 Piecewise constant PDFs of β derived from the MLE and the EVR

weak or corroded blades, which are then replaced. Considering the failure distribution of the blade set, a replacement of a single blade may be considered as minimal repair.

Data is obtained from five blade sets operating for varying periods of time. In the particular case considered, a time limit of 560 days is set. In case a failure is identified before that limit, the corroded blade is replaced. If the failure is identified beyond that limit, with some probability, the whole blade set is replaced, yielding full repair. A study of the replacements indicated that the probability of a full repair is of the order of 20%. The following data is collected for the five sets:

Blade set 1: Operating 170 days. No failures.
Blade set 2: Operating 815 days. Failures at 281, 304, 487, 537, 599, 712. All with minimal repair.
Blade set 3: Operating 365 days. Failures at 35 and 197. All with minimal repair.
Blade set 4: Operating 410 days. Failures at 79, 234 and 378 days. All with minimal repair.
Blade set 5: Operating 740 days. Failures at 232, 479 with minimal repair and a full repair failure at 707 days.

The Monte Carlo procedure for sampling a single history for the above scenario is outlined in Figure 5.2.12. There are 14 event points: 281, 304, 487, 537, 599, 712, 35, 197, 79, 234, 378, 232, 479, 707. The zero age points are: 170, 815, 365, 410, 707, 33. The MLE estimates are: $\beta_{MLE} = 1.42$, $\lambda_{MLE} = 7.726 \times 10^{-5}$ and $\eta_{MLE} = 3.226 \times 10^{-3}$. The SMVR takes the value $\eta_s = 1.79$. Figures 5.2.13(a,b) display the bias factor and the average SMVR as functions of β, respectively. The solution of the equation $\beta B(\beta) = \beta_{MLE}$ yields $\beta_{ube} = 1.341$, $\lambda_{ube} = 6.468 \times 10^{-4}$ and $\eta_{ube} = 4.186 \times 10^{-3}$. From the average SMVR curve, and using the observed value of 1.79, one obtains $\beta_{evr} = 1.343$, $\lambda_{evr} = 6.386 \times 10^{-4}$ and $\eta_{evr} = 4.147 \times 10^{-3}$. The point estimates of the MLE and the variance ratio are very close. The interval estimates are shown in Figure 5.2.14. Both methods provide a coherent and comprehensive data analysis. It can be seen, however, that in the above examples the MLE observable yields smaller confidence intervals then the variance ratio observable.

Inference for the scale parameter λ In the previous section the use of the Monte Carlo method in the inference of the shape parameter, for the Weibull distribution, was

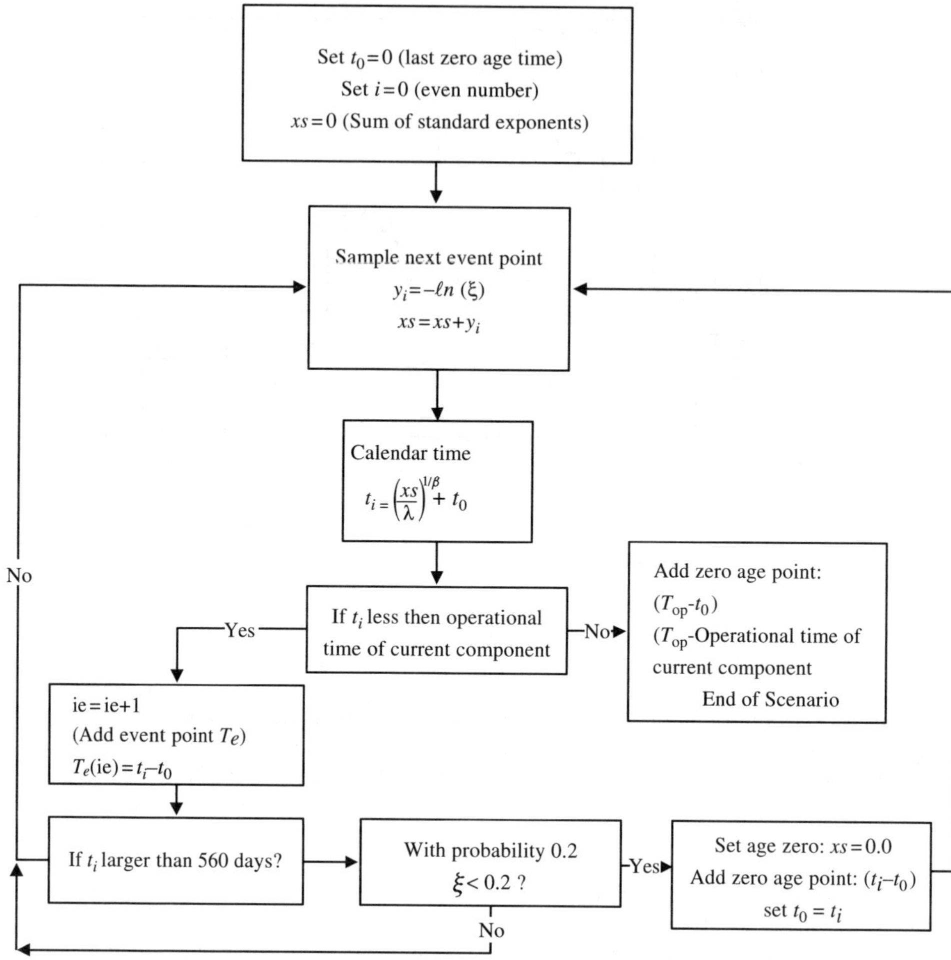

Figure 5.2.12 Sampling event and zero age points for the scenario of Example 5.2.4

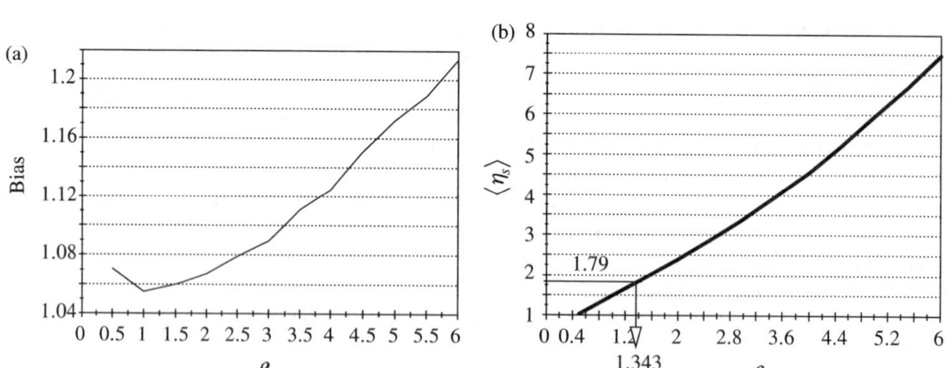

Figure 5.2.13 (a) The bias factor as a function of β, (b) average SMVR as a function of β

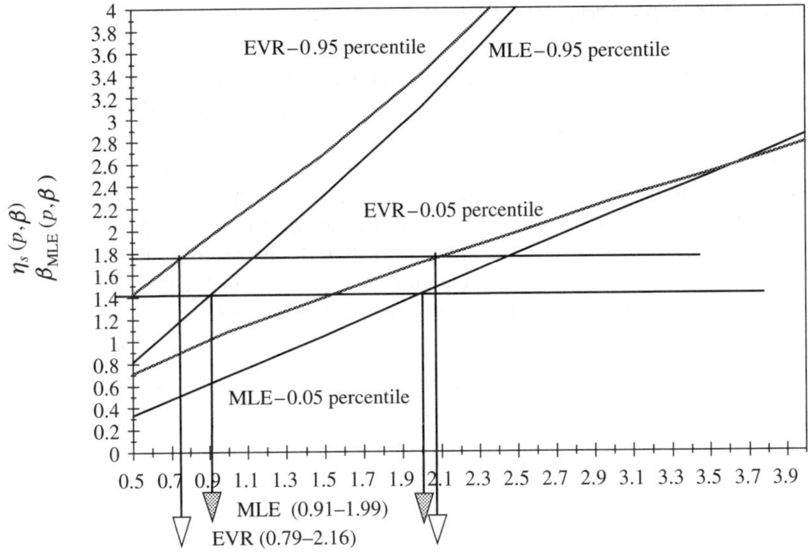

Figure 5.2.14 Interval estimates for the MLE and variance ratio

discussed. The scale parameter was assumed to be a function of the shape parameter and the observed data, such that it is the MLE value calculated by equation (5.1.9) for the given shape parameter and zero age points. It is, however, possible to define an observable for inference on the scale parameter assuming that the shape parameter is given.

Let $t_1, \ldots, t_n$ be a set of observed event points as defined for equation (5.1.8). Consider the observable

$$T = \sum_{i=1}^{r} t_i. \tag{5.2.16}$$

This is simply the sum of the event points. Each event point is given either in the form of $t_i = \{\theta \times [-\ln(\xi)]\}^{1/\beta}$ (with $\theta = 1/\lambda$), where ξ is a uniform r.v. in [0, 1] in the case of full repair, or $t_i = \{\theta \times [-\ln(z_i)]\}^{1/\beta}$, where z_i is a product of i such variables in the case of minimal repair (see equation (5.1.15b)). This can be written as $t_i = \{\theta \times [-\ln(y)]\}^{1/\beta}$, where y is a random variable known to be in the range [0, 1]. Substituting the last expression into (5.2.16) yields

$$T = \sum_{i=1}^{r} t_i = \theta^{1/\beta} \times \sum_{i=1}^{n} [-\ln(y_i)]^{1/\beta}, \tag{5.2.17}$$

indicating that T is an increasing function of θ and is thus a legitimate observable for inference. Furthermore, the variable

$$g_n = \frac{T}{\theta^{1/\beta}} = \sum_{i=1}^{n} [-\ln(y_i)]^{1/\beta} \tag{5.2.18}$$

is a pivot variable.

Example 5.2.5: Confidence interval for the scale parameter

A pump has a Weibull failure distribution with known shape parameter $\beta = 1.75$. Five independent failure times were measured: 295 h, 357 h, 530 h, 626 h and 780 h. A 0.95 confidence interval is required for the scale parameter λ.

To obtain the confidence interval the percentiles of the pivot variable g_n of equation (5.2.18) are required. A histogram of this variable can be easily generated by sampling N sets of five independent failure points from the Weibull distribution with unit scale parameter. The sum of these points comprises a value of g_n. Using $N = 10\,000$, the percentiles of g_n, namely $g_n(p)$ as a function of p, are shown in Figure 5.2.15(a) and the histogram in Figure 5.2.15(b).

The 0.025 and 0.975 percentiles are, respectively, 2.375 and 6.919, thus

$$P_r\{2.375 \le g_n \le 6.91\} = \left\{2.375 \le \frac{T_r}{\theta^{1/1.75}} \le 6.91\right\} = 0.95,$$

where T_r is the realization of the observable which, in this example, is the sum of the observed event times, namely 2588 h, hence:

$$P_r\left\{\frac{2588}{6.91} \le \theta^{1/1.75} \le \frac{2588}{2.375}\right\} = 0.95,$$

or finally $P_r\{31886 \le \theta \le 206670\} = P_r\{4.83 \times 10^{-6} \le \lambda \le 3.13 \times 10^{-5}\} = 0.95$.

Distribution edge effects and zero events: The construction of the histograms of an observable involves the sampling of N realizations of the observable. However, since N is finite, one will always obtain a smallest and a largest realization. In Example 5.2.5 these are 1.0404 and 9.732, respectively. Clearly, one cannot claim that these are the upper and lower bounds of the domain of the observable. Increasing the number of samples will change these boundaries. Furthermore, with regard to the observable of Example 5.2.5, it is obvious that it may obtain values throughout the real axis from zero to infinity. Thus, limiting the distribution between 1.0404 and 9.732 is unjustified. It is then reasonable to inquire as to the probability that a realization will appear outside these boundaries. A similar question arises in general Monte Carlo calculations involving event probabilities. For all the estimators described in Chapters 3 and 4, it is essential that they obtain non-zero values. If the evaluated event does not occur in N

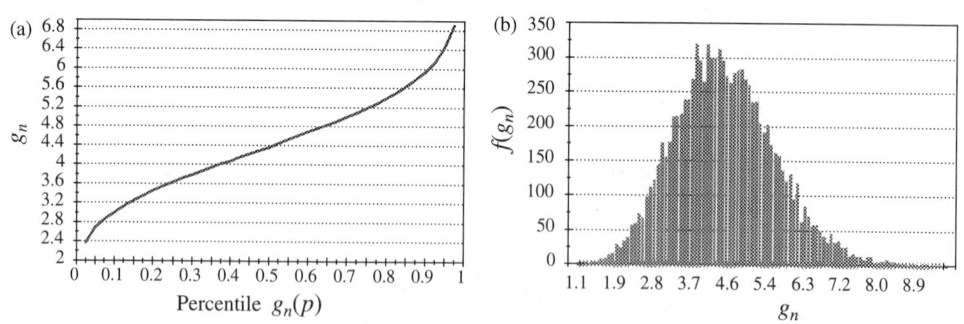

Figure 5.2.15 (a) Percentiles of the observable; (b) histogram of the observable

histories, then no estimate is obtained. This question was considered in discussing rare events and biasing in previous chapters. Indeed, if the evaluated event never occurs, then no estimate and no confidence interval (statistical error) is obtained (see equations (3.3.3)–(3.3.6) for example). Still, if one conducts, say, $N = 1\,000\,000$ case histories for the evaluation of the failure probability of a system and no failure event appears, then something may be said about the event probability. This general question is discussed next.

Let the unknown probability that an event will occur be denoted by p. Consider the binomial probability $P(k, N, p) = \binom{N}{k} p^k (1-p)^{N-k}$. This is the probability for k successes in N experiments, with success probability p. Let $n_{L\alpha}$ be defined as the largest integer for which $P_r[k \geq n_{L\alpha}] \geq 1 - \alpha$, where $\alpha < 1$, i.e. the largest solution of the inequality $\sum_{k=n_{L\alpha}}^{N} P(k, N, p) \geq 1 - \alpha$.

For any given α, $n_{L\alpha}$ is an increasing function of p with the observation that for $p = 0$ $n_{L\alpha} = 0$, and for $p = 1$ $n_{L\alpha} = 1\ \forall \alpha > 0$. Also, as p increases from zero, $n_{L\alpha}$ remains zero until $p = p_1 = 1 - \sqrt[n]{\alpha}$, at which point $n_{L\alpha} = 1$, since $\sum_{k=1}^{N} P(k, N, p) = 1 - (1 - p_1)^N \equiv 1 - \alpha$. $n_{L\alpha}$ will remain 1 in the range $[p_1, p_2)$, where p_2 is the solution of $1 - (1 - p_2)^N - Np_2(1 - p_2)^{N-1} = 1 - \alpha$, at which point $n_{L\alpha} = 2$, and so on. Clearly, $n_{L\alpha}(p)$ is a discontinuous function of p. Let K_e be a realization of the number of successes in N experiments. Let p_e be the minimal value of p for which $n_{L\alpha}(p_e) = K_e$. Thus, if and only if $p \leq p_e$, then $K_e \in [n_{L\alpha}(p), N]$. Thus, the event $A = \{K_e \in [n_{L\alpha}, N]\}$ is identical to the event $B = \{p \leq p_e\}$.

The dependence of $n_{L\alpha}$ on p and the determination of p_e are shown in Figure 5.2.16 for $N = 12$ and $\alpha = 0.07$. Now, since $P_r(A) = P_r\{K_e \in [n_{L\alpha}, N]\} \geq 1 - \alpha$, it may be concluded that

$$P_r(B) = P_r\{p \leq p_e\} \geq 1 - \alpha. \tag{5.2.19}$$

This establishes a one-sided confidence interval for p. In particular, if $K_e = 0$, which is the case of zero events, and since $p_e(K_e = 0) = 1 - \alpha^{1/N}$, one obtains the confidence interval

$$P_r(p \leq 1 - \alpha^{1/N}) \geq 1 - \alpha. \tag{5.2.20}$$

For example, in the case of Figure 5.2.16, $N = 12$. Thus for $\alpha = 0.07$, if in 12 case histories the event does not appear, one may state that with a probability larger than 0.93, the event probability is smaller than 0.1987. This can also be addressed in the following way: suppose that the failure of a system has a probability smaller than $p_f = 10^{-5}$, and one wishes to make the statement 'With probability larger than 0.95 the failure probability is less than 10^{-5}'. This means that $\alpha = 0.05$ and $1 - \alpha^{1/N} = 10^{-5}$, or

$$N = \frac{\ln(\alpha)}{\ln(1 - p_f)} = \frac{\ln(\alpha)}{\ln(1 - 10^{-5})} = 74892. \tag{5.2.21}$$

Thus, performing a Monte Carlo calculation with 74 892 case histories or more in which the failure event never occurs justifies the required statement.

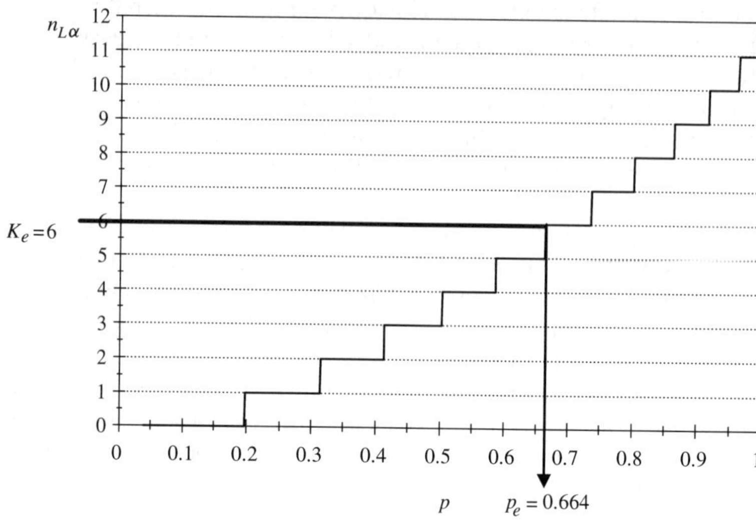

Figure 5.2.16 Dependence of $n_{L\alpha}$ on the event probability

5.3 Basic Use of Data Uncertainty in Systems Analysis

In the previous section, the estimation of failure distribution parameters was discussed. It becomes evident that, even with a small sample of experimental data and with the use of the scenario leading to the data, meaningful confidence intervals and approximate distributions may be obtained. The use of these distributions in the calculations is straightforward; it simply means that in addition to averaging over the phase space of the problem, an additional averaging is done over the parameters of the distributions involved in the calculation.

Let $C(\bar{\nu})$ be any system related quantity. $\bar{\nu}$ is a general notation for all the distribution parameters upon which the quantity depends. Let $g(\bar{\nu})$ represent an approximated joint PDF of the distribution parameters. Such PDF arise when field data is translated into a distribution of possible values of the parameters or confidence intervals. The uncertainty in the data propagates into any quantity calculated, in the sense that $C(\bar{\nu})$ becomes a random variable. The average value of $C(\bar{\nu})$ can be written as

$$\langle C(\bar{\nu})\rangle = \int_{\bar{\nu}} C(\bar{\nu}) g(\bar{\nu})\, d\bar{\nu}. \tag{5.3.1}$$

In some cases, the detailed distribution of $C(\bar{\nu})$ is of importance. In both cases, since equation (5.3.1) is often a complex multidimensional integral, the use of Monte Carlo methods may facilitate the calculation.

Consider a single component with exponential failure and repair distributions in a steady state. The steady state availability is given by $A = \mu/(\lambda + \mu)$, where λ and μ are the failure and repair rates, respectively. The failure rate λ is inferred from measurements, and may be given in the form of a distribution, such as the χ^2 distribution (see

Table 5.3.1 Piecewise uniform distribution of the failure rate

p_1	p_2	χ^2_{14}, p_1	Range: $\left[\frac{\chi^2_{14}, p_1}{18292}, \frac{\chi^2_{14}, p_2}{18292}\right]$	Probability (divided by 0.98 for normalization)
0.01	0.05	4.66	$2.54 \times 10^{-4} - 3.59 \times 10^{-4}$	0.04088
0.05	0.1	6.571	$3.59 \times 10^{-4} - 4.25 \times 10^{-4}$	0.05112
0.1	0.3	7.79	$4.25 \times 10^{-4} - 5.91 \times 10^{-4}$	0.204
0.3	0.5	10.821	$5.91 \times 10^{-4} - 7.29 \times 10^{-4}$	0.204
0.5	0.7	13.339	$7.29 \times 10^{-4} - 8.868 \times 10^{-4}$	0.204
0.7	0.9	16.222	$8.868 \times 10^{-4} - 1.151 \times 10^{-3}$	0.204
0.9	0.95	21.064	$1.151 \times 10^{-3} - 1.294 \times 10^{-3}$	0.05112
0.95	0.99	23.685	$1.294 \times 10^{-3} - 1.593 \times 10^{-3}$	0.04088
		29.14 (99%)		

expression (5.2.7)), or in the form of a piecewise uniform distribution. Assume that eight independent failure times are known: 760, 860, 1020, 1230, 1416, 1760 and 2100. Using equation (5.2.7), one obtains

$$P_r\left[\frac{\chi^2_{14,p_1}}{18292} \leq \lambda \leq \frac{\chi^2_{14,p_2}}{18292}\right] = p_2 - p_1.$$

Setting p_1 and p_2 successively increasing and using the percentiles of the χ^2 distribution, a piecewise constant approximate distribution for λ is obtained, and is shown in Table 5.3.1.

In the distribution, the range below the 1% percentile and above the 99% percentile are neglected as part of the approximation. The average failure rate resulting from the above distribution is $\langle\lambda\rangle = 7.649 \times 10^{-4}$ also $\mu = 0.0116$. The steady state availability with this average and with $\mu = 0.0116$ is $A = \mu/(\langle\lambda\rangle + \mu) = 0.9381$. However, this value is not necessarily identical to the average value of the availability as defined by equation (5.3.1). The PDF of λ can be written as

$$g(\lambda) = \frac{P_i}{\lambda_i - \lambda_{i-1}}$$

for the ith interval, where P_i is the probability for each interval and the denominator is the size of the interval. The average steady state availability is then

$$\langle A\rangle = \int_{\lambda_0}^{\lambda_8} \frac{\mu}{\lambda+\mu} g(\lambda)\, d\lambda = \sum_{i=1}^{8} \frac{P_i}{\lambda_i - \lambda_{i-1}} \int_{\lambda_{i-1}}^{\lambda_i} \frac{\mu}{\lambda+\mu}\, d\lambda = \sum_{i=1}^{8} \frac{\mu P_i}{\lambda_i - \lambda_{i-1}} \ln\left(\frac{\lambda_i+\mu}{\lambda_{i-1}+\mu}\right). \tag{5.3.2}$$

This expression yields $\langle A\rangle = 0.8647$, which is considerably different to the value obtained when the average failure rate is used. Thus, using average values of distribution parameters in non-linear system related quantities (and most quantities are indeed non-linear) may yield incorrect results.

Table 5.3.2 Field data from seven gas turbines

Turbine number	Failure times	Operating time (hours)
1	1670, 2020, 5330	16 000
2	No failures	6000
3	2340	4200
4	2700, 2780, 4220, 3450	16 000
5	2910, 4400	8700
6	No failures	11 000
7	2870, 3090	8700

Example 5.3.1: Uncertainty and warranty in power production

A gas turbine facility involves three identical turbines, each producing 50 MGW. Two stochastic elements affect the expected total production. First, the nominal production is known to have fluctuations described by a normal distribution around the average of 50 MGW with a standard deviation assumed to be 5%. Also, the failure rate is uncertain. Field data contains 12 independent full repair failure time points with the scenario described in Table 5.3.2.

The failure time distribution is assumed to be a Weibull function. The average repair time is 32 hours. The target of the analysis is the probability that the average production will be below 136 MGW, because a warranty is given such that if the average production goes beyond this limit, a penalty must be paid to recover for loss of production.

In this scenario there are 12 event points and 19 zero age points. The zero age points include the event points and the operation times. Thus, for the first component the last event point occurs at 9020 hours (sum of the first event times), hence the zero age point, related to the end of observation, is 6980 h. The MLE estimate is $\beta_{\mathrm{MLE}} = 1.692$, $\lambda_{\mathrm{MLE}} = 7.602 \times 10^{-7}$ and $\eta_{\mathrm{MLE}} = 2.418 \times 10^{-4}$. The bias factor as function of β is shown in Figure 5.3.1 in the form of the relation between $\langle\beta_{\mathrm{MLE}}\rangle$ and β_{true}, resulting in $\beta_{\mathrm{ube}} = 1.535$, $\lambda_{\mathrm{ube}} = 2.918 \times 10^{-7}$ and $\eta_{\mathrm{ube}} = 2.478 \times 10^{-4}$. The variance ratio observable yields the estimate $\beta_{\mathrm{evr}} = 3.17$, which is seriously inconsistent with the MLE estimate. However, this inconsistency becomes apparent recalling that the variance ratio is based only on the event points. The zero age points, and particularly periods in which no failures were observed, are neglected. In the present example such periods are important. Most of the event points are in the range 1500–4500 hours, while there is a period of 11 000 hours in which no failure occurred. This period tends to reduce the value of β. Indeed, the MLE estimate based only on the event points will produce $\beta_{\mathrm{MLE}} = 3.319$, which is very close to the variance ratio estimate. This is an example of a case in which the use of any observable which does not account for the zero age points may be misleading. The analysis then proceeds using only the MLE observable.

The next step is to establish a distribution for β. To this end, the procedure discussed in Section 5.2 can be executed with the following steps. A sample of N sets of event points and zero age points is generated according to the scenario of Table 5.3.1 for each given value of β, with the value of λ calculated according to the MLE formula with the observed zero age points of Table 5.3.1. For the ith sample, the value of β^i_{MLE} is calculated. The average value

$$\langle\beta_{\mathrm{MLE}}(\beta)\rangle = \frac{1}{N}\sum_{i=1}^{N}\beta^i_{\mathrm{MLE}}$$

is calculated for the estimation of the bias (see Figure 5.3.1), and the percentiles of the distribution of β_{MLE} are also estimated. The 0.05 and 0.95 percentiles as a function of β and the determination of the 90% confidence interval, yielding (1.057, 2.239), are shown in Figure 5.3.2.

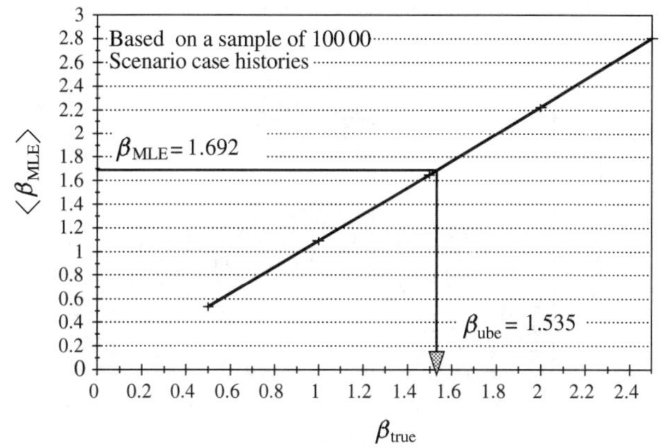

Figure 5.3.1 The bias of the MLE estimate

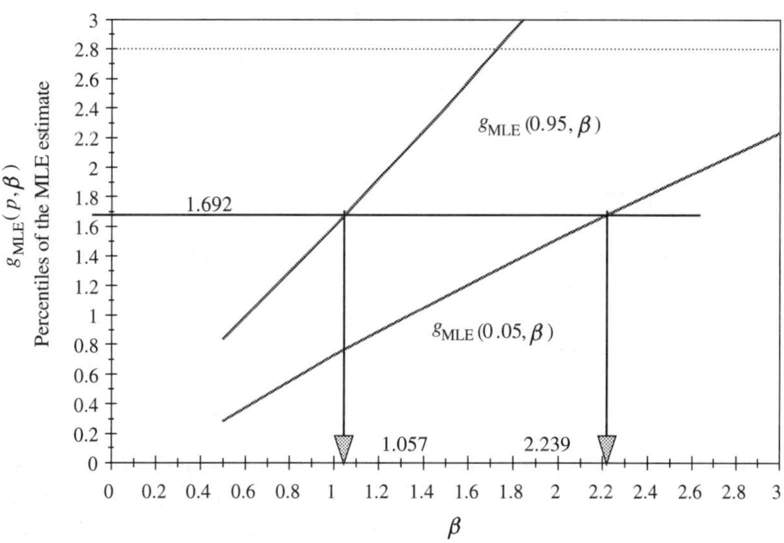

Figure 5.3.2 A 90% confidence interval determination for β

Using a number of percentiles, an approximate distribution for β is obtained and displayed in Figure 5.3.3. The distribution is expressed in terms of intervals, and the probability of each interval. This form is convenient for sampling. First, the interval is sampled, and then the value of β is sampled uniformly within the interval. A tabular form of the same distribution is also presented in Table 5.3.3 for convenience. It should be noted that the 5% edges of the distribution are neglected and a normalization factor is applied to each interval.

This completes the inference of the shape parameter. At this point one may proceed to the actual calculation with the scale parameter evaluated as the MLE for each value of β, using the zero age points of the scenario. It would, however, be useful to add independent inference for λ, at least for the purposes of comparison. To that end, the method suggested in equation (5.2.16) and (5.2.17) may be deployed. However, that method is based only on the event points. This may cause a serious overestimation of λ, because relatively long periods in which no events occurred

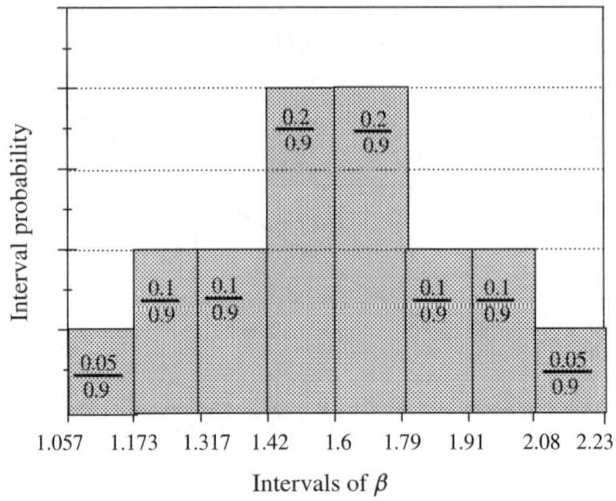

Figure 5.3.3 Interval probabilities for the values of β

Table 5.3.3 The approximate distribution of β

Interval	Probability y
1.057–1.173	0.05555
1.173–1.317	0.11111
1.317–1.42	0.11111
1.420–1.60	0.22222
1.60–1.79	0.22222
1.79–1.91	0.11111
1.91–2.08	0.11111
2.08–2.23	0.05555

are neglected. To overcome this, the zero age points which do not coincide with event points and which are larger than the largest event point are added to the event points. In the present case, this includes three points: 6980 h, 6000 h and 11 000 h. These points arise for the first, second and sixth components, respectively. Note that the sixth component, for example, had 11 000 hours of operation with no events. Thus, an event point for this component will occur, and will definitely be larger than 11 000 h. Taking 11 000, hours is an underestimation of this event point, while still adding an event point which is larger than any of the observed event points. Inference for λ is done for a given value of β. Equation (5.2.17) is then used with the two extreme values of β, namely 1.05 and 2.23. The inference procedure follows that of Example 5.2.5 with 15 event points. The sum of the event points (with the three added points) is 61760 h. For $\beta = 1.05$, the 0.025 percentile is 8.48 and the 0.975 percentile is 22.7; this yields the confidence interval

$$P_r\left\{\frac{61760}{22.7} \le \theta^{1/1.05} \le \frac{61760}{8.48}\right\} = 0.95, \quad \left(\theta = \frac{1}{\lambda}\right) \quad \text{or}$$

$$P_r\{8.802 \times 10^{-5} \le \lambda \le 2.475 \times 10^{-4}\} = 0.95, \tag{5.3.3a}$$

leading to

$$P_r\{1.373 \times 10^{-4} \le \eta = \lambda^{1/\beta} \le 3.765 \times 10^{-4}\} = 0.95.$$

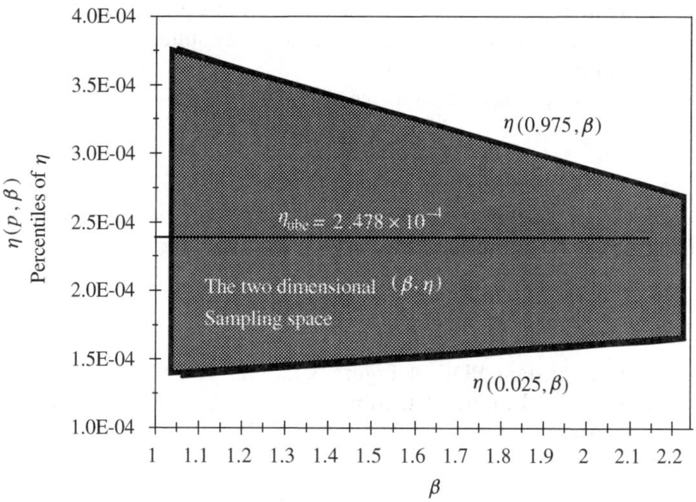

Figure 5.3.4 Shifting 95% confidence intervals for η as a function of the shape parameter

For $\beta = 2.23$ the 0.025 percentile is 10.19 and the 0.975 percentile is 16.53; this yields the confidence interval

$$P_r\left\{\frac{61760}{16.52} \leq \theta^{1/2.23} \leq \frac{61760}{10.19}\right\} = 0.95 \quad \text{or} \quad P_r\{3.672 \times 10^{-9} \leq \lambda \leq 1.078 \times 10^{-8}\} = 0.95, \tag{5.3.3b}$$

leading to

$$P_r\left\{1.649 \times 10^{-4} \leq \eta = \lambda^{1/(\beta=1.05)} \leq 2.675 \times 10^{-4}\right\} = 0.95.$$

As β varies from 1.05 to 2.23, the 0.025 percentile of η varies from 1.373×10^{-4} to 1.649×10^{-4}, and the 0.975 percentile from 3.765×10^{-4} to 2.675×10^{-4}. The boundaries for each value of β can be obtained by assuming that the percentile is a linear function of β. This is shown in Figure 5.3.4.

The 0.025 curve has the form $\eta_L(\beta) = 2.33 \times 0510^{-5}\beta + 1.129 \times 10^{-4}$ and that of the 0.975 by $\eta_U(\beta) = -9.2203 \times 10^{-5}\beta + 4.733 \times 10^{-4}$. To sample values of β and η, the value of β is sampled first using the distribution of Table 5.3.2; once β is defined, η_L and η_U are determined and η is sampled uniformly in the interval $[\eta_L(\beta), \eta_U(\beta)]$.

This completes the first step of preparing estimates for the distribution of the unknown failure distribution parameters. The next step is to evaluate the impact of the stochastic and uncertain elements on the distribution of the power generation. The power production at time t can be described as

$$W(t) = \sum_{i=1}^{3} \omega_i \times A_i(\lambda, \beta, t), \tag{5.3.4}$$

where ω_i is the production of the ith turbo generator. ω_i is a random variable having a normal distribution with mean $\mu = 50_{\text{MGW}}$ and standard deviation $\mu = 2.5_{\text{MGW}}$, reflecting the fluctuation in the nominal throughput. $A_i(\lambda, \beta, t)$ is the availability. At this point, it is assumed that the three generators are identical, independent and operate in a steady state. The steady state assumption renders the production to be independent of time.

In a steady state the availability is given by

$$A = \frac{\text{MTTF}}{\text{MTTF} + \text{MTTR}}.$$

This expression holds for a component with two states for any failure and repair distribution. Substituting the expression for the MTTF of a Weibull distribution yields

$$A = \frac{\Gamma(1 + 1/\beta)}{\Gamma(1 + 1/\beta) + 32 \times \lambda^{1/\beta}}. \tag{5.3.5}$$

Denoting the piecewise constant PDF of Figure 5.3.3 and Table 5.3.2 for β by $f_\beta(\beta)$, the distribution of η by $f_\eta(\eta|\beta)$ and the normal distribution of ω, $N(\mu, \sigma, \omega)$, the average power production per unit time is given by

$$\langle W \rangle = \int_0^\infty d\omega_1 \int_0^\infty d\omega_2 \int_0^\infty d\omega_3 \int_{\beta_1}\int_{\beta_2}\int_{\beta_3}\int_{\eta_1}\int_{\eta_2}\int_{\eta_3} \left[\sum_{i=1}^{3} \omega_i \times A_i(\lambda_i, \beta_i)\right] \times \prod_{j=1}^{3} \big(N(\mu, \sigma, \omega_j) f_\beta(\beta_j) f_\eta(\eta_j|\beta_j)\big)\, d\beta_1\, d\beta_2\, d\beta_3\, d\eta_1\, d\eta_2\, d\eta_3. \tag{5.3.6}$$

Since this expression is independent of time, it represents the average power production per unit time averaged over any period of time. This nine-dimensional integral is suitable for Monte Carlo estimation.

The distributions are used to sample variables, and the estimator is $\left[\sum_{i=1}^{3} \omega_i \times A_i(\lambda_i, \beta_i)\right]$. ω_i is sampled from the standard normal distribution and then transformed using the mean and standard deviation [`Program gpower.for`]. In each sample a value of W is obtained; by counting the frequency of appearance of such values in intervals, the histogram of W is obtained. The power generation histogram is shown in Figure 5.3.5.

Three histograms are depicted in Figure 5.3.5. The light gray line describes the histogram of the normal distribution without taking into account the unavailability of the system, i.e. it is the

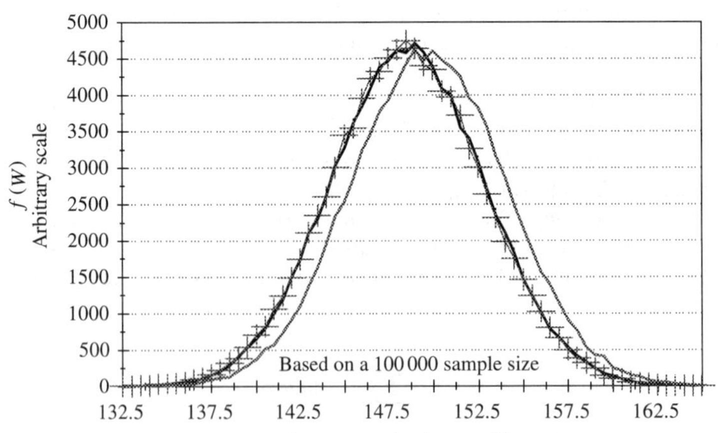

Figure 5.3.5 Histogram of the power production

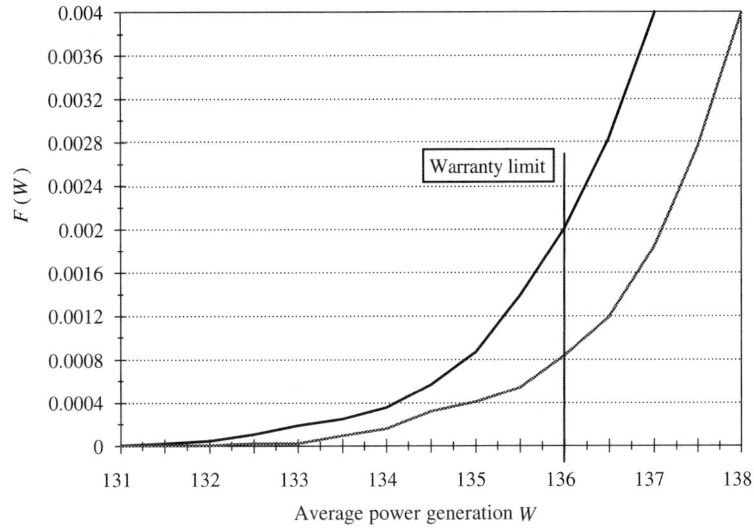

Figure 5.3.6 Cumulative distribution of the average power production

histogram of $(w_1 + w_2 + w_3)$. The average of this histogram is at 150 MGW. The solid black curve indicates the histogram with the effect of unavailability, i.e. the histogram of $\left[\sum_{i=1}^{3} \omega_i \times A_i(\lambda_i, \beta_i)\right]$ when the distribution of η is the shifting intervals described in Figure 5.3.4. The histogram is seen to be shifted to the left with a reduced average being 148.793 MGW. The plus markers line indicate the histogram obtained when λ is derived as the MLE for each β, and it can be seen that the histograms are almost identical.

To evaluate the probability that average production will get below 136 MW, the CDF of the histogram is calculated. The CDF in the range of interest (131–138) MGW is shown in Figure 5.3.6. From this graph it can be seen that the probability of having average power generation smaller than 136 MGW is 0.00201 (solid black curve). The probability based on the nominal normal dispersion is only 0.00084 (gray curve). It should be noted that it is possible to add redundancies such as auxiliary burners that will back up partial loss of power, and thereby reduce the probability of crossing the warranty limit to a negligible level. Denoting the cost of such redundancy added to the design by $C_d(W_0)$ and the penalty cost by $C_p(W_0)$, one can look for the relation $C_p(W_0) \times F(W_0) \geq C_d(W_0)$ upon which the design of such redundancies becomes worthwhile.

6
System Transport

It has already been hinted that the most general approach to the calculation of systems behavior is through the Monte Carlo method utilizing the concept of the 'history' of the system. A history is denoted by $c_k = (\vec{\mathbf{B}}_1, t_1; \vec{\mathbf{B}}_2, t_2; \ldots; \vec{\mathbf{B}}_k, t_k)$, where $\vec{\mathbf{B}}_i$ is the state vector of the system and t_i is the time point at which the system enters that state. Every aspect of the system and every quantity related to its operation can be accounted for and calculated by observing the system's histories. In this chapter, the transport of systems through the phase space of time and states will be discussed. In previous chapters, basic concepts of system engineering and the Monte Carlo method were introduced, and specific problems and Monte Carlo solutions were presented. The Monte Carlo method was presented mainly through its efficiency in addressing multiple dimensional integrals. In the following, the concept of random walk will be introduced and used as an approach to general system engineering problems.

Definition of continuous and discrete random walks Consider a system of any sort defined by its capacity to be in a set of states in some phase space. A general point in the phase space is denoted by a general phase space vector $\vec{\mathbf{P}}$. In general, the coordinates of $\vec{\mathbf{P}}$ may be either discrete or continuous. The system starts at a specific state $\vec{\mathbf{P}}_1$; it then moves to the next state $\vec{\mathbf{P}}_2$, and so on until it reaches a state $\vec{\mathbf{P}}_k$, at which point it is terminated in some sense. The sequence of states $(\vec{\mathbf{P}}_1, \vec{\mathbf{P}}_2, \ldots, \vec{\mathbf{P}}_k)$ is called a *random walk*. It is a random walk process if the following distributions and probabilities are attached to it.

$f^1(\vec{\mathbf{P}}_1)$ – first state probability. This is the probability (or probability density if the space is continuous) that the first state of the system will be $\vec{\mathbf{P}}_1$. In the continuous case $f^1(\vec{\mathbf{P}})\,d\vec{\mathbf{P}}$ is the probability that the initial state will be in a phase space element $d\vec{\mathbf{P}}$ around $\vec{\mathbf{P}}$. In the discrete case, $f^1(\vec{\mathbf{P}}_1)$ is replaced by $p^1(j_1)$, where j_1 is a discrete state of the system.

$K(\vec{\mathbf{P}}_i, \vec{\mathbf{P}}_{i+1})$ – state transition kernel. This is the probability density that, given a system which entered state $\vec{\mathbf{P}}_i$, the next state will be a particular state $\vec{\mathbf{P}}_{i+1}$ (per unit phase space volume).

$p(\vec{\mathbf{P}}_k)$ – probability that the walk is terminated in state $\vec{\mathbf{P}}_k$.

In the discrete case, $k(\vec{\mathbf{P}}_i, \vec{\mathbf{P}}_{i+1})$ is replaced by $p(j_i, j_{i+1})$ or $p_{j_i, j_{i+1}}$ and $p(\vec{\mathbf{P}}_k)$ is replaced by $p(j_k)$. The following conditions are fulfilled by the process:

$$\int_\Gamma f^1(\vec{\mathbf{P}}_1)\, d\vec{\mathbf{P}}_1 = 1, \tag{6.1}$$

where Γ denotes the phase space, and the normalization condition indicates that the probability that the system will have an initial state is unity. The discrete analog of (6.1) is $\sum_{m=1}^{n} p^1(j_m) = 1$, where n is the number of states in the space.

$$\int_\Gamma K(\vec{\mathbf{P}}_i, \vec{\mathbf{P}}_{i+1})\, d\vec{\mathbf{P}}_{i+1} = 1 - p(\vec{\mathbf{P}}_i) = q(\vec{\mathbf{P}}_i). \tag{6.2}$$

This condition merely indicates that the system may either transfer into another phase space state or be absorbed in its current state. $q(\vec{\mathbf{P}})$ is the survival probability, or the probability that the random walk will not be terminated. The discrete analog of equation (6.2) is

$$\sum_{m=1}^{n} p(j_i, m) = 1 - p(j_i) = q(j_i). \tag{6.2a}$$

It should be noted that the definition of the process is not limited to Markovian processes. The transition kernel $K(\vec{\mathbf{P}}_i, \vec{\mathbf{P}}_{i+1})$ may be dependent not only on the current state, but on any other previous state. Previous states are not explicitly entered for simplicity of notation. The transition kernel may also be presented in the normalized form

$$V(\vec{\mathbf{P}}_i, \vec{\mathbf{P}}_{i+1}) = \frac{K(\vec{\mathbf{P}}_i, \vec{\mathbf{P}}_{i+1})}{\int_\Gamma K(\vec{\mathbf{P}}_i, \vec{\mathbf{P}})\, d\vec{\mathbf{P}}} = \frac{K(\vec{\mathbf{P}}_i, \vec{\mathbf{P}}_{i+1})}{q(\vec{\mathbf{P}}_i)}. \tag{6.3}$$

Let $P^k(j)$ denote the probability that the kth state of a discrete random walk is j. Then, from the definition of the random walk, one may write

$$P^k(j) = \sum_{i=1}^{n} P^{k-1}(i) p(i, j). \tag{6.4}$$

The number of states is inclusive of the case $n = \infty$, with the continuous parallel

$$P^k(\vec{\mathbf{P}}) = \int_\Gamma P^{k-1}(\vec{\mathbf{P}}') K(\vec{\mathbf{P}}', \vec{\mathbf{P}})\, d\vec{\mathbf{P}}'. \tag{6.4a}$$

Equations (6.4) and (6.4a) present a single step transition in a random walk process. The average number of entries into state j, $E(j)$, is given by

$$E(j) = \sum_{k=1}^{\infty} P^k(j) = P^1(j) + \sum_{k=2}^{\infty} P^k(j).$$

Noting that $P^1(j) = p^1(j)$ and substituting equation (6.4) for the terms in the sum yields

$$E(j) = \sum_{k=1}^{\infty} P^k(j) = p^1(j) + \sum_{k=2}^{\infty}\sum_{i=1}^{n} P^{k-1}(i)p(i,j);$$

rearranging the summation over $(k-1)$ instead of k and interchanging the sums yields

$$E(j) = p^1(j) + \sum_{i=1}^{n}\left(\sum_{k=1}^{\infty} P^k(i)\right)p(i,j),$$

and again using the definition of $E(i)$ yields the final result

$$E(j) = p^1(j) + \sum_{i=1}^{n} E(i)p(i,j). \tag{6.5}$$

Likewise for the continuous case, replace the notation $E(j)$ by $\psi(\vec{\mathbf{P}})$ to denote the average number density of entries into state $\vec{\mathbf{P}}$, and obtain

$$\begin{aligned}\psi(\vec{\mathbf{P}}) &= \sum_{k=1}^{\infty} P^k(\vec{\mathbf{P}}) = f^1(\vec{\mathbf{P}}) + \sum_{k=2}^{\infty} P^k(\vec{\mathbf{P}}) \\ &= f^1(\vec{\mathbf{P}}) + \sum_{k=2}^{\infty}\int_{\Gamma} K(\vec{\mathbf{P}}',\vec{\mathbf{P}})P^{k-1}(\vec{\mathbf{P}}')\,d\vec{\mathbf{P}}' = f^1(\vec{\mathbf{P}}) + \int_T\left(\sum_{k=1}^{\infty} P^k(\vec{\mathbf{P}}')\right)K(\vec{\mathbf{P}}',\vec{\mathbf{P}})\,d\vec{\mathbf{P}}',\end{aligned}$$

and finally,

$$\psi(\vec{\mathbf{P}}) = f^1(\vec{\mathbf{P}}) + \int_T \psi(\vec{\mathbf{P}}')K(\vec{\mathbf{P}}',\vec{\mathbf{P}})\,d\vec{\mathbf{P}}'. \tag{6.5a}$$

Equations (6.5) and (6.5a) are the basic relations of the random walk process. They relate the transfer between two consecutive states in the space to the event densities (event density, or rates, are the number density of entries into a given state). They are called the *transport equations* of the process, or the phase space. A random walk process is a transport process in a phase space, and the equations that control this process are the transport equations. Thus, any process that can be described as a Markovian or non-Markovian random walk process automatically fulfils a transport equation (as for the phase space of time and states of system engineering). A simple form of the equation was introduced in Section 2.2.2. Equations (2.2.23) and (2.2.40a,b) are also special cases of the transport equation.

6.1 Discrete Random Walks

In this section a number of applications of discrete random walk processes are discussed. The first is the solution of linear equations. It is presented both to demonstrate the insensitivity of the Monte Carlo method to the dimension of the space, and to indicate how the method can be applied in areas that have no apparent connection to statistics and probability.

6.1.1 Solution of linear equations

Let

$$\vec{\mathbf{x}} = \mathbf{H}\vec{\mathbf{x}} + \vec{\mathbf{a}} \tag{6.1.1}$$

denote a set of n linear equations, where $\mathbf{H}$ is a $n \times n$ real matrix, $\vec{\mathbf{a}}$ is a known vector and $\vec{\mathbf{x}}$ is the vector of solutions of the equation. The solution of this equation can be written in the form

$$\vec{\mathbf{x}} = (I - \mathbf{H})^{-1}\vec{\mathbf{a}},$$

which can be expanded such that the jth element of the solution is expressed as an infinite series expansion

$$x_j = a_j + [\mathbf{H}\vec{\mathbf{a}}]_j + [\mathbf{H}^2\vec{\mathbf{a}}]_j + \sum_{n=3}^{\infty}[\mathbf{H}^n\vec{\mathbf{a}}]_j. \tag{6.1.2}$$

This expansion is called the *Neumann expansion*, and will hold if the series converges absolutely. Absolute convergence of the series is guaranteed if the largest modulus of the eigenvalues λ of the matrix $\mathbf{H}$ is smaller than unity. The largest modulus is also named the spectral radius of $\mathbf{H}$, $\rho(\mathbf{H})$, thus the convergence criterion is $\rho(\mathbf{H}) < 1$. The eigenvalues of the matrix are real or complex values for which the equation $\mathbf{H}\vec{\mathbf{g}} = \lambda\vec{\mathbf{g}}$, is satisfied. The vectors $\vec{\mathbf{g}}$ are called the *eigenvectors* of the matrix. An equivalent condition to the eigenvalue equation is that the determinant of the matrix $\mathbf{H} - \lambda\mathbf{I}$ is zero, i.e. $\mathrm{Det}(\mathbf{H} - \lambda\mathbf{I}) = |\mathbf{H} - \lambda\mathbf{I}| = 0$, where $\mathbf{I}$ is the unit diagonal matrix. The set of the eigenvalues of a matrix, or any other linear operator, is called the *spectrum* of the operator. The determination of the eigenvalues, as can be easily seen from the condition on the determinant, involves a solution of an nth order polynomial. (Recipes for eigenvalue and eigenvector determination can be found in Press, W. H., Teukolsky, S. A., Vetterling, W. T. and Flannery, B. P. (1992). *Numerical Recipes in Fortran*, Cambridge University Press). For general reading on convergence of linear operators expansion, see Kantorovich L. V. and Akilov, G. P. (1964) *Functional Analysis in Normed Spaces*, Pergamon Press. Also, see Spanier, J. and Gelbard, E. M. (1969). *Monte Carlo Principles and Neutron Transport Problems*, Addison-Wesley and the references therein).

For any set of linear equations, for which the series (6.1.2) converges, define a general random walk process as follows: Let there be n states with the following properties:

1. $p_1(i)$ $i = 1, \ldots, n$ be subject to the condition $a_i \neq 0 \Rightarrow p^1(i) \neq 0$, i.e. if the ith element of the vector $\vec{\mathbf{a}}$ is non-zero, then the probability that the initial state is i must also be non-zero.
2. Let $p(i, j)$ be defined such that $p(i, i) = 0$ and $h_{j,i} \neq 0 \Rightarrow p(i, j) \neq 0$, i.e. if the matrix element is non-zero, so also is the transfer matrix element. Also, the matrix $p(i, j)$ is chosen such that its spectrum is contained in the unit circle. To guarantee this property it is enough to choose the matrix such that every state is accessible and, in at least one state, the absorption probability is non-zero. This also means that the probability for a random walk with an infinite number of states is zero.
3. For the calculation of a particular element x_{j_0} of the solution vector $\vec{\mathbf{x}}$ chose the absorption probability $p(j_0) = 1 - \sum_{i=1}^{n} p(j_0, j) > 0$.

With the above definition the random walk, or random history, takes the form $c_k = (i_1, i_2, \ldots, i_k)$, with the probability for the random walk being

$$P(c_k) = p^1(i_1)p(i_1, i_2)p(i_2, i_3)\cdots p(i_{k-1}, i_k)p(i_k). \tag{6.1.3}$$

If $\eta(c_k)$ is an estimator defined on the random walk process, then the mean value of the estimator is obtained by averaging it over all possible histories, namely

$$\langle\eta(c_k)\rangle = \sum_{k=1}^{\infty}\sum_{i_1=1}^{n}\sum_{i_2=1}^{n}\sum_{i_3=1}^{n}\cdots\sum_{i_k=1}^{n}\eta(c_k)p^1(i_1)p(i_1, i_2)p(i_2, i_3)\cdots p(i_{k-1}, i_k)p(i_k). \tag{6.1.4}$$

Let Λ_k denote the subset of all the histories that have exactly k steps; then the averaging can be written in the concise form

$$\langle\eta(c_k)\rangle = \sum_{k=1}^{\infty}\sum_{c\in\Lambda_k} P(c)\eta(c).$$

Now define the estimator of the random walk $c_k = (i_1, i_2, \ldots, i_k)$ beginning at i_1 and terminating at i_k in the form

$$\eta_{j_0}(c_k) = \frac{a_{i_1}}{p^1(i_1)}\frac{h_{i_2,i_1}}{p(i_1, i_2)}\frac{h_{i_3,i_2}}{p(i_2, i_3)}\cdots\frac{h_{i_k,i_{k-1}}}{p(i_{k-1}, i_k)}\frac{\delta_{j_0,i_k}}{p(i_k)}. \tag{6.1.5}$$

η_{j_0} is an unbiased estimator for the j_0 element of the solution vector $\vec{\mathbf{x}}$. $\delta_{j,i}$ is the Kronneker delta function (the discrete analog of the dirac delta function), being 1 if $i = j$ and zero otherwise. To show that η_{j_0} is an unbiased estimator, expression (6.1.5) is substituted in equation (6.1.4), yielding

$$\langle\eta(c_k)\rangle = \sum_{k=1}^{\infty}\sum_{i_1=1}^{n}\sum_{i_2=1}^{n}\sum_{i_3=1}^{n}\cdots\sum_{i_k=1}^{n} a_{i_1}h_{i_2,i_1}h_{i_3,i_2}\cdots h_{i_k,i_{k-1}}\delta_{i_k,j_0}. \tag{6.1.6}$$

The term $k = 1$ yields $\sum_{i_1=1}^{n} a_{i_1}\delta_{i_1,j_0} = a_{j_0}$. Likewise, for $k = 2$,

$$\sum_{i_1=1}^{n}\sum_{i_2=1}^{n} a_{i_1}h_{i_2,i_1}\delta_{i_2,j_0} = \sum_{i_1=1}^{n} a_{i_1}h_{j_0,i_1} = [\mathbf{H}\vec{\mathbf{a}}]_{j_0},$$

which is the second term in the Neumann expansion. Similarly, the $k = 3$ term yields the third term

$$\sum_{i_1=1}^{n}\sum_{i_2=1}^{n}\sum_{i_3=1}^{n} a_{i_1}h_{i_2,i_1}h_{i_3,i_2}\delta_{i_3,j_0} = \sum_{i_1=1}^{n}\sum_{i_2=1}^{n} a_{i_1}h_{i_2,i_1}h_{j_0,i_2} = \sum_{i_1=1}^{n} a_{i_1}h^2_{j_0,i_1} = [\mathbf{H}^2\vec{\mathbf{a}}]_{j_0},$$

and similarly, the nth term in the sum yields $[\mathbf{H}^n\vec{\mathbf{a}}]_{j_0}$, concluding that (6.1.6) is identical to (6.1.2), and hence proving that η_{j_0} is an unbiased estimator. The method used in the above proof is common to linear operators in general. The solution of a non-homogeneous equation involving a linear operator can always be expanded as a Neumann series, and unbiased estimators often estimate these expansions term-by-term. This means that a history containing n events, with an estimator that obtains a non-zero value upon the last event, contributes directly to the nth term of the Neumann series expansion. This has interesting consequences regarding optimization. The relative magnitude of a Neumann expansion term could serve as a measure of the true importance, and optimization may follow the idea of matching the relative number of histories 'invested' in estimating a given term with its importance.

Example 6.1.1

The matrix **H** and the constant vector $\vec{\mathbf{a}}$ shown below (the notation 3.274^{-1} is a shortened notation for 3.274×10^{-1}) are part of an equation of the form $\vec{\mathbf{x}} = \mathbf{H}\vec{\mathbf{x}} + \vec{\mathbf{a}}$ with an unknown solution vector **H**.

$$\mathbf{H} = \begin{pmatrix} -3.274^{-1} & -1.826^{-1} & -1.366^{-1} & 0.0 & 1.211^{-1} & -1.169^{-1} & 4.131^{-2} & 0.0 & 0.0 & 1.526^{-1} \\ -9.288^{-18} & -3.288^{-2} & 1.277^{-1} & 0.0 & 0.0 & 2.660^{-2} & 0.0 & -3.802^{-2} & 1.67^{-2} & -4.912^{-2} \\ -6.4248^{-18} & -6.391^{-2} & -7.588^{-2} & 6.260^{-2} & 7.088^{-2} & -3.56^{-2} & 2.816^{-3} & 0.0 & 6.86^{-2} & -1.249^{-1} \\ 0.0 & -5.477^{-15} & -2.541^{-21} & 9.086^{-2} & 4.301^{-2} & 1.997^{-2} & -9.472^{-3} & 0.0 & -3.025^{-3} & 1.148^{-2} \\ 0.0 & 2.914^{-16} & 0.0 & -1.048^{-2} & 9.684^{-2} & 3.9^{-3} & 1.607^{-2} & -3.093^{-2} & 0.0 & 3.248^{-2} \\ 2.32^{-1} & 4.701^{-2} & 3.698^{-21} & 3.513^{-15} & 0.0 & -3.241^{-2} & 6.129^{-3} & 0.0 & -1.856^{-2} & -6.453^{-2} \\ 2.41^{-1} & 2.281^{-1} & 4.097^{-3} & 0.0 & 6.338^{-18} & -6.865^{-20} & 1.821^{-2} & 3.284^{-2} & 3.472^{-2} & 4.315^{-3} \\ 2.819^{-28} & -5.509^{-2} & -1.83^{-1} & -5.961^{-3} & -7.209^{-17} & 2.541^{-21} & 0.0 & -2.934^{-2} & 0.0 & 0.0 \\ 0.0 & 0.0 & 7.501^{-2} & -1.448^{-1} & 1.247^{-1} & -6.377^{-18} & 0.0 & 0.0 & 5.675^{-2} & -1.783^{-2} \\ 0.0 & -1.229^{-1} & 8.813^{-3} & -3.195^{-2} & -1.913^{-1} & 1.044^{-2} & 0.0 & 0.0 & 0.0 & 4.401^{-2} \end{pmatrix}$$

$$\vec{\mathbf{a}} = \begin{pmatrix} 1.2 \\ 0.5 \\ 4.8 \\ 1.2 \\ 0.5 \\ 4.2 \\ 1.8 \\ 0.2 \\ 2.1 \\ 4.0 \end{pmatrix}$$

For the corresponding random walk process, the following choice is often useful. $p^1(i)$ is chosen relative to $a(i)$ according to

$$p^1(i) = \frac{a(i)}{\sum_{j=1}^{n} a(j)}. \tag{6.1.7}$$

$p(i,j)$ is set to zero if $h(j,i) = 0$, and otherwise it is

$$p(i,j) = \frac{1 - p_a(i)}{n_0(i)}, \tag{6.1.7a}$$

where $n_0(i)$ is the number of non-zero elements in the ith column of $h(j,i)$, and $p_a(i)$ is the termination probability set as a parameter. For the above case, $p_a(i) = 0.05\ \forall i$, and the solution vector obtained with 10 million histories is the vector $\vec{\mathbf{x}}$ shown below. The PRSD of each element is smaller then 0.6%.

$$\vec{\mathbf{x}} = \begin{pmatrix} 0.489 \\ 1.06 \\ 5.0 \\ 1.46 \\ 0.764 \\ 3.94 \\ 2.27 \\ -0.756 \\ 2.41 \\ 3.93 \end{pmatrix}.$$

The realization of this transport process, for a single history, requires the following steps:

1. Set $k = 1$. With probability $p^1(i)$ sample (using discrete sampling) the initial state, i_k.
2. With probability $p_a(i_k)$, terminate the history. If the history is terminated, go to step 4. Otherwise proceed to step 3.
3. History continues: set $k = k + 1$. With probability $p(i_{k-1}, j)$, sample the next state j, and set $i_k = j$. Proceed to step 2.
4. History terminated: calculate the value of the estimator using expression (6.1.5), and start a new history.

The sample average

$$\langle \eta \rangle_N = \frac{1}{N} \sum_{i=1}^{N} \eta_{j_0,i}(c_k),$$

where $\eta_{j_0,i}(c_k)$ is the value of the estimator obtained from the ith history, will converge, by virtue of (6.1.6), to the solution x_{j_0}. The second moment is estimated by

$$\langle S^2 \rangle_N = \frac{1}{N} \sum_{i=1}^{N} \eta^2_{j_0,i}(c_k)$$

[program lineq.for].

Example 6.1.2: A discrete state reliability model

Consider a system with n possible states. Numbering the states by $i = 1, \ldots, n$, state 1 represents a new system and each higher state represents a degraded state of operation. There is an additional, absorbing state in which the system is destroyed and discarded. Such a model can be adopted to a situation in which many identical systems are operating and the rate of events per system in state i is λ_i. Time is divided into discrete equal intervals, ΔT, such that the number of events in an interval, per system, is $\lambda_i \Delta T$. Given an event in state i, with probability Σ^{ji} the system passes into state j. With the above definitions,

$$\Sigma^{ji} = \frac{\lambda_{ji}}{\lambda_i},$$

where λ_{ji} is the rate of events at which the system passes from i to j and $\lambda_i = \sum_{j=1}^{n} \lambda_{ji}$. Also, Σ^{ji} for $j \leq i$ may be non-zero, i.e. the system may either remain in the same state or transfer into a less degraded state. This event models a detected malfunction that results in an immediate repair operation that improves the state of the system. The probability of a system being discarded (absorbed) in an event in state i is $p_a(i) = 1 - \sum_{j=1}^{n} \Sigma^{ji}$. Assuming that at least for one state, that can be, ultimately, reached from any other state, the probability of discard is non-zero, it is obvious that the number of operating systems, will decrease in time until all systems are discarded. Systems are also produced at a rate of $a(i)$ per unit time in state i. The time unit is the interval ΔT. Newly produced systems can already emerge in a degraded state either because of manufacturing defects or because they are rebuilt from discarded systems that are 'renewed'.

Denoting the number of events in a steady state in state i by $x(i)$ and the transfer matrix by

$$\mathbf{H} = \begin{pmatrix} \Sigma^{1,1} & \Sigma^{1,2} & \cdots & \Sigma^{1,n} \\ \cdot & & & \\ \cdot & & & \\ \cdot & & & \\ \cdot & & & \\ \Sigma^{n,1} & \Sigma^{n,2} & \cdots & \Sigma^{n,n} \end{pmatrix}.$$

It is clear that

$$\mathbf{H}\vec{\mathbf{x}} = \begin{pmatrix} \Sigma^{1,1} & \Sigma^{1,2} & \dots & \Sigma^{1,n} \\ \cdot & & & \\ \cdot & & & \\ \cdot & & & \\ \Sigma^{n,1} & \Sigma^{n,2} & \dots & \Sigma^{n,n} \end{pmatrix} \begin{pmatrix} x(1) \\ \cdot \\ \cdot \\ \cdot \\ \cdot \\ x(n) \end{pmatrix}$$

represents a vector containing the average number of transfers into each state, due to *events in the field*. Adding the vector $\vec{\mathbf{a}}$ to $\mathbf{H}\vec{\mathbf{x}}$ yields the total number of systems that are entering each state either due to events, or due to being produced in the corresponding state. This must equal $\vec{\mathbf{x}}$, which represents the number of systems leaving each state, since the number of systems entering a state must equal the number of systems leaving the state, per unit time. Indeed, if a system has an event and remains in the same state, it is counted in $\vec{\mathbf{x}}$ as leaving the state and in $\mathbf{H}\vec{\mathbf{x}}$ as entering the state. Hence, the above model fulfils equation (6.1.1).

Recall that, to use Monte Carlo estimation of $\vec{\mathbf{x}}$, it is required that the spectral radius of $\mathbf{H}$ will be bounded by unity. This property has a rather simple physical interpretation. Convergence of the series (6.1.2) means that the terms of the series decrease at a rate that guarantees convergence (for example, a rate of $1/n^{\alpha}$, where $\alpha > 1$). The terms of the series represent the repeated operation of the operator (the matrix in this case) on the initial vector. If one starts with a single system in state 1, then the average number of systems transferring into any state, following the first event, is $\mathbf{H}\vec{\mathbf{a}}$, and after the kth event it is $\mathbf{H}^k\vec{\mathbf{a}}$. Thus, convergence is guaranteed if the number of units occupying each state is diminishing at an appropriate rate. This also means that, starting with a single system, the system will be discarded (absorbed) with a probability 1 after a finite number of steps. This is guaranteed if the following conditions are fulfilled:

1. Every state can be ultimately reached from any other state, not necessarily in a single step. This property is called *irreducibility*.
2. At least one state exists for which $\sum_{j=1}^{n} \Sigma^{ji} < 1$, i.e. the sum of the elements in at least one matrix column is less than unity, leading to a non-zero probability for termination (absorption) in at least one state.

(for further discussion on the subject, see Varga, R. S. (1962). *Matrix Iterative Analysis*, Prentice Hall, Englewood Cliffs, NJ). This basically means that looking at a physical process, in which there are discrete states, and the transfer from one state to another is represented by an operation of a linear operator on a state vector, it is evident that the probability that the process will proceed an infinite number of steps without reaching an absorbing state is zero. It is then guaranteed that the process will terminate after a finite number of steps, and that a Monte Carlo simulation of such a process will terminate accordingly. Clearly, if there is a non-zero probability for a infinite process, then on the one hand the Neumann series will not converge and, from the point of view of a Monte Carlo process, there is a probability for an infinite history which will require an infinite amount of computer time.

For the above process the random walk can be defined such that

$$p^1(i) = \frac{a(i)}{\sum_{j=1}^{n} a(j)},$$

as in (6.1.7), $p(i,j) = \Sigma^{j,i}$ and $p(i) = 1 - \sum_{j=1}^{n} \Sigma^{j,i}$. Substituting this random walk process into the estimator (6.1.5) yields

$$\eta_{j_0}(c_k) = \frac{a_{i_1}}{p^1(i_1)} \frac{h_{i_2,i_1}}{p(i_1,i_2)} \frac{h_{i_3,i_2}}{p(i_2,i_3)} \cdots \frac{h_{i_k,i_{k-1}}}{p(i_{k-1},i_k)} \frac{\delta_{j_0,i_k}}{p(i_k)} = A\frac{\delta_{j_0,j_k}}{p(i_k)}$$

where

$$A = \sum_{j=1}^{n} a(j)$$

is simply the source strength which is a constant multiplier. Assuming a unit source strength, the estimator takes the simple form

$$\eta_{j_0}(c_k) = \frac{\delta_{j_0, j_k}}{p(i_k)}. \tag{6.1.8}$$

The average value of this estimator is the average number of entries into state j_0 per system starting in the field. This estimator is called a 'last event estimator', as it scores information only upon the last event, the absorption event. Consider the state transfer matrix

$$\mathbf{H} = \begin{pmatrix} 0.08 & 0.001 & 0.002 & 0 & 0 & 0.02 & 0.05 & 0.07 \\ 0.2 & 0.08 & 0.001 & 0 & 0 & 0.04 & 0.07 & 0.09 \\ 0.05 & 0.3 & 0.1 & 0.05 & 0.07 & 0.03 & 0 & 0.09 \\ 0 & 0.02 & 0.31 & 0.05 & 0.07 & 0.04 & 0.02 & 0 \\ 0.07 & 0 & 0.033 & 0.25 & 0.1 & 0.115 & 0.02 & 0 \\ 0 & 0 & 0.025 & 0.07 & 0.6 & 0.05 & 0.02 & 0.02 \\ 0 & 0.04 & 0.025 & 0.05 & 0.06 & 0.65 & 0.32 & 0 \\ 0 & 0.06 & 0 & 0.05 & 0 & 0.05 & 0.12 & 0.12 \end{pmatrix}$$

Summing the columns yields the absorption vector

$$\vec{\mathbf{p}} = \begin{pmatrix} 0.6 \\ 0.499 \\ 0.504 \\ 0.475 \\ 0.1 \\ 0.005 \\ 0.38 \\ 0.53 \end{pmatrix}.$$

The source vector is taken to have one unit generated in state 1, per time interval, and zero in all other states. The calculation results with 200 000 histories are shown in Table 6.1.1. It is worth noting that the estimates for state 5, and especially state 6, are distinctly worse than all the others. This is a result of the small absorption probability in these states of 0.1 and 0.005, respectively. Since information is obtained only upon absorption in a state, a small absorption probability results in a small number of histories contributing to the estimate, and thereby a high variance. This is another reflection of the rare events phenomenon; in this case the rare event is being absorbed in the state.

This rare event phenomenon can be improved by an estimator which accumulates information upon *entry into* a state rather than upon being *absorbed* in it. Consider the partial weight quantity

$$W_m = \frac{a_{i_1}}{p^1(i_1)} \frac{h_{i_2,i_1}}{p(i_1, i_2)} \frac{h_{i_3,i_2}}{p(i_2, i_3)} \cdots \frac{h_{i_m,i_{m-1}}}{p(i_{m-1}, i_m)}$$

Table 6.1.1 Event rates for the 8 state system

State	Event rate	PRSD (%)
1	1.1016	0.16
2	0.2573	0.58
3	0.1705	0.73
4	0.07805	1.1
5	0.1265	1.97
6	0.088	10.6
7	0.1284	0.98
8	0.0496	1.4

defined on the mth event in the history, and define the estimator

$$\eta_{j_0} = \sum_{m=1}^{k} W_m(i_1, \ldots, i_m)\delta_{i_m, j_0}, \tag{6.1.9}$$

with k being the size of the history. The average value of this estimator is given by

$$\langle \eta_{j_0} \rangle = \sum_{k=1}^{\infty}\sum_{i_1=1}^{n}\sum_{i_2=1}^{n}\sum_{i_3=1}^{n}\cdots\sum_{i_k=1}^{n}\sum_{m=1}^{k} W_m(i_1, \ldots, i_m)\delta_{i_m, j_0} p^1(i_1)p(i_1, i_2)p(i_2, i_3)\cdots p(i_{k-1}, i_k)p(i_k).$$

The summation over m and k may be interchanged to yield

$$\langle \eta_{j_0} \rangle = \sum_{m=1}^{\infty}\sum_{k=m}^{\infty}\sum_{i_1=1}^{n}\sum_{i_2=1}^{n}\sum_{i_3=1}^{n}\cdots\sum_{i_k=1}^{n} W_m(i_1, \ldots, i_m)\delta_{i_m, j_0} p^1(i_1)p(i_1, i_2)p(i_2, i_3)\cdots p(i_{k-1}, i_k)p(i_k)$$

$$= \sum_{m=1}^{\infty}\sum_{i_1=1}^{n}\sum_{i_2=1}^{n}\sum_{i_3=1}^{n}\cdots\sum_{i_k=1}^{n} W_m(i_1, \ldots, i_m)\delta_{i_m, j_0} p^1(i_1)p(i_1, i_2)\cdots p(i_{m-1}, i_m)$$

$$\times \left\{ p_{i_m} + \sum_{i_{m+1}=1}^{n} p_{i_m, i_{m+1}} p_{i_{m+1}} + \sum_{i_{m+1}=1}^{n}\sum_{i_{m+2}=1}^{n} p_{i_m, i_{m+1}} p_{i_{m+1}, i_{m+2}} p_{i_{m+2}} + \cdots \right\}.$$

The term in the curly brackets is the probability that the system will be absorbed in any step following and including the mth step. This, however, is guaranteed to be unity by virtue of the fact that the spectral radius of the matrix $\mathbf{H}$, and therefore also the matrix $p(i, j)$, is smaller than 1. Thus, replacing the term in the curly brackets by unity yields

$$\langle \eta_{j_0} \rangle = \sum_{m=1}^{\infty}\sum_{i_1=1}^{n}\sum_{i_2=1}^{n}\sum_{i_3=1}^{n}\cdots\sum_{i_k=1}^{n} W_m(i_1, \ldots, i_m)\delta_{i_m, j_0} p^1(i_1)p(i_1, i_2)p(i_2, i_3)\cdots p(i_{m-1}, i_m).$$

Substituting W_m into the above sum yields an expression identical to (6.1.6), and hence the estimator of expression (6.1.9) is an unbiased estimator of the solution. For the present case, and assuming a unit source strength, it takes the simple form

$$\eta_{j_0} = \sum_{m=1}^{k} \delta_{i_m, j_0}. \tag{6.1.10}$$

Thus, accumulating unity each time, an event resulting in entry into the state j_0 occurs. The estimator (6.1.9) differs from the last event estimator in that it collects information

Table 6.1.2 Results of the event type estimator for the event rates

State	Event rate	PRSD (%)
1	1.10	0.067
2	0.2575	0.43
3	0.1705	0.59
4	0.0771	0.89
5	0.1276	0.77
6	0.094	0.87
7	0.1280	0.87
8	0.0496	1.2

not only in the last event, but at every event. It is thus called an *event* (or collision) *type estimator*. This estimator is expected to produce better results in 'rare event' situations when the probability of absorption in a given state is very small. Results obtained with the event estimator (6.1.10) for the above problem are shown in Table 6.1.2.

Comparison with Table 6.1.1 indicates that the event estimator yields a smaller variance for all states. Particularly impressive are the results for state 6, which now has the same PRSD as other states. The computer time required to process a history is about the same for the two estimators, so that the FOMs can be compared on the basis of the PRSD. The benefit ratio for the sixth state is

$$B = \frac{\text{FOM}_{\text{evt}}}{\text{FOT}_{\text{lst}}} = \left(\frac{\text{PRSD}_{\text{lst}}}{\text{PRSD}_{\text{evt}}}\right)^2 = 148.4.$$

The collision type estimator is about 150 times more efficient than the last event estimator. (For further reading about the origin of estimators (6.1.5) and (6.1.9) see Wasow, W. (1952). A note on the inversion of matrices. *Math. Tables Aids to Comp.*, **6**, 78 and Forsyth, G. E. and Leibler, R. A. (1950). Matrix inversion by a Monte Carlo method. *Math. Tables Aids to Comp.*, **4**, 127.) These estimators were, in fact, sought in connection with the motion of a neutral particle (a neutron) in the energy domain. Dividing the energy domain into a set of intervals (energy groups), neutrons are generated from a neutron source into any one of these groups with a known distribution. Upon collision (event) with a nucleus, the neutron may lose energy, thereby transferring into a lower energy group. It may remain in its energy group, and sometimes even gain energy from the interaction, and go into a higher group. It may also be absorbed and disappear in the event. Thus, a set of equations of the form of equation (6.1.1), with a matrix similar to that of Example 6.1.2, can be established (see Spanier, J. and Gelbard, E. M. (1969). *Monte Carlo Principles and Neutron Transport Problems*, Addison-Wesley). *The transport of the neutron in the energy space and that of the system in the degraded states space are analogous.*

6.2 The General Random Walk and Transport Equation of a System

The situation in system engineering bears close resemblance to the situation in the transport of neutral particles. This point was already mentioned, and will now be further discussed. It is quite natural to look at a neutron as a particle moving in its space from event to event. It will be very useful to take a similar view of systems. The transport theory of neutrons is well established, and was developed over the past 60 years with impressive success. Understanding the similarities (as well as the differences) beween the two phenomena may make it possible to draw on the many ideas developed for neutral particle transport and 'borrow' them for use in system engineering. Example 6.1.2 is a demonstration of this process. A neutral particle moves in space in straight lines until it hits a nucleus. The collision point of the particle is identified by the coordinates of the collision point, namely the position vector $\vec{\mathbf{r}}$, the direction with which the particle enters the collision $\vec{\Omega}$ and the energy of the particle, E. It is then said that the particle entered a collision at the phase space point $\vec{\mathbf{P}} = (\vec{\mathbf{r}}, \vec{\Omega}, E)$. As a result of the collision, a nuclear reaction occurs, resulting in the particle either being absorbed (which is the analog of the termination of a system in Example 6.1.2), or being scattered and exiting the reaction at the same spatial point with a different direction and energy. Thus, the particle exits the event with a phase space vector $\vec{\mathbf{P}}' = (\vec{\mathbf{r}}, \vec{\Omega}', E')$. Following the collision, the particle proceeds in a straight line from the point $\vec{\mathbf{r}}$ in direction $\vec{\Omega}'$ until the next event occurs. That event will happen at a spatial point $\vec{\mathbf{r}}'$, which is readily expressed as $\vec{\mathbf{r}}' = \vec{\mathbf{r}} + \ell\vec{\Omega}$, where ℓ is the spatial distance between the events. The particle enters the next event at a phase space position $\vec{\mathbf{P}}' = (\vec{\mathbf{r}}', \vec{\Omega}', E')$. This process continues until the particle is either absorbed or escapes the boundaries of the medium. The process that controls the transfer from one event point $\vec{\mathbf{P}}$ to the next one, $\vec{\mathbf{P}}'$, encapsulates the theory of neutral particles transport. This process can be separated into two distinct parts. One is the collision itself, namely, the relationship between the incoming and outgoing direction and energy. For this purpose, the collision kernel $C(\vec{\mathbf{r}}; \vec{\Omega}, E \to \Omega', E')$ is defined as the probability density that a particle entering a collision at point $\vec{\mathbf{r}}$ with direction $\vec{\Omega}$ and energy E will emerge from the collision in direction $\vec{\Omega}'$ and energy E'. Let Σ_j be the probability for a reaction of type j to occur (per unit length in the specific media in which the particle is traveling). Σ_j is called the *reaction cross-section*. Let Σ_t be the sum of the cross-sections for all reactions, also referred to as the *total* cross-section. Let $f_j(\theta', \phi', E')$ be the PDF of the polar angle, the azimuthal angle and the energy of the particle emerging from a reaction of this type. Then the collision kernel can be expressed in the form

$$C(\vec{\mathbf{r}}; \vec{\Omega}, E \to \vec{\Omega}', E') = \sum_j \frac{\Sigma_j}{\Sigma_t} f(\theta', \phi', E'). \tag{6.2.1}$$

Also, note that the integral $q(\vec{\mathbf{P}}) = \int_{\vec{\Omega}} \int_{E'} C(\vec{\mathbf{r}}; \vec{\Omega}, E \to \vec{\Omega}', E') \sin\theta' d\theta' d\phi' dE'$ yields the number of particles emerging from a single reaction at $\vec{\mathbf{P}}$, thus $p(\vec{\mathbf{P}}) = 1 - q(\vec{\mathbf{P}})$ is the absorption probability in a reaction at $\vec{\mathbf{P}}$. θ' and ϕ' are the polar and azimuthal angles in a coordinate system relative to the direction of the incoming particle, and $d\vec{\Omega}' = \sin\theta' d\theta' d\phi'$ is an infinitesimal solid angle element.

The second kernel defines the free flight of the particle emerging from the collision onto the next event. This kernel is denoted by $T(\vec{\Omega}', E'; \vec{\mathbf{r}} \to \vec{\mathbf{r}}')$. It is defined as the probability density that a particle emerging at point $\vec{\mathbf{r}}$ in direction $\vec{\Omega}'$ with energy E' will have a next collision at $\vec{\mathbf{r}}'$ (per unit volume). The product of collision kernel and the free flight kernel

$$K(\vec{\mathbf{P}} \to \vec{\mathbf{P}}') = C(\vec{\mathbf{r}}; \vec{\Omega}, E \to \vec{\Omega}', E') \times T(\vec{\Omega}', E'; \vec{\mathbf{r}} \to \vec{\mathbf{r}}') \tag{6.2.2}$$

is called the transport kernel. It defines the probability density that a particle entering a collision at $\vec{\mathbf{P}}$ will enter its next collision at $\vec{\mathbf{P}}'$. The transport kernel contains all the physical knowledge available on the process. The collision density, or event density, $\psi(\vec{\mathbf{P}})$ is the number of particles (systems) entering a collision at $\vec{\mathbf{P}}$ per unit volume in the phase space (per unit solid angle per unit volume and unit energy).

Let $d\vec{\mathbf{P}}'$ be an infinitesimal volume element in phase space. The contribution of collisions in this element to the collision density at $\vec{\mathbf{P}}$ is given, by the above definitions, as $\psi(\vec{\mathbf{P}}')K(\vec{\mathbf{r}}', \vec{\Omega}', E' \to \vec{\mathbf{r}}, \vec{\Omega}, E)$. Integrating over the contributions from all points in the phase space yields

$$\psi(\vec{\mathbf{P}}) = S(\vec{\mathbf{P}}') + \int_{\vec{\mathbf{P}}'} \psi(\vec{\mathbf{P}}')K(\vec{\mathbf{r}}', \vec{\Omega}', E' \to \vec{\mathbf{r}}, \vec{\Omega}, E)\, d\vec{\mathbf{P}}', \tag{6.2.3}$$

where $S(\vec{\mathbf{P}})$ is a source term of first collisions in the medium. This is the Boltzmann Transport equation, and it is identical to equation (6.5a). It is an equation common to all random walk processes. It has a simple meaning of a balance equation between events in consecutive points in space, and is the basic, and only, equation to be solved for the analysis of neutral particles behavior in medium (for further details on the neutron transport in this context see, for example, Dubi, A. (1986). Monte Carlo calculations for nuclear reactors. In: Y. Ronen (ed), *Handbook of Nuclear Reactors Calculations*, CRC Press). An analytic solution for the above equation only exists for the trivial case of a single isotropic point source in an infinite homogeneous media (see Case, K. M., De Hoffmann, F. and Placzek, G. (1953). *Introduction to the Theory of Neutron Diffusion, Vol. 1*, Los Alamos Scientific Laboratory, Los Alamos, NM). Numerical solutions exist for one and two dimensional approximations, and a full heterogeneous six dimensional solution requires the application of the Monte Carlo method.

The behavior of systems will now be discussed in a form analogous to the above description, to enhance and introduce the perception of a system as a particle moving in a phase space.

Consider a system with n components. Let b_i indicate the state of the ith component and τ_i indicate the time point at which the component entered its corresponding state. The vectors $\vec{\mathbf{B}} = (b_1, b_2, \ldots, b_n)$ and $\vec{\ } = (\tau_1, \tau_2, \ldots, \tau_n)$ are the state vector and time vector of the system, and $\vec{\mathbf{P}} = (\vec{\mathbf{B}}, \vec{\ }; t)$ is the most general phase space vector of the system. It indicates that the system has entered the state $\vec{\mathbf{B}}$ at time t, where $t = \max(\tau_1, \tau_2, \ldots, \tau_n)$. The time of entry of the system into the state is the time at which the last component entered its corresponding state. It is also convenient to denote the phase space vector by the simple notation $\vec{\mathbf{P}} = (\vec{\mathbf{B}}, t)$, where the vector $\vec{\ }$ is left out for simplicity. In models in which the future depends only upon the point of entry of the system into a state (and not on the details of the past entries of other components into their states) the simple notation is sufficient for all purposes. The event density $\psi(\vec{\mathbf{B}}, t)$ is the number of entries, per unit time, into state $\vec{\mathbf{B}}$ at time t. If the time of entry of every component is of importance, then the very detailed density $\psi(\vec{\mathbf{B}}, \vec{\ }; t)$ must be considered and $\psi(\vec{\mathbf{B}}, t)$ is obtained by integration of $\psi(\vec{\mathbf{B}}, \vec{\ }; t)$ over the range of possible past entry times. The kernel $T(\vec{\mathbf{B}}'; t' \to t)$ is the free flight kernel. It is defined as the probability density that a system entering state $\vec{\mathbf{B}}'$ at time t' will enter the next event at time t per unit time. This kernel defines the transfer of system through time with its state unchanged, just as the neutron kernel defines the transfer of the particle through the spatial coordinate with its angle and energy unchanged. The collision kernel is now denoted by $C(t; \vec{\mathbf{B}}' \rightarrow \vec{\mathbf{B}})$ and defined as the probability that upon entering an event at t in state $\vec{\mathbf{B}}'$ the system will transfer into a new state $\vec{\mathbf{B}}$.

The transport kernel is now $K(\vec{\mathbf{B}}'; t' \rightarrow \vec{\mathbf{B}}; t) = T(\vec{\mathbf{B}}'; t' \rightarrow t)C(t; \vec{\mathbf{B}}' \rightarrow \vec{\mathbf{B}})$, namely, the probability density that a system that entered state $\vec{\mathbf{B}}'$ at time t' will enter in the next event into state $\vec{\mathbf{B}}$ at time t. It should be noted that the inclusion of the time vector $\vec{\ }$ in the kernels definition is similar to the statement that the transport kernel $K(\vec{\mathbf{P}} \rightarrow \vec{\mathbf{P}}')$ is not only a function of the present state $\vec{\mathbf{P}}$, but also of the states incurred by the system in the past. In models in which the transport kernel depends only upon the time of entry into a state the transport equation takes the form

$$\psi(\vec{\mathbf{B}}, t) = S(\vec{\mathbf{B}}, t) + \sum_{\vec{\mathbf{B}}'} \int_{t'} \psi(\vec{\mathbf{B}}', t')K(\vec{\mathbf{B}}', t' \rightarrow \vec{\mathbf{B}}, t)\, dt'. \tag{6.2.4}$$

This equation can be derived with similar arguments as equation (6.2.3), or directly from equation (6.5a). The explicit form of this equation for the case of a single component with n states was discussed in some detail in Section 2.2.2, where a simple explicit form of the equation was derived from first principles (equation (2.2.23)). In Section 2.2.3 it was shown that the simple n state equation reduces to the Markov equation if all the transition rates are constants, and in Section 2.2.4 the concept of age was considered for the two state equation. It is through the above transport process and random walk that the general form of the equation can be obtained. In the more general case where the transport kernel depends upon past events, the transport equation takes the, rather implicit, form

$$\psi(\vec{\mathbf{B}}, \vec{\ }; t) = S(\vec{\mathbf{B}}, \vec{\ }; t) + \sum_{\vec{\mathbf{B}}'} \int_{\vec{\ }} \psi(\vec{\mathbf{B}}', \vec{\ }; t')K(\vec{\mathbf{B}}', \vec{\ }; t' \rightarrow \vec{\mathbf{B}}, \vec{\ }; t)\, d\vec{\ }. \tag{6.2.5}$$

The integration over the space $\vec{\ }$ is model dependent, and depends upon the states $\vec{\mathbf{B}}'$ and $\vec{\mathbf{B}}$.

The system may now be viewed as a particle moving in a phase space $\vec{\mathbf{P}} = (\vec{\mathbf{B}}, t)$. Its transport is composed of a sequence of events in which it enters a state $\vec{\mathbf{P}}' = (\vec{\mathbf{B}}', t')$ followed by a segment of free flight on the time axis up to the next event. Just as the neutron may cross the spatial boundaries of the medium and leak out, so may the system cross the *time* of interest and leak out of the range. The similarities with the neutron transport are obvious. There are, however, a number of differences that make the system transport problem different, and more complicated.

In the case of a neutral particle, the free flight kernel is inherently based on an exponential distribution. The probability that a neutron will have a collision at distance x from its point of origin is $f(x) = \Sigma_t \mathrm{e}^{-\Sigma_t x}$, where Σ_t is the total cross-section, which depends only upon the medium and the energy. Neutrons in this sense are Markovian systems and are ageless, just like two state components with exponential failure distribution. The free flight distribution of a system, on the other hand, may be a Weibull distribution or any other complex composition of distributions related to the components comprising the system. Components may age (see Section 2.2.4) and the failure distribution may depend upon the type of repair applied. Also, as is shown in expression (6.2.1), the collision kernel of a neutron is universal; it depends only upon the nuclei of the medium and the energy. Once measured in a nuclear experiment, it can be applied whenever the same medium is considered. There is a limited number of possible nuclear reactions, such as elastic and non-elastic scattering, absorption, fission, etc., and the data may be compiled and stored for general use. The dynamics of a collision for a system are vastly different. They are partially of a physical nature (such as the formation of minute fractures due to fatigue and stress), and even then they are considerably less coherent than nuclear reactions, and partially composed of man-made rules. The contingencies applied in the case of a failure, the use of backup systems, shipment of a failed unit, type of repair applied, maintenance, etc., are part of the dynamics of the collision – it is impossible to compile a finite set of collision rules. This is another reflection of the complexity of reality. This means that, unlike equation (6.2.1), which is a general expression for the neutron collision kernel, collision kernels for systems must be built for each model separately. It is an open world situation. A neutron reaction in Oak-Ridge, Moscow, and Galaxy M41 obey exactly the same rules, the exact same system may obey completely different rules in two organizations in the same town.

Furthermore, the neutron phase space is a six-dimensional continuous space (still requiring the MC method to obtain full transport solutions). The system phase space is, in the general case, a $2 \times N$ dimensional space, N being the number of components with N continuous time coordinates and N discrete dimensions.

6.2.1 Some explicit forms of the system transport equation

In this subsection, a number of explicit forms of the transport equation are presented. In the case of a single system with n states, it has already been noted that the notation can be simplified so that $\psi_i(t)$ indicates the event rate of the ith state. The free flight kernel takes the form

$$T(\vec{\mathbf{B}}; t' \to t) = T(j; t' \to t) = f_j(t', t) = R_j(t', t) z_j(t', t), \tag{6.2.6a}$$

where $R_j(t', t)$ is the 'state reliability', i.e. the probability that a system entering state j at t' will not exit this state up to time t, and $z_j(t', t)$ is the hazard in this state. The collision kernel takes the form

$$C(t; \vec{\mathbf{B}}' \to \vec{\mathbf{B}}) = C(t;\ j \to i) = \frac{z_{j,i}(t', t)}{z_j(t', t)}, \tag{6.2.6b}$$

where $z_{j,i}(t', t)$ is the partial hazard for transfer from state j into state i. Substituting these kernels into equation (6.2.4) yields equation (2.2.23):

$$\psi_i(t) = P_{i0}\delta(t) + \sum_{\substack{j=1 \\ j \neq 1}}^{n} \int_0^t \psi_j(t') R_j(t - t') z_{ji}(t - t')\, dt'. \tag{6.2.7}$$

P_{i0} is the probability that the system starts in state i at $t = 0$.

Fixed time failure and exponential repair

In the case of two state systems, the transport equations take the form (2.2.24a) and (2.2.24b):

$$\psi_1(t) = \delta(t) + \int_0^t \psi_0(t')f_0(t-t')\,dt'; \quad \psi_0(t) = \int_0^t \psi_1(t')f_1(t-t')\,dt'.$$

Substituting $\psi_0(t)$ into $\psi_1(t)$ yields

$$\psi_1(t) = \delta(t) + \int_0^t \psi_1(t')\left[\int_{t'}^t f_1(x-t')f_0(t-x)\,dx\right]dt' = \delta(t) + \int_0^t \psi_1(t')K(t,t')\,dt'. \quad (6.2.8)$$

Consider the case in which failure occurs in fixed time intervals Δ, and repair is exponential. Thus, the failure distribution is a Dirac delta function: $f_1(t-t') = \delta[(t-t')-\Delta]$ and the repair distribution is $f_0(t-t') = \mu\,\mathrm{e}^{-\mu(t-t')}$. The kernel $K(t,t')$ takes the form

$$K(t,t') = \int_{t'}^t \delta(x-t'-\Delta)\mu\,\mathrm{e}^{-\mu(t-x)}\,dx = \begin{cases} 0 & \text{if } t-t' < \Delta \\ \mu\,\mathrm{e}^{-\mu(t-t'-\Delta)} & \text{if } t-t' < \Delta. \end{cases}$$

Substituting this result into (6.2.8), one may note that for $t \le \Delta$ the kernel is zero, and hence $\psi_1(t) = \delta(t)$. This is plausible since no failure can occur before Δ. Also, for $t > \Delta$, in the range $t' > t - \Delta$ the kernel vanishes, hence the integral yields

$$\psi_1(t) = \int_0^{t-\Delta} \psi_1(t')\mu\,\mathrm{e}^{-\mu(t-t'-\Delta)}\,dt'. \quad (6.2.8a)$$

One can now look at values of t in intervals of size Δ. In the range $0 \le t \le \Delta$, $\psi_1(t) = \delta(t)$. In the range $\Delta \le t \le 2\Delta$, the integration variable is always smaller then Δ, hence $\psi_1(t') = \delta(t')$, and the integration for that range yields $\psi_1(t) = \mu\,\mathrm{e}^{-\mu(t-\Delta)}$.

For the range $2\Delta \le t \le 3\Delta$, the integration variable ranges from $0 \le t' \le 2\Delta$, hence the integral may be written as

$$\psi_1(t) = \int_0^{\Delta} \delta(t')\mu\,\mathrm{e}^{-\mu(t-t'-\Delta)}\,dt' + \int_{\Delta}^{t-\Delta} \mu\,\mathrm{e}^{-\mu(t'-\Delta)}\,\mu\,\mathrm{e}^{-\mu(t-t'-\Delta)}\,dt',$$

where the known values of $\psi_1(t')$ are introduced at each range, yielding $\psi_1(t) = \mu\,\mathrm{e}^{-\mu(t-\Delta)} + \mu^2(t-2\Delta)\,\mathrm{e}^{-\mu(t-2\Delta)}$. Proceeding in this manner and using induction, one can readily verify that, for the general range $n\Delta \le t \le (n+1)\Delta$, the event rate takes the form

$$\psi_1(t) = \sum_{k=1}^{n} \frac{\mu^k(t-k\Delta)^{k-1}}{(k-1)!}\mathrm{e}^{-\mu(t-k\Delta)}. \quad (6.2.8b)$$

Now using equation (2.2.21), the availability can be obtained as $P_1(t) = \int_0^t \psi_1(t')R_1(t-t')\,dt'$. Now $R_1(x) = 1 - F_1(x)$, where $F_1(x)$ is the CDF of the delta function, and is thus a step function being zero for $x < \Delta$ and unity otherwise. The availability then takes the form

$$P_1(t) = \int_{t-\Delta}^{t} \sum_{k=1}^{n} \frac{\mu^k (t' - k\Delta)^{k-1}}{(k-1)!} \mathrm{e}^{-\mu(t'-k\Delta)}\, dt'.$$

Given that t is in the range $n\Delta \leq t \leq (n+1)\Delta$, the integral can be divided into two parts:

$$P_1(t) = \sum_{k=1}^{n-1} \int_{t-\Delta}^{n\Delta} \frac{\mu^k (t' - k\Delta)^{k-1}}{(k-1)!} \mathrm{e}^{-\mu(t'-k\Delta)}\, dt' + \sum_{k=1}^{n} \int_{n\Delta}^{t} \frac{\mu^k (t' - k\Delta)^{k-1}}{(k-1)!} \mathrm{e}^{-\mu(t'-k\Delta)}\, dt'.$$

Using now the transformation $x = \mu(t' - k\Delta)$ yields

$$P_1(t) = \sum_{k=1}^{n-1} \frac{1}{(k-1)!} \int_{\mu[t-(k+1)\Delta]}^{\mu(n-k)\Delta} x^{k-1}\, \mathrm{e}^{-x}\, dx + \sum_{k=1}^{n} \frac{1}{(k-1)!} \int_{\mu(n-k)\Delta}^{\mu(t-k\Delta)} x^{k-1}\, \mathrm{e}^{-x}\, dx.$$

Noting that

$$H(x, k-1) = \frac{1}{(k-1)!} \int x^{k-1}\, \mathrm{e}^{-x}\, dx = -\,\mathrm{e}^{-x} \sum_{m=0}^{k-1} \frac{x^m}{m!}$$

leads to the form

$$\begin{aligned}
P_1(t) &= \sum_{k=1}^{n-1} \{H[\mu(n-k)\Delta, k-1] - H[\mu(t-(k+1)\Delta), k-1]\} \\
&\quad + \sum_{k=1}^{n} \{H[\mu(t-k\Delta), k-1] - H[\mu(n-k\Delta), k-1]\} \\
&= \sum_{k=1}^{n-1} \{H[\mu(t-k\Delta), k-1] - H[\mu(t-(k+1)\Delta), k-1]\} \\
&\quad + H[\mu(t-n\Delta), n-1] + H[0, n-1] \\
&= \sum_{k=1}^{n-1} \{H[\mu(t-k\Delta), k-1] - H[\mu(t-(k+1)\Delta), k-1]\} \\
&\quad - \mathrm{e}^{-\mu(t-n\Delta)} \sum_{m=0}^{n-1} \frac{[\mu(t-n\Delta)]^m}{m!} + 1 \\
&= \sum_{k=0}^{n-2} H[\mu(t-(k+1)\Delta), k] - \sum_{k=1}^{n-1} H[\mu(t-(k+1)\Delta), k-1] + 1 \\
&\quad - \mathrm{e}^{-\mu(t-n\Delta)} \sum_{m=0}^{n-1} \frac{[\mu(t-n\Delta)]^m}{m!}.
\end{aligned}$$

Substituting $H(x, j)$ explicitly, and separating the term $k = 0$ in the first sum and $k = n - 1$ in the second, yields

$$P_1(t) = -\,e^{-\mu(t-\Delta)} - \sum_{k=1}^{n-2} e^{-\mu(t-(k+1))\Delta} \sum_{m=0}^{k} \frac{[\mu(t-(k+1)\Delta)]^m}{m!}$$
$$+ \sum_{k=1}^{n-2} e^{-\mu(t-(k+1))\Delta} \sum_{m=0}^{k-1} \frac{[\mu(t-(k+1)\Delta)]^m}{m!}$$
$$+ e^{-\mu(t-n\Delta)} \sum_{m=0}^{n-2} \frac{[\mu(t-n\Delta)]^m}{m!} + 1 - e^{-\mu(t-n\Delta)} \sum_{m=0}^{n-1} \frac{[\mu(t-n\Delta)]^m}{m!}.$$

Note that the second and third terms cancel each other, except for the term $m = k$ in the second term. Also, the fourth and sixth terms subtract and leave only the $m = n - 1$ term, thus the expression reads

$$P_1(t) = -\,e^{-\mu(t-\Delta)} - \sum_{k=1}^{n-2} e^{-\mu(t-(k+1))\Delta} \frac{[\mu(t-(k+1)\Delta)]^k}{k!} + 1 - e^{-\mu(t-n\Delta)} \frac{[\mu(t-n\Delta)]^{n-1}}{(n-1)!},$$

which yields the final result

$$P_1(t) = 1 - \sum_{k=0}^{n-1} e^{-\mu(t-(k+1))\Delta} \frac{[\mu(t-(k+1)\Delta)]^k}{k!}. \tag{6.2.9}$$

The case of exponential failure time and fixed repair time can be easily derived from (6.2.9) by shifting time $t \to t + \Delta$ and replacing μ by λ, hence

$$P_1(t) = 1 - \sum_{k=0}^{n-1} e^{-\lambda(t-k\Delta)} \frac{[\lambda(t-k\Delta)]^k}{k!} \tag{6.2.9a}$$

(For a derivation of a similar relation through a Laplace transform from a relation not stemming from the transport equation, see Barlow, R. E. and Proschan, F. (1965). *Mathematical Theory of Reliability*, John Wiley & Sons).

6.2.2 Systems with interacting components

In the next section the explicit form of the transport equation for a more elaborate case is discussed.

Two components with repair dependence Two pumps are constructed in the cooling cycle of a process. The first one is the main pump, and the second is for support and backup. Both pumps are active in regular operation of the system. Structure and piping considerations force a situation in which the first pump can be repaired while the second one is operating, but the second pump cannot be repaired unless the first one is non-operational. Thus, upon failure of the second pump no repair is applied and it remains failed (state -2) until a failure occurs in the first pump (or until periodic maintenance is applied, at which time both pumps are shut). Also, if both pumps are failed the repair process is sequential, and the second pump is always repaired first. The states and state transitions in the system are portrayed in the Figure 6.2.1. Unlike a Markov diagram, this diagram does not include transition rates, but it includes the detailed relation between the time of entries of components to their respective states.

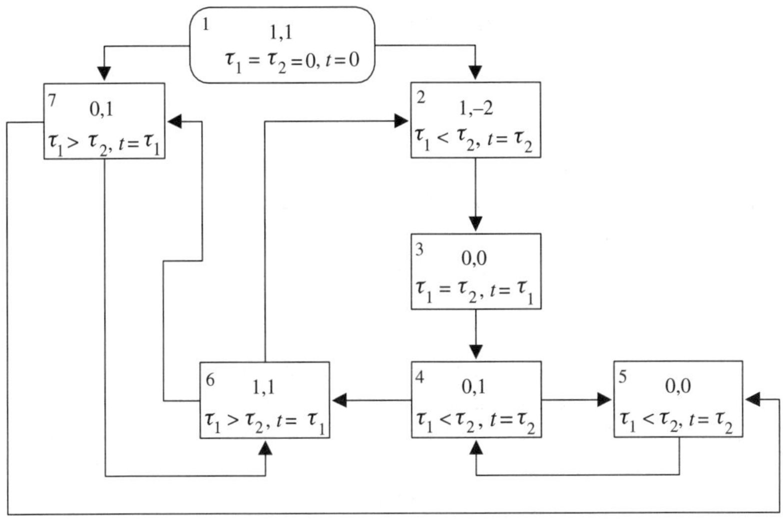

Figure 6.2.1 Non-Markovian state transition diagram

At $t = 0$ both components are operational. If component 2 fails before 1 the system transfers into state 2; the state of component 2 is denoted by -2, indicating a failed and non-repairable state. From state 2 the system can move only to state 3, in which both components are failed and repairable. Note that as component 1 moves into state 0, component 2 moves from state -2 into state 0, thus the time of entry into the current state is identical for both components, although component 1 triggers the event. In this state, components are repaired sequentially and component 2 is always repaired first, since 1 cannot be made operational while 2 is under repair, thus moving into state 4. From this state the system can go either into state 5, in which component 2 fails before 1 is repaired. State 5 differs from state 3 in the order of entry times (which makes a difference in the free flight kernel). State 6 is reached if component 1 is repaired. From state 5 one can move only back into state 4 by a repair of component 2. From state 6 a failure of component 2 will cause a transfer back into state 2, while a failure of component 1 will transfer the system into state 7, in which component 1 is in a failed repairable state. Failure of component 2 in state 7 will cause a transfer into state 5, in which again component 2 must be repaired first. Figure 6.2.1 is a necessary step in the process of inducing the transport equations of the system.

The general transport kernel Identifying the system transport equations requires some general considerations of the transport kernel. Let $h_k(b_k, t)$ denote the PDF of component k in state b_k, $H_k(b_k, t)$ the CDF and $z_k(b_k, t)$ the corresponding hazard. Let the state be denoted by $(\vec{\mathbf{B}}'; \vec{\tau}'; t') = (b'_1, \ldots, b'_n; \tau'_1, \ldots, \tau'_n; t')$, and such that $t' = \tau'_m$, i.e. the system entered into that state at time t' by an entry of the mth component into its respective state. The free flight kernel is then given by

$$T(\vec{\mathbf{B}}'; \vec{\tau}'; t' \to t) = \left[\prod_{j=1}^{n} \frac{1 - H(b'_j, t - \tau'_j)}{1 - H(b'_j, t' - \tau'_j)} \right] \times Z_s(t), \tag{6.2.10}$$

where $Z_s(t)$ is the sum of the hazard functions of the components, namely

$$Z_s(t) = \sum_{j=1}^{n} z_j(b_j'; t - \tau_j'). \tag{6.2.11}$$

$z_j(b_j'; t - \tau_j')$ is the hazard or rate function of the jth component in state b_j'. The hazard is in turn the sum of the partial hazard of all the states that can be reached from b_j' (see expression (2.1.1)), namely

$$z_j(b_j'; t - \tau_j') = \sum_{b_j''} z_j(b_j' \to b_j''; t - \tau_j'). \tag{6.2.11a}$$

The collision kernel assumes that a system state transfer involves a single transfer of one component; other immediate changes may follow, but it is a single component transfer that triggers the change. The collision kernel for a transfer from state $\vec{\mathbf{B}}'$ into state $\vec{\mathbf{B}}$ in which the ith component has transferred from state b_i' into state b_i is given by the ratio of the partial hazard for this transfer to the total hazard of the system. The collision kernel must also reflect the transfer of the vector $\vec{\tau}'$ into $\vec{\tau}$. For all the components that do not change their state, the entry time does not change, i.e. $\tau_j = \tau_j'$. This is reflected by introducing a delta function of the form $\delta(\tau_j - \tau_j')$ which, upon integration over τ_j', will merely replace τ_j' by τ_j. For the component that undergoes the transfer, $\tau_i = t$, hence a delta function $\delta(\tau_i - t)$ is introduced in the collision kernel. The same applies to any other component that enters a new state as a result of interaction with the 'trigger' component (such as component 2 in state 3 in Figure 6.2.1). This delta function will remain in the event density function. With these assumptions, the collision kernel takes the form

$$C(t; \vec{\mathbf{B}}' \to \vec{\mathbf{B}}) = \frac{z_i(b_i' \to b_i; t - \tau_i')\delta(\tau_i - t)}{Z_s(t)} \prod_{\substack{j=1 \\ j \neq i}}^{n} \delta(\tau_j' - \tau_j). \tag{6.2.12}$$

If the component transfer can be made only into a single state, then $z_i(b_i' \to b_i; t - \tau_i') = z_i(b_i'; t - \tau_i')$ and the transport kernel takes the form

$$\begin{aligned} K(\vec{\mathbf{B}}'; \vec{\tau}'; t' \to \vec{\mathbf{B}}; \vec{\tau}; t) &= \left[\prod_{j=1}^{n} \frac{1 - H_j(b_j', t - \tau_j')}{1 - H_j(b_j', t' - \tau_j')}\right] \left[\prod_{\substack{j=1 \\ j \neq i}}^{n} \delta(\tau_j - \tau_j')\right] \\ &\quad \times \delta(\tau_i - t) \times z_i(b_i', t - \tau_i') \\ &= \left[\prod_{\substack{j=1 \\ j \neq i}}^{n} \frac{1 - H_j(b_j', t - \tau_j')}{1 - H_j(b_j', t' - \tau_j')} \delta(\tau_j - \tau_j')\right] \\ &\quad \times \delta(\tau_i - t) \frac{h_i(b_i', t - \tau_i')}{1 - H_i(b_i', t' - \tau_i')}. \end{aligned} \tag{6.2.13}$$

In (6.2.13) it is assumed that only one component (the trigger component) changes its state in the event. If, however, the change in component i results in an immediate change in the state of another component, say m, then the delta function $\delta(\tau_m - \tau_m')$ should be replaced by $\delta(\tau_m - t)$.

With the above definitions, the transport equations for the system of Figure 6.2.1 take the following explicit form. The event density for the initial state, (1), is simply a delta function of the entry times of the two components, namely

$$\psi_1(1,1;\tau_1=\tau_2=t)=\delta(\tau_1)\delta(\tau_2). \tag{6.2.14a}$$

The equation for the event density of the second state is constructed as the sum of the contributions from state (1) and state (6). Substituting the explicit expressions of the transport kernel of expression (6.2.12) into equation (6.2.5) yields

$$\begin{aligned}
&\psi_2(1,-2;\tau_1<\tau_2=t)\\
&=\int_0^t\int_0^t d\tau_1'\,d\tau_2'\,\delta(\tau_1')\delta(\tau_2')h_2(1,t-\tau_2')\times[1-H_1(1,t-\tau_1')]\delta(\tau_1'-\tau_1)\delta(\tau_2-t)\\
&\quad+\int_0^t d\tau_1'\int_0^{\tau_1'} d\tau_2'\,\psi_6(1,1;\tau_1'>\tau_2',t')\delta(\tau_1'-\tau_1)\delta(\tau_2-t)[1-H_1(1,t-\tau_1')]\\
&\quad\times\frac{h_2(1,t-\tau_2')}{[1-H_2(1,\tau_1'-\tau_2')]}\\
&=\delta(\tau_1)\delta(\tau_2-t)h_2(1,t)[1-H_1(1,t)]+\delta(\tau_2-t)[1-H_1(1,t-\tau_1)]\\
&\quad\times\int_0^{\tau_1} d\tau_2'\,\psi_6(1,1;\tau_1>\tau_2',t')\frac{h_2(1,t-\tau_2')}{[1-H_2(1,\tau_1-\tau_2')]}.
\end{aligned} \tag{6.2.14b}$$

The transfer process into state number 3 from state 2 is demonstrated in Figure 6.2.2. This transfer yields the relation

$$\begin{aligned}
\psi_3(0,0;\tau_1=\tau_2=t)&=\delta(\tau_1-t)\delta(\tau_2-t)\int_0^t d\tau_2'\int_0^{\tau_2'} d\tau_1'\,\psi_2(1,-2;\tau_1'<\tau_2',t')\\
&\quad\times[1-H_2(-2,t-\tau_2')]\frac{h_1(1,t-\tau_1')}{[1-H_1(1,\tau_2'-\tau_1')]}.
\end{aligned} \tag{6.2.14c}$$

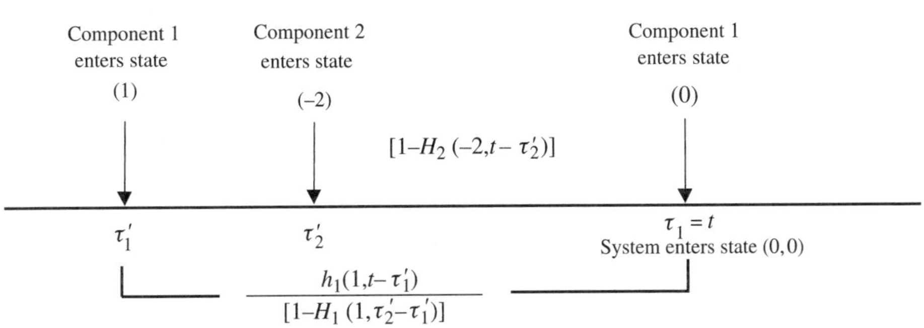

Figure 6.2.2 Transfer from state 2 into state 3

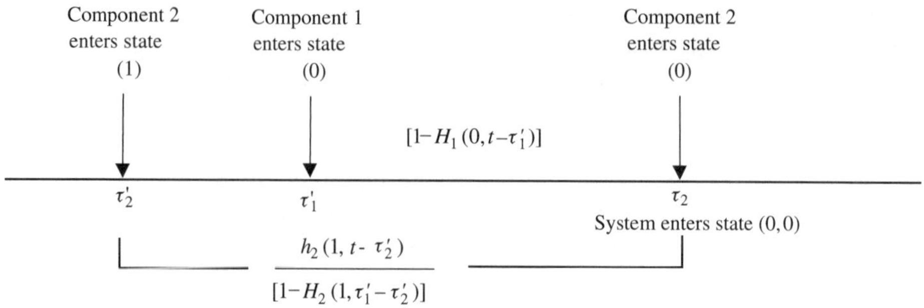

Figure 6.2.3 Transfer from state 7 into state 5

In the above equation, $[1 - H_2(-2, t - \tau_2')] \equiv 1$, since the failed pump remains in that state until the first pump fails. The relation between state 4 and states 3 and 5 is given by

$$\psi_4(0,1;\tau_1 < \tau_2, t) = \delta(\tau_2 - t)\psi_3(0,0;\tau_1,\tau_2)h_2(1, t - \tau_1)$$
$$+ \delta(\tau_2 - t)\int_0^t d\tau'_2 \psi_5(0,0,\tau_1 < \tau_2', t' = \tau_2')h_2(0, t - \tau_2')\frac{1 - H_1(0, t - \tau_1)}{1 - H_1(0, \tau_2' - \tau_1)}.$$

Consider now the contributions to state 5. The transfer from state 7 into state 5 is displayed in Figure 6.2.3. This results in the contribution

$$\int_0^t d\tau_1' \int_0^{\tau_1'} d\tau_2' \, \psi_7(0,1;\; \tau_1' > \tau_2', t')\delta(\tau_1' - \tau_1)\delta(\tau_2 - t)[1 - H_1(0, t - \tau_1')]\frac{h_2(1, t - \tau_2')}{[1 - H_2(1, \tau_1' - \tau_2')]},$$

which, upon integration over τ_1' and taking account of the delta function, yields

$$\int_0^{\tau_1} d\tau_2' \, \psi_7(0,1;\; \tau_1, \tau_2')\delta(\tau_2 - t)[1 - H_1(0, t - \tau_1)]\frac{h_2(1, t - \tau_2')}{[1 - H_2(\tau_1 - \tau_2')]},$$

where the inequality sign in the arguments of ψ_7 has been removed, since the integration limits guarantee that $\tau_2' < \tau_1$. Similarly, the contribution from state 4 to state 5 is schematically shown in Figure 6.2.4.

$$\int_0^t d\tau_1' \int_{\tau_1'}^t d\tau_2' \, \psi_4(0,1;\tau_1' < \tau_2', t')\delta(\tau_1' - \tau_1)\delta(\tau_2 - t)h_2(1, t - \tau_2')\frac{[1 - H_1(0, t - \tau_1')]}{[1 - H_1(0, \tau_2' - \tau_1')]},$$

which, upon integration over τ_1' and taking account of the delta function, yields

$$\int_{\tau_1}^t d\tau_2' \, \psi_4(0,1;\; \tau_1, \tau_2')\delta(\tau_2 - t)h_2(1, t - \tau_2')\frac{[1 - H_1(0, t - \tau_1)]}{[1 - H_1(\tau_2' - \tau_1)]}.$$

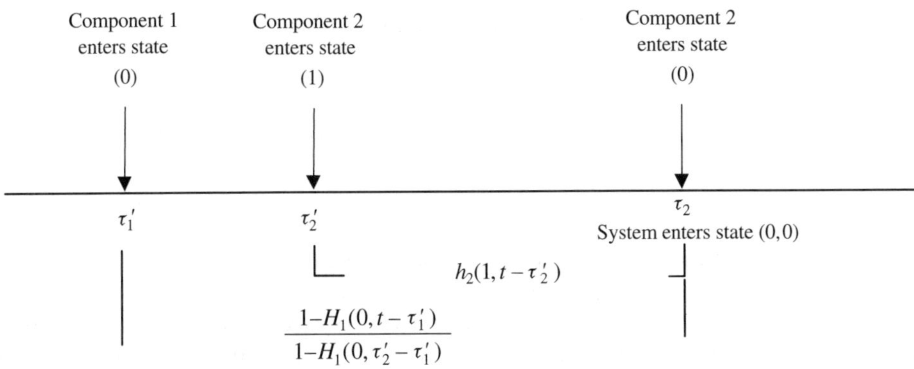

Figure 6.2.4 Transfer from state 4 into state 5

The equation for state 5 is the sum of the above two contributions. Likewise, the equations for states 6 and 7 take the form

$$\psi_6(1,1,\tau_1>\tau_2,t=\tau_1)$$
$$=\left\{\int_0^{\tau_2} d\tau_1'\,\psi_4(0,1,\tau_1'<\tau_2)[1-H_2(1,t-\tau_2)]\frac{h_1(1,t-\tau_1')}{[1-H_1(1,\tau_2-\tau_1')]}\right.$$
$$\left.+\int_{\tau_2}^{t} d\tau_1'\,\psi_7(0,1,\tau_1'>\tau_2)h_1(1,t-\tau_1')\frac{1-H_2(1,t-\tau_2)}{[1-H_2(1,\tau_1'-\tau_2)]}\right\}\delta(\tau_1-t), \qquad (6.2.14e)$$

and

$$\psi_7(0,1,\tau_1>\tau_2,t=\tau_1)=\delta(\tau_2)\delta(\tau_1-t)h_1(1,t)[1-H_2(1,t)]$$
$$+\int_{\tau_2}^{t} d\tau_1'\,\psi_6(1,1,\tau_1'>\tau_2)h_1(1,t-\tau_1')\frac{[1-H_2(1,t-\tau_2)]}{[1-H_2(1,\tau_1'-\tau_2)]}\delta(\tau_1-t). \qquad (6.2.14f)$$

The availability of the system is obtained by extension of equation (2.2.21a), namely, integrating over past entries into all the operational states folded with their probability to remain in the state up to the time of interest. This takes the general form

$$A(t)=\sum_{\Gamma(n)}\int_0^{\chi(\tau_1)} d\tau_1\int_0^{\chi(\tau_2)} d\tau_2\,\psi_n(k_1,k_2,\tau_1,\tau_2)\frac{[1-H_1(k_1,t-\tau_1)]}{[1-H_1(k_1,\tau_L-\tau_1)]}\frac{[1-H_2(k_2,t-\tau_2)]}{[1-H_2(k_2,\tau_L-\tau_2)]}, \qquad (6.2.15)$$

where $\Gamma(n)$ is the set of all values of n for which the system is operational, k_i is the state of the ith component, τ_L is the time of entry into the state, and

$$\chi(\tau)=\begin{cases}\tau_j & \text{if } \tau_i<\tau_L,\ j=1 \text{ if } i=2;\ j=2 \text{ if } i=1\\ t & \text{if } \tau_i=\tau_L.\end{cases}$$

It is worth noting that the first term in (6.3.15) involves the state at $t = 0$. Using (6.2.14a), this takes the form

$$\int_0^t d\tau_1 \int_0^t d\tau_2\, \psi_1(1,1,\tau_1,\tau_2)[1 - H_1(1, t-\tau_1)][1 - H_2(1, t-\tau_2)]$$
$$= [1 - H_1(1,t)][1 - H_2(1,t)].$$

This term can be compared to the zero term of expression (2.2.2).

The above example and, in particular, equations (6.2.10)–(6.2.13) together with the non-Markovian diagram form, provide a recipe for the explicit construction of general system transport equations. It is difficult not to express some delight at the beauty in which the Dirac delta functions set order in the above equations so that everything falls into place. It may be no coincidence that it was Dirac to have been quoted as saying "It is so beautiful, I wish it was also correct".

6.2.3 The random walk process and estimators

The realization of the system random walk process takes the following steps:

History generation procedure – 6.2.1

1. Set $i = 0$ (step number), set initial phase space state
$$\vec{\mathbf{P}}_i = (b_1^i, b_2^i, \ldots, b_n^i, \tau_1^i, \tau_2^i, \ldots, \tau_n^i) \quad (\tau_j^i = 0; b_j^i = 1\ \forall j),$$
with $\tau_j^i = 0$ and $b_j^i = 1$ for $j = 1, \ldots, n$. This is the state of the system at $t = 0$.
2. Sample next event time t_{i+1} from the free flight distribution (6.2.10)
$$T(\vec{\mathbf{B}}_i; \vec{\tau}_i; t_i \to t_{i+1}) = \left[\prod_{j=1}^{n} \frac{1 - H(b_j^i, t_{i+1} - \tau_j^i)}{1 - H(b_j^i, t_i - \tau_j^i)}\right] \times Z_s(t_{i+1}).$$
3. If $t_{i+1} > T_{\max}$, terminate the history; otherwise proceed to step 4.
4. Sample trigger component, i.e. the component to which the current change occurs. The probability that component j is the trigger is given by (see (6.2.12)).
$$P_j = \frac{z_j(b_j^i; t - \tau_j^i)}{Z_s(t)}.$$
5. If the sampled component has more than one state into which it may transfer, sample the next trigger state b_j^i with probability
$$P(b_j^i) = \frac{z_j(b_j' \to b_j; t - \tau_j')}{z_j(b_j'; t - \tau_j')}$$
(see equation (6.2.11a)).
6. Apply immediate transfer to all components that are affected by the trigger change. Increase the step value $i = i + 1$ so that a next phase space state $\vec{\mathbf{P}}_i = (b_1^i, b_2^i, \ldots, b_n^i, \tau_1^i, \tau_2^i, \ldots, \tau_n^i; t_i = \tau_j^i)$ is established.
7. Go to step 2.

As a result of this process, a single random walk, or a history, is generated of the form:

$$c_k = (\vec{\mathbf{P}}_1, \vec{\mathbf{P}}_2, \ldots, \vec{\mathbf{P}}_k) = (\vec{\mathbf{B}}_1, \vec{1}; \vec{\mathbf{B}}_2, \vec{2}; \ldots; \vec{\mathbf{B}}_k, \vec{k}),$$

where there are k events such that the $(k+1)$th event time is beyond the time range limit T_m.

The above process is somewhat different, perceptually, from the way in which one would observe changes in systems. A typical report of such change would first specify the nature of the change, such as a short circuit which has occurred in circuit breaker number 6. The time at which the event happened will be added (if at all) as additional, and often secondary, information. In observing events, we usually see what happens and only then glance at the clock. The modeling of systems transport indicates that things are quite different. In the above process, the question answered first is 'when?', namely, when did a change occur (not yet specifying the nature of the change)? Only after the time of change is known does one answer the question 'What?' This is an essential point since, in fact, *the nature of the event cannot be specified before the elapsed time is known.* The hazard functions in the collision kernel may be dependent on the sojourn time in each state of each component. In comparison, it is similar to asking for the nuclear reaction occurring to a neutron before knowing the distance of travel, hence the material in which it has the collision. This further emphasizes the 'particle like' nature of the system.

In sampling the next event time, it is to be noted that $T(\vec{\mathbf{B}}';\vec{*}; t' \to t)$ is the PDF for the next event time. Consider the CDF of the next event point

$$F_s(t', t) = \int_{t'}^{t} T(\vec{\mathbf{B}}';\vec{*}; t' \to t)\, dt,$$

where for simplicity the notation of the state and time vectors are omitted from the CDF. From the form of the free flight kernel and the notion that the PDF is the product of the hazard and the state reliability, it is clear that one may write

$$F_s(t', t) = 1 - \left[\prod_{j=1}^{n} \frac{1 - H(b_j', t - \tau_j')}{1 - H(b_j', t' - \tau_j')}\right] = 1 - \left[\prod_{j=1}^{n} 1 - \frac{H(b_j', t - \tau_j') - H(b_j', t' - \tau_j')}{1 - H(b_j', t' - \tau_j')}\right]. \tag{6.2.16}$$

Hence, the sampling equation for the next event time should be

$$F_s(t' - t) = \xi. \tag{6.2.17}$$

Taking now the time derivative of (6.2.16) yields

$$\frac{dF_s(t', t)}{dt} = \sum_{i=1}^{n} \left[\prod_{\substack{j=1 \\ j \neq 1}}^{n} \frac{1 - H(b_j', t - \tau_j')}{1 - H(b_j', t' - \tau_j')}\right] \left(\frac{dH(b_i', t - \tau_i')/dt}{1 - H(b_i', t' - \tau_i')}\right)$$

yet $dH(b_i', t-\tau_i')/dt$ is the PDF of the ith component, $h(b_i', t-\tau_i')$, hence

$$\frac{dF_s(t',t)}{dt} = \sum_{i=1}^{n}\left[\prod_{\substack{j=1\\ j\neq 1}}^{n}\frac{1-H(b_j', t-\tau_j')}{1-H(b_j', t'-\tau_j')}\right]\left(\frac{h(b_i', t-\tau_i')}{1-H(b_i', t'-\tau_i')}\right).$$

By definition, the PDF can be written as $h(b_i', t-\tau_i') = \left[1-H(b_j', t-\tau_j')\right] \times z(b_j', t-\tau_j')$ which is the state reliability multiplied by the total hazard (6.2.11a). Substituting back yields

$$\frac{dF_s(t',t)}{dt} = \left[\prod_{j=1}^{n}\frac{1-H(b_j', t-\tau_j')}{1-H(b_j', t'-\tau_j')}\right]\sum_{i=1}^{n} z(b_i', t-\tau_i'),$$

which is identical to the definition of the free flight kernel (6.2.10). Furthermore, substituting the exponential form for each component's CDF yields

$$F_s(t',t) = \left[\prod_{j=1}^{n}\frac{\mathrm{e}^{-\int_{\tau_j'}^{t} z(b_j', y-\tau_j')\,dy}}{\mathrm{e}^{-\int_{\tau_j'}^{t'} z(b_j', y-\tau_j')\,dy}}\right]\sum_{i=1}^{n} z(b_i', t-\tau_i') = \mathrm{e}^{-\int_{t'}^{t}\sum_{j=1}^{n} z(b_j', y-\tau_j')\,dy}\sum_{i=1}^{n} z(b_i', t-\tau_i'),$$

which reduces to the exponential form

$$F_s(t',t) = Z_s(t)\,\mathrm{e}^{-\int_{t'}^{t} Z_s(y)\,dy}, \tag{6.2.18}$$

with the hazard of the system taken from expression (6.2.11). The derivation of equation (6.2.18) is, in fact, a proof that the hazard of the system, at any state, is simply the sum of the hazards of its components.

Estimators The above procedure describes how the analog random walk of a system may be sampled and a history constructed. To obtain specific information concerning the system, estimators must be used. In the following a simplified notation, $T(\vec{\mathbf{B}}_i; t_i, t)$ for the next event time PDF (the free flight kernel) and $C(t; \vec{\mathbf{B}}', \vec{\mathbf{B}})$ for the collision kernel are used. The states entries time vector $\vec{\ }$ is omitted, as this can cause no ambiguity. The probability density of a history is then given by

$$\begin{aligned} f(c_k) &= f(\vec{\mathbf{B}}_1, \vec{1};\ \vec{\mathbf{B}}_2, \vec{2}; \ldots; \vec{\mathbf{B}}_k, \vec{k}) \\ &= Q(\vec{\mathbf{B}}_0)T(\vec{\mathbf{B}}_0, t_0;\ t_1)C(t_1;\ \vec{\mathbf{B}}_0, \vec{\mathbf{B}}_1)T(\vec{\mathbf{B}}_1, t_1;\ t_2)C(t_2;\ \vec{\mathbf{B}}_1, \vec{\mathbf{B}}_2)\cdots \\ &\quad \times T(\vec{\mathbf{B}}_{k-1}, t_{k-1};\ t_k)C(t_k;\ \vec{\mathbf{B}}_{k-1}, \vec{\mathbf{B}}_k)[1-F_s(\vec{\mathbf{B}}_k, t_k, T_m)]. \end{aligned}$$

$Q(\vec{\mathbf{B}}_0)$ is the probability that the initial state of the system at $t_0 = 0$ is $\vec{\mathbf{B}}_0$.

Let Λ_k denote the subset of all histories with k events; then the probability to have a history in this subset is

$$P(\Lambda_k) = \sum_{\vec{\mathbf{B}}_1}\sum_{\vec{\mathbf{B}}_2}\sum_{\vec{\mathbf{B}}_3}\cdots\sum_{\vec{\mathbf{B}}_k}\int_{t_1}\cdots\int_{t_{k-1}}\int_{0}^{T} dt_k f(\vec{\mathbf{B}}_1, \vec{1}, t_1; \vec{\mathbf{B}}_2, \vec{2}, t_2; \ldots; \vec{\mathbf{B}}_k, \vec{k}, t_k)\,dt_1\cdots dt_{k-1}. \tag{6.2.19}$$

In the above, a shorthand integration format is used. The integral $\int_{t_i} dt_i$ indicates the integral $\int_{\vec{\tau}_i} d\vec{\tau}_i \int_{t_i} dt_i$ (i.e. integrating over the entry time vector of all the components and then integration over the entry time into the state). It should be noted that if the jth component entry time in the ith state, τ_i^j, equals the system entry time t_i into the state, then the event kernel includes a delta function of the form $\delta(\tau_i^j - t_i)$ (see equation (6.2.12)). Hence, integrating over τ_i^j merely replaces τ_i^j by t_i. This convention holds also in the following discussion.

The set of all histories is called the *sample space*. Let $\xi(c_k)$ be any estimator applied to the history c_k; the expected value of the estimator over the sample space is given by

$$
\langle \xi \rangle = \sum_{k=0}^{\infty} \sum_{\vec{\mathbf{B}}_0} \sum_{\vec{\mathbf{B}}_1} \sum_{\vec{\mathbf{B}}_2} \sum_{\vec{\mathbf{B}}_3} \cdots \sum_{\vec{\mathbf{B}}_k} \int_{t_0} \int_{t_1} \cdots
\times \int_{t_{k-1}} \xi(c_k) f(\vec{\mathbf{B}}_1, \vec{\tau}_1, t_1; \vec{\mathbf{B}}_2, \vec{\tau}_2, t_2; \ldots; \vec{\mathbf{B}}_k, \vec{\tau}_k, t_k)\, dt_0\, dt_1 \cdots dt_k. \tag{6.2.20}
$$

The integration over each entry time extends from zero to the next event entry time, and for the last event time, from zero to the calculation time T. Considering the history generation procedure (6.2.1), let N histories be sampled and let c_i denote the ith history and ξ_i the estimator value obtained in this history, then

$$
\tilde{\xi}_N = \frac{1}{N} \sum_{i=1}^{N} \xi_i \tag{6.2.21}
$$

is the sample average which converges to $\langle \xi \rangle$ at a rate of $\sigma/\sqrt{N}$, where σ is the standard deviation of $\xi(c_k)$.

To prove that $\xi(c_k)$ is an unbiased estimator of any quantity of interest I, it is necessary and sufficient to prove that $\langle \xi \rangle = I$. Since I is an expression involving the event densities, the proof means that $\langle \xi \rangle$ must equal a quantity obtained from the solution of the system transport equation. Thus, from the point of view of the application of the Monte Carlo technique, the existence of a transport equation is critical for the establishment of unbiased estimators. This concept will become obvious in the following, where a number of general unbiased estimators will be presented.

Consider the transport equation (6.2.5):

$$
\psi(\vec{\mathbf{B}}, \vec{\tau}, t) = S(\vec{\mathbf{B}}, \vec{\tau}, t) + \sum_{\vec{\mathbf{B}}'} \int_{\vec{\tau}'} \int_{t'} \psi(\vec{\mathbf{B}}', \vec{\tau}', t') K(\vec{\mathbf{B}}', \vec{\tau}', t' \to \vec{\mathbf{B}}, \vec{\tau}, t)\, d\vec{\tau}'\, dt'.
$$

Let the source term be

$$
S(\vec{\mathbf{B}}, \vec{\tau}, t) = \delta_{\vec{\mathbf{B}}, \vec{\mathbf{B}}_0} \prod_{i=1}^{n} \delta(\tau_i - t)\, \delta(t),
$$

indicating that at $t = 0$ the system is in state $\vec{\mathbf{B}}_0$ and all the entry times of the components are at t. Also, denote the zero order approximation to the solution $\psi^0(\vec{\mathbf{B}}, \vec{\tau}, t) = S(\vec{\mathbf{B}}, \vec{\tau}, t)$. Next define

$$
\psi^{n+1}(\vec{\mathbf{B}}, \vec{\tau}, t) = S(\vec{\mathbf{B}}, \vec{\tau}, t) + \sum_{\vec{\mathbf{B}}'} \int_{\vec{\tau}'} \int_0^t dt'\, \psi^n(\vec{\mathbf{B}}', \vec{\tau}', t') K(\vec{\mathbf{B}}', \vec{\tau}', t' \to \vec{\mathbf{B}}, \vec{\tau}, t)\, d\vec{\tau}' \tag{6.2.22}
$$

such that

$$\psi^1(\vec{\mathbf{B}},\vec{;}t) = S(\vec{\mathbf{B}},\vec{;}t) + \sum_{\vec{\mathbf{B}}'}\int_{\vec{\ }}\int_0^t dt'\, S(\vec{\mathbf{B}}',\vec{;}t')K(\vec{\mathbf{B}}',\vec{;}t' \to \vec{\mathbf{B}},\vec{;}t)\,d\vec{\ }$$

$$= S(\vec{\mathbf{B}},\vec{;}t) + K(\vec{\mathbf{B}}_0, t_0 \to \vec{\mathbf{B}},\vec{;}t). \tag{6.2.22a}$$

The second order approximation will take the form:

$$\psi^2(\vec{\mathbf{B}},\vec{;}t) = S(\vec{\mathbf{B}},\vec{;}t) + \sum_{\vec{\mathbf{B}}_0}\int_{\vec{0}}\int_0^t dt_0\, S(\vec{\mathbf{B}}_0,\vec{_0},t_0)K(\vec{\mathbf{B}}_0,\vec{_0},t_0 \to \vec{\mathbf{B}},\vec{;}t)\,d\vec{_0}$$

$$+\sum_{\vec{\mathbf{B}}_0}\sum_{\vec{\mathbf{B}}_1}\int_{\vec{0}}\int_{\vec{1}}\int_{t_0} dt_0 \int_{t_1} dt_1\, S(\vec{\mathbf{B}}_0,\vec{_0},t_0)K(\vec{\mathbf{B}}_0,\vec{_0},t_0 \to \vec{\mathbf{B}}_1,\vec{_1},t_1)$$

$$\times K(\vec{\mathbf{B}}_1,\vec{_1},t_1 \to \vec{\mathbf{B}},\vec{;}t)\,d\vec{_0}\,d\vec{_1}. \tag{6.2.22b}$$

Proceeding this process of successively inserting increasing order approximations yields the nth order approximation as

$$\psi^n(\vec{\mathbf{B}},\vec{;}t) = \sum_{k=0}^{n}\sum_{\vec{\mathbf{B}}_0}\cdots\sum_{\vec{\mathbf{B}}_{k-1}}\int_{t_0}\cdots\int_{t_{k-1}} S(\vec{\mathbf{B}}_0,\vec{_0},t_0)K(\vec{\mathbf{B}}_0,\vec{_0},t_0 \to \vec{\mathbf{B}}_1,\vec{_1},t_1)$$

$$\times K(\vec{\mathbf{B}}_1,\vec{_1},t_1 \to \vec{\mathbf{B}}_2,\vec{_2},t_2)\cdots K(\vec{\mathbf{B}}_{k-1},\vec{_{k-1}},t_{k-1} \to \vec{\mathbf{B}},\vec{;}t)\,dt_0\,dt_1\cdots dt_{k-1},$$

where the convention of the time domain integration of equation (6.2.19) is used. As n increases, the solution of the transport equation is obtained in the form

$$\psi(\vec{\mathbf{B}},\vec{;}t) = \lim_{n\to\infty}\psi^n(\vec{\mathbf{B}},\vec{;}t)$$

$$= \sum_{k=0}^{\infty}\sum_{\vec{\mathbf{B}}_0}\cdots\sum_{\vec{\mathbf{B}}_{k-1}}\int_{t_0}\cdots\int_{t_{k-1}} S(\vec{\mathbf{B}}_0,\vec{_0},t_0)K(\vec{\mathbf{B}}_0,\vec{_0},t_0 \to \vec{\mathbf{B}}_1,\vec{_1},t_1)$$

$$\times K(\vec{\mathbf{B}}_1,\vec{_1},t_1 \to \vec{\mathbf{B}}_2,\vec{_2},t_2)\cdots K(\vec{\mathbf{B}}_{k-1},\vec{_{k-1}},t_{k-1} \to \vec{\mathbf{B}},\vec{;}t)\,dt_0\,dt_1\cdots dt_{k-1}. \tag{6.2.23}$$

This expression is the Neumann series expansion of the solution of the transport equation. Equation (6.2.23) is a generalization of expression (2.2.26h) discussed in Section 2.2.3. The convergence of the approximation and the uniqueness of the solution are guaranteed by the fact that for any finite time range, the transport kernels are probability density functions and their integrals are bounded by unity (see Spanier, J. (1967). Some results on transport theory and their application to Monte Carlo methods. *J. Math. Anal. Appl.* **17**, 27 and Spanier, J. and Gelbard, E. M. (1969). *Monte Carlo Principles and Neutron Transport Problems*, Addison-Wesley). In simple terms, this means that for any given finite time the probability that a history will have an infinite number of events is zero. The only exception to this rule would occur if there is a set of closed states (such that once entered into that set, no exit is possible), with delta functions relating to the transfers between the states. This will result in an infinite number of events on any finite time interval. However, such situations are highly unrealistic in any system.

Event type estimators Any time-dependent quantity of interest related to the system behavior can be expressed as an integral of the form

$$G(T) = \sum_{\vec{\mathbf{B}}\in\Sigma} \int_0^T \psi(\vec{\mathbf{B}}, \vec{;} t) D(\vec{\mathbf{B}}, \vec{;} t)\, dt = \sum_{\vec{\mathbf{B}}} \int_0^T \psi(\vec{\mathbf{B}}, \vec{;} t) D(\vec{\mathbf{B}}, \vec{;} t) \chi(\vec{\mathbf{B}})\, dt, \qquad (6.2.24)$$

where Σ is a subset of states of interest and $D(\vec{\mathbf{B}}, \vec{;} t)$ is a general response function or a 'detector' function indicating the score accumulated for the quantity of interest upon entry of the system into the state. In most cases, the detector function is independent of the entries time vector, and depends only upon the state entered and the time of entry into the state. Equation (6.2.24) is a generalization of expression (2.2.21). $\chi(\vec{\mathbf{B}})$ is a range function introduced to allow summation over all the state space, and defined as

$$\chi(\vec{\mathbf{B}}) = \begin{cases} 1 & \text{if } \vec{\mathbf{B}} \in \Sigma \\ 0 & \text{otherwise.} \end{cases}$$

Recall that, by extension of equation (2.2.21) and using the definition of the system CDF (6.2.16), the availability can be expressed as

$$A(T) = \sum_{\vec{\mathbf{B}}\in\Sigma} \int_0^T \psi(\vec{\mathbf{B}}, \vec{;} t)[1 - F_s(\vec{\mathbf{B}}, \vec{;} t, T)\, dt = \sum_{\vec{\mathbf{B}}} \int_0^T \psi(\vec{\mathbf{B}}, \vec{;} t)[1 - F_s(\vec{\mathbf{B}}, \vec{;} t)] S(\vec{\mathbf{B}})\, dt, \qquad (6.2.25)$$

where Σ is the subset of states for which $S(\vec{\mathbf{B}}) = 1$. Thus for the availability, one obtains

$$D(\vec{\mathbf{B}}, t) = S(\vec{\mathbf{B}})[1 - F_s(\vec{\mathbf{B}}, t, T)]. \qquad (6.2.25a)$$

Similarly, if $Y(\vec{\mathbf{B}})$ is a throughput function describing the production rate of the system, then the average production is given by replacing $S(\vec{\mathbf{B}})$ by $Y(\vec{\mathbf{B}})$ in expression (6.2.25). Hence, for the throughput, $D(\vec{\mathbf{B}}, t') = Y(\vec{\mathbf{B}})[1 - F_s(\vec{\mathbf{B}}, t', t)]$. For the reliability, equation (6.2.25) holds, provided that a failed system state is an absorbing state. In expression (6.2.24), quantities that depend upon the system being in a given state at time T are presented with the aid of the event density by multiplying the quantity by the probability of staying in the state from the time point of entry and up to T. Cost functions can be divided into two general categories: costs of type C_1, that incur upon entry into a state, such as costs of failure, spare parts, repair, inspection, etc.; and other costs of type C_2, which are dependent upon the time of residing in given states, such as the cost of unavailability (system down-time). A general cost detector function will then take the form

$$D(\vec{\mathbf{B}}, t) = C_1(\vec{\mathbf{B}}, t) + \int_t^T C_2(\vec{\mathbf{B}}, t')(t' - t)[f_s(\vec{\mathbf{B}}, t, t')]\, dt'. \qquad (6.2.25b)$$

$G(T)$ of equation (6.2.24) is obtained by folding (6.2.25a) with the event density and summing over all possible states such that it yields the accumulated cost up to time T. Substituting $\psi(\vec{\mathbf{B}}, \vec{;}, t)$ from equation (6.2.23) into equation (6.2.24) yields

$$
\begin{aligned}
G(T) = \sum_{k=0}^{\infty} \sum_{\vec{\mathbf{B}}} \int_0^T dt \sum_{\vec{\mathbf{B}}_0} \cdots \sum_{\vec{\mathbf{B}}_{k-1}} \int_{t_0} \cdots \int_{t_{k-1}} S(\vec{\mathbf{B}}_0, \vec{;}_0, t_0) \\
\times K(\vec{\mathbf{B}}_0, \vec{;}_0, t_0 \to \vec{\mathbf{B}}_1, \vec{;}_1, t_1) K(\vec{\mathbf{B}}_1, \vec{;}_1, t_1 \to \vec{\mathbf{B}}_2, \vec{;}_2, t_2) \cdots \\
\times K(\vec{\mathbf{B}}_{k-1}, \vec{;}_{k-1}, t_{k-1} \to \vec{\mathbf{B}}, \vec{;}, t) D(\vec{\mathbf{B}}, \vec{;}, t) \chi(\vec{\mathbf{B}})\, dt_0\, dt_1 \cdots dt_{k-1}.
\end{aligned}
\tag{6.2.26}
$$

This is the Neumann series expansion of the quantity of interest related to the problem. Expression (6.2.26) is important, as it provides a general framework for introducing unbiased estimators. To prove that any estimator $\xi(c_k)$ is an unbiased estimator of $G(T)$, it is necessary and sufficient to prove that expression (6.2.20) is equal to (6.2.26).

Let an estimator be defined in the form

$$
\xi(c_k) = \sum_{i=0}^{k} D(\vec{\mathbf{B}}_i, t_i) \chi(\vec{\mathbf{B}}_i). \tag{6.2.27}
$$

This estimator accumulates information at each event. Consider the expected value of this estimator. Substituting (6.2.27) into (6.2.20) yields

$$
\begin{aligned}
\langle \xi(T) \rangle &= \sum_{k=0}^{\infty} \sum_{\vec{\mathbf{B}}_0} \sum_{\vec{\mathbf{B}}_1} \sum_{\vec{\mathbf{B}}_2} \cdots \sum_{\vec{\mathbf{B}}_k} \int_{t_0} \cdots \int_{t_k} \left[\sum_{i=1}^{k} D(\vec{\mathbf{B}}_i, t_i) \chi(\vec{\mathbf{B}}_i) \right] \\
&\quad \times f(\vec{\mathbf{B}}_1, \vec{;}_1, t_1; \vec{\mathbf{B}}_2, \vec{;}_2, t_2; \ldots; \vec{\mathbf{B}}_k, \vec{;}_k, t_k)\, dt_0 \cdots dt_k \\
&= \sum_{k=0}^{\infty} \sum_{i=1}^{k} \sum_{\vec{\mathbf{B}}_0} \sum_{\vec{\mathbf{B}}_1} \sum_{\vec{\mathbf{B}}_2} \cdots \sum_{\vec{\mathbf{B}}_k} \int_{t_0} \cdots \int_{t_k} D(\vec{\mathbf{B}}_i, t_i) \chi(\vec{\mathbf{B}}_i) \\
&\quad \times f(\vec{\mathbf{B}}_1, \vec{;}_1, t_1; \vec{\mathbf{B}}_2, \vec{;}_2, t_2; \ldots; \vec{\mathbf{B}}_k, \vec{;}_k, t_k)\, dt_0 \cdots dt_k.
\end{aligned}
$$

Interchanging the order of summation from $\sum_{k=0}^{\infty} \sum_{i=0}^{k}$ to $\sum_{i=0}^{\infty} \sum_{k=i}^{\infty}$ and substituting the explicit form of the history PDF yields

$$
\begin{aligned}
&= \sum_{i=0}^{\infty} \sum_{k=i}^{\infty} \sum_{\vec{\mathbf{B}}_0} \sum_{\vec{\mathbf{B}}_1} \sum_{\vec{\mathbf{B}}_2} \cdots \sum_{\vec{\mathbf{B}}_k} \int_{t_0} \cdots \int_{t_k} S(\vec{\mathbf{B}}_0, \vec{;}_0, t_0) \\
&\times K(\vec{\mathbf{B}}_0, \vec{;}_0, t_0 \to \vec{\mathbf{B}}_1, \vec{;}_1, t_1) \cdots K(\vec{\mathbf{B}}_{i-1}, \vec{;}_{i-1}, t_{i-1} \to \vec{\mathbf{B}}_i, \vec{;}_i, t_i) \\
&\times D(\vec{\mathbf{B}}_i, t_i) \chi(\vec{\mathbf{B}}_i) K(\vec{\mathbf{B}}_i, \vec{;}_i, t_i \to \vec{\mathbf{B}}_{i+1}, \vec{;}_{i+1}, t_{i+1}) \cdots K(\vec{\mathbf{B}}_{k-1}, \vec{;}_{k-1}, t_{k-1} \to \vec{\mathbf{B}}_k, \vec{;}_k, t_k) \\
&\times [1 - F_s(\vec{\mathbf{B}}_k, t_k, T)]\, dt_1 \cdots dt_k.
\end{aligned}
$$

The terms in the sum following the ith term can be separated in the form

$$= \sum_{t=0}^{\infty} \sum_{\vec{\mathbf{B}}_0} \sum_{\vec{\mathbf{B}}_1} \sum_{\vec{\mathbf{B}}_2} \cdots \sum_{\vec{\mathbf{B}}_l} \int_{\tau_0} \cdots \int_{\tau_l} S(\vec{\mathbf{B}}_0, \vec{\mathbf{t}}_0, t_0) K(\vec{\mathbf{B}}_0, \vec{\mathbf{t}}_0, t_0 \to \vec{\mathbf{B}}_1, \vec{\mathbf{t}}_1, t_1) \cdots$$

$$\times K(\vec{\mathbf{B}}_{i-1}, \vec{\mathbf{t}}_{i-1}, t_{i-1} \to \vec{\mathbf{B}}_i, \vec{\mathbf{t}}_i, t_i) D(\vec{\mathbf{B}}_i, t_i) \chi(\vec{\mathbf{B}}_i)\, dt_1 \cdots dt_i$$

$$\times \left\{ \left[1 - F_s(\vec{\mathbf{B}}_i, t_i, T)\right] + \sum_{\vec{\mathbf{B}}_{i+1}} \int_{t_{i+1}} K(\vec{\mathbf{B}}_i, t_i \to \vec{\mathbf{B}}_{i+1}, t_{i+1})\, dt_{i+1} \right.$$

$$\left. + \sum_{\vec{\mathbf{B}}_{i+1}} \sum_{\vec{\mathbf{B}}_{i+2}} \int_{t_{i+1}} \int_{t_{i+2}} K(\vec{\mathbf{B}}_i, t_i \to \vec{\mathbf{B}}_{i+1}, t_{i+1}) K(\vec{\mathbf{B}}_{i+1}, t_{i+1} \to \vec{\mathbf{B}}_{i+2}, t_{i+2})\, dt_{i+1} dt_{i+2} + \cdots \right\},$$

which can be written as

$$\langle \xi(T) \rangle = \sum_{i=0}^{\infty} \sum_{\vec{\mathbf{B}}_0} \sum_{\vec{\mathbf{B}}_1} \sum_{\vec{\mathbf{B}}_2} \cdots \sum_{\vec{\mathbf{B}}_i} \int_{t_0} \cdots \int_{t_i} S(\vec{\mathbf{B}}_0, \vec{0}, t_0) K(\vec{\mathbf{B}}_0, \vec{0}, t_0 \to \vec{\mathbf{B}}_1, \vec{1}, t_1) \cdots$$

$$\times K(\vec{\mathbf{B}}_{i-1}, \vec{i-1}, t_{i-1} \to \vec{\mathbf{B}}_i, \vec{i}, t_i) D(\vec{\mathbf{B}}_i, t_i) \chi(\vec{\mathbf{B}}_i)\, dt_1 \cdots dt_i$$

$$\times \left\{ \sum_{m=0}^{\infty} \sum_{\vec{\mathbf{B}}'_1} \sum_{\vec{\mathbf{B}}'_2} \sum_{\vec{\mathbf{B}}'_3} \cdots \sum_{\vec{\mathbf{B}}'_m} \int_{t_0} \cdots \int_{t_i} f(\vec{\mathbf{B}}_i, t_i, \vec{\mathbf{B}}'_1, t'_1; \vec{\mathbf{B}}'_2, t'_2 \ldots \vec{\mathbf{B}}_m, t_m)\, dt'_1 \cdots dt'_m \right\}.$$

The infinite sum in the curly brackets is a sum of probabilities. The first term is the probability that the ith state remain unchanged up to time T. The second term is the probability that a single event will occur in which the ith state $(\vec{\mathbf{B}}_i, \vec{i}, t_i)$ transfers into state $(\vec{\mathbf{B}}'_1, \vec{1}, t'_1)$, which then remains unchanged up to time T, and so on. Thus, the infinite sum is the probability of any future continuation of the history starting from the ith state. This sum is unity, and the average takes the form

$$\langle \xi(T) \rangle = \sum_{i=0}^{\infty} \sum_{\vec{\mathbf{B}}_0} \sum_{\vec{\mathbf{B}}_1} \cdots \sum_{\vec{\mathbf{B}}_{i-1}} \sum_{\vec{\mathbf{B}}} \int_{t_0} \cdots \int_{t} S(\vec{\mathbf{B}}_0, \vec{0}, t_0) K(\vec{\mathbf{B}}_0, \vec{0}, t_0 \to \vec{\mathbf{B}}_1, \vec{1}, t_1) \cdots$$

$$\times K(\vec{\mathbf{B}}_{i-1}, \vec{i-1}, t_{i-1} \to \vec{\mathbf{B}}, \vec{;} t) D(\vec{\mathbf{B}}, t) \chi(\vec{\mathbf{B}})\, dt_1 \cdots dt_{i-1} dt. \tag{6.2.28}$$

Upon comparison, it can be seen that (6.2.28) is identical with the average (6.2.26), hence $\langle \xi(T) \rangle = G(T)$, and the event type estimator is an unbiased estimator.

The basic advantage of the event type estimator is that it scores information at every event. Consider the calculation of the unavailability at time T (substituting $[1 - S(\vec{\mathbf{B}})]$ into (6.2.25a) instead of $S(\vec{\mathbf{B}})$). It is intuitively plausible (and will be formally confirmed shortly) that a natural estimator would sample the history and yield information if, upon crossing the time T, the system is unavailable. If, however, the system enters an unavailable state at time $t_1 < T$ and exits this state before T, no information is scored. The event type estimator will also provide a score from this event. Upon entry to the state, the estimate $[1 - F_s(\vec{\mathbf{B}}, t_1, T)]$ is scored. This estimate expresses the probability that this unavailable state will persist up to T.

Surface crossing estimators Event type estimators are tallied upon the entry of a system into a state. Many quantities of interest are dependent on the system *being* in a state at time T rather than *entering* into it. Such is the unavailability at time t and the production rate. Referring to a time point as a surface in the time domain, one may define the function

$$\theta(\vec{\mathbf{B}}_i, t_i, t_{i+1}, t) = \begin{cases} 1 & \text{if } t_i \le t < t_{i+1} \\ 0 & \text{otherwise.} \end{cases}$$

This function, in the time domain, serves the same purpose that the function $\chi(\vec{\mathbf{B}})$ serves in the states space. With this function, the estimator

$$\xi(c_k) = \sum_{i=1}^{k} D(\vec{\mathbf{B}}_i)\theta(\vec{\mathbf{B}}_i, t_i, t_{i+1}, t) \tag{6.2.29}$$

is an unbiased estimator of the quantity

$$I(t) = \sum_{\vec{\mathbf{B}}} P(\vec{\mathbf{B}}, t)D(\vec{\mathbf{B}}, t), \tag{6.2.30}$$

where ($P(\vec{\mathbf{B}}, t)$ is the probability of being in state $\vec{\mathbf{B}}$ at time t), and $I(t)$ is the average value of $D(\vec{\mathbf{B}})$ at t. Equation (6.2.30) represents the quantity as an average over all possible states at the t. Setting $D(\vec{\mathbf{B}}, t) = 1 - S(\vec{\mathbf{B}}, t)$ would yield the unavailability at t. To prove that (6.2.29) is an unbiased estimator of (6.2.30), substitute (6.2.29) in (6.2.20), yielding for every $t \le T$

$$\langle \xi \rangle = \sum_{k=0}^{\infty} \sum_{\vec{\mathbf{B}}_0} \sum_{\vec{\mathbf{B}}_1} \sum_{\vec{\mathbf{B}}_2} \sum_{\vec{\mathbf{B}}_3} \cdots \sum_{\vec{\mathbf{B}}_k} \int_{t_0} \int_{t_1} \cdots \int_{t_k} \left[\sum_{i=1}^{k} D(\vec{\mathbf{B}}_i)\theta(t_i, t_{i+1}, t) \right]$$
$$\times f(\vec{\mathbf{B}}_1, \vec{1}, t_1; \vec{\mathbf{B}}_2, \vec{2}, t_2; \ldots; \vec{\mathbf{B}}_k, \vec{k}, t_k)\, dt_0\, dt_1 \cdots dt_k.$$

T is the time limit of the calculation and t is any point in the range. The case in which $t = T$ is simple. By the proof that the event type estimator as an unbiased estimator, it follows that the estimator of (6.2.29) is an unbiased estimator of the average of $D(\vec{\mathbf{B}}_i)\theta(\vec{\mathbf{B}}_i, t_i, t_{i+1}, T)$ over the sample space of a system's histories. This estimator scores the value of $D(\vec{\mathbf{B}}, T)$, i.e. the value of the response function for the last state of the system in the history, upon crossing T. Defining, for example, $D(\vec{\mathbf{B}}, T) = 1 - S(\vec{\mathbf{B}}, T)$ will yield the average value of $1 - S(\vec{\mathbf{B}}, T)$, hence the unavailability at T.

For $t < T$, the surface crossing estimator is zero in each interval between two successive events, except for an interval satisfying $t_i \le t < t_{i+1}$, in which case it equals $D(\vec{\mathbf{B}}_i)$. Notably, i can take any value from 0 to k. The sample space can be organized according to a division, by which $t_0 \le t < t_1$, $t_1 \le t < t_2$ and up to $t_k \le t < T$. First substitute the explicit form of the history PDF to obtain

$$\langle\xi\rangle = \sum_{k=0}^{\infty}\sum_{\vec{\mathbf{B}}_0}\sum_{\vec{\mathbf{B}}_1}\sum_{\vec{\mathbf{B}}_2}\sum_{\vec{\mathbf{B}}_3}\cdots\sum_{\vec{\mathbf{B}}_k}\int_{t_0}\int_{t_1}\cdots\int_{t_k}\left[\sum_{i=1}^{k} D(\vec{\mathbf{B}}_i)\theta(t_i, t_{i+1}, t)\right] S(\vec{\mathbf{B}}_0, t_0)$$
$$\times\, T(\vec{\mathbf{B}}_0, t_0 \to t_1)C(t_1; \vec{\mathbf{B}}_0 \to \vec{\mathbf{B}}_1)\cdots T(\vec{\mathbf{B}}_{k-1}, t_{k-1} \to t_k)C(t_k; \vec{\mathbf{B}}_{k-1} \to \vec{\mathbf{B}}_k)$$
$$\times\, [1 - F_s(B_k, t_k, T)]\, dt_0\, dt_1 \cdots dt_k.$$

The above division suggests that for the first case the time integration over t_0 will range from $0 \le t_0 < t$, and for t_1 the integration range is $t \le t_1 \le T$. Also, for each t_i, with $i > 1$, the integration range is $t_{i-1} \le t_i < T$. The integration of the time domain will then take the form

$$\int_0^t dt_0 \int_t^T dt_1 \int_{t_1}^T dt_2 \cdots \int_{t_{k-1}}^T dt_k.$$

It should be recalled that the integration over t_0 is formal, since a delta function $\delta(t_0)$ in the source term fixes $t_0 \equiv 0$. Also, recall that each integration over t_i involves integration over the components entry time vector $\vec{\ }_i$. For the second group, the integration of the time domain takes the form

$$\int_0^t dt_0 \int_{t_0}^t dt_1 \int_t^T dt_2 \int_{t_2}^T dt_3 \cdots \int_{t_{k-1}}^T dt_k.$$

For the third group $t_2 \le t \le t_3$, so that the integration limits are

$$\int_0^t dt_0 \int_{t_0}^t dt_1 \int_{t_1}^t dt_2 \int_t^T dt_3 \int_{t_3}^T dt_4 \cdots \int_{t_{k-1}}^T dt_k$$

and so on, up to the last group $t_k \le t \le T$, resulting in the integration limits

$$\int_0^t dt_0 \int_{t_0}^t dt_1 \int_{t_1}^t dt_2 \cdots \int_{t_{k-1}}^t dt_k.$$

This division is applied to the averaging process, resulting in the sum

$$\langle\xi(T)\rangle = \sum_{k=0}^{\infty}\sum_{\vec{\mathbf{B}}_0}\sum_{\vec{\mathbf{B}}_1}\sum_{\vec{\mathbf{B}}_2}\sum_{\vec{\mathbf{B}}_3}\cdots\sum_{\vec{\mathbf{B}}_k}\int_0^t dt_0\int_t^T dt_1\int_{t_1}^T dt_2\cdots\int_{t_{k-1}}^T dt_k\, D(\vec{\mathbf{B}}_0)S(\vec{\mathbf{B}}_0,t_0)$$
$$\times\, T(\vec{\mathbf{B}}_0,t_0\to t_1)C(t_1;\vec{\mathbf{B}}_0\to\vec{\mathbf{B}}_1)\cdots T(\vec{\mathbf{B}}_{k-1},t_{k-1}\to t_k)C(t_k;\vec{\mathbf{B}}_{k-1}\to\vec{\mathbf{B}}_k)$$
$$\times\,[1-F_s(\vec{\mathbf{B}}_k,t_k,T)]$$
$$+\sum_{k=0}^{\infty}\sum_{\vec{\mathbf{B}}_0}\sum_{\vec{\mathbf{B}}_1}\sum_{\vec{\mathbf{B}}_2}\sum_{\vec{\mathbf{B}}_3}\cdots\sum_{\vec{\mathbf{B}}_k}\int_0^t dt_0\int_{t_0}^t dt_1\int_t^T dt_2\int_{t_2}^T dt_3\cdots\int_{t_{k-1}}^T dt_k\, D(\vec{\mathbf{B}}_1)S(\vec{\mathbf{B}}_0,t_0)$$
$$\times\, T(\vec{\mathbf{B}}_0,t_0\to t_1)C(t_1;\vec{\mathbf{B}}_0\to\vec{\mathbf{B}}_1)\cdots T(\vec{\mathbf{B}}_{k-1},t_{k-1}\to t_k)C(t_k;\vec{\mathbf{B}}_{k-1}\to\vec{\mathbf{B}}_k)$$
$$\times\,[1-F_s(\vec{\mathbf{B}}_k,t_k,T)]$$
$$+\sum_{k=0}^{\infty}\sum_{\vec{\mathbf{B}}_0}\sum_{\vec{\mathbf{B}}_1}\sum_{\vec{\mathbf{B}}_2}\sum_{\vec{\mathbf{B}}_3}\cdots\sum_{\vec{\mathbf{B}}_k}\int_0^t dt_0\int_{t_0}^t dt_1\int_{t_1}^t dt_2\int_t^T dt_3\int_{t_3}^T dt_4\cdots\int_{t_{k-1}}^T dt_k\, D(\vec{\mathbf{B}}_2)S(\vec{\mathbf{B}}_0,t_0)$$
$$\times\, T(\vec{\mathbf{B}}_0,t_0\to t_1)C(t_1;\vec{\mathbf{B}}_0\to\vec{\mathbf{B}}_1)\cdots T(\vec{\mathbf{B}}_{k-1},t_{k-1}\to t_k)C(t_k;\vec{\mathbf{B}}_{k-1}\to\vec{\mathbf{B}}_k)$$
$$\times\,[1-F_s(\vec{\mathbf{B}}_k,t_k,T)]$$
$$+\cdots$$
$$+\cdots$$
$$+\cdots$$
$$+\sum_{k=0}^{\infty}\sum_{\vec{\mathbf{B}}_0}\sum_{\vec{\mathbf{B}}_1}\sum_{\vec{\mathbf{B}}_2}\sum_{\vec{\mathbf{B}}_3}\cdots\sum_{\vec{\mathbf{B}}_k}\int_0^t dt_0\int_{t_0}^t dt_1\int_{t_1}^t dt_2\int_{t_2}^t dt_3\cdots\int_{t_{k-1}}^t dt_k\, D(\vec{\mathbf{B}}_k)S(\vec{\mathbf{B}}_0,t_0)$$
$$\times\, T(\vec{\mathbf{B}}_0,t_0\to t_1)C(t_1;\vec{\mathbf{B}}_0\to\vec{\mathbf{B}}_1)\cdots T(\vec{\mathbf{B}}_{k-1},t_{k-1}\to t_k)C(t_k;\vec{\mathbf{B}}_{k-1}\to\vec{\mathbf{B}}_k)$$
$$\times\,[1-F_s(\vec{\mathbf{B}}_k,t_k,T)].$$

This can be written in a concise form

$$\langle\xi(T)\rangle = \sum_{k=0}^{\infty}\sum_{\vec{\mathbf{B}}_0}\sum_{\vec{\mathbf{B}}_1}\sum_{\vec{\mathbf{B}}_2}\sum_{\vec{\mathbf{B}}_3}\cdots\sum_{\vec{\mathbf{B}}_k}\sum_{i=1}^{k}\int_0^t dt_0\int_{t_0}^t dt_1\cdots\int_{t_{i-1}}^t dt_i\int_t^T dt_{i+1}\int_{t_{i+1}}^T dt_{i+2}\cdots\int_{t_{k-1}}^T dt_k$$
$$\times\, D(\vec{\mathbf{B}}_i)S(\vec{\mathbf{B}}_0,t_0)$$
$$\times\, T(\vec{\mathbf{B}}_0,t_0\to t_1)C(t_1;\vec{\mathbf{B}}_0\to\vec{\mathbf{B}}_1)\cdots T(\vec{\mathbf{B}}_{k-1},t_{k-1}\to t_k)C(t_k;\vec{\mathbf{B}}_{k-1}\to\vec{\mathbf{B}}_k)$$
$$\times\,[1-F_s(\vec{\mathbf{B}}_k,t_k,T)].$$

Interchanging the order of summation from $\sum_{k=0}^{\infty}\sum_{i=0}^{k}$ to $\sum_{i=0}^{\infty}\sum_{k=i}^{\infty}$ is applied, as was done for the event type estimator. Furthermore, the free flight kernel of the ith event, $T(\vec{\mathbf{B}}_i,t_i\to t_{i+1})$ can be written in the form

$$[1-F_s(\vec{\mathbf{B}}_i,t_i,t)]\frac{T(\vec{\mathbf{B}}_i,t_i\to t_{i+1})}{[1-F_s(B_i,t_i,t)]} = [1-F_s(\vec{\mathbf{B}}_i,t_i,t)]T^*(\vec{\mathbf{B}}_i,t_i\to t_{i+1}|t),$$

namely, the PDF for the next event at t_{i+1} is divided into the product of the probability to have no event up to time t and the conditional probability density to have a next event at t_{i+1} given no event up to t, $T^*(\vec{\mathbf{B}}_i, t_i \to t_{i+1}|t)$.

The expected value then takes the form

$$\langle\xi(T)\rangle = \sum_{i=0}^{\infty}\sum_{k=i}^{\infty}\sum_{\vec{\mathbf{B}}_0}\sum_{\vec{\mathbf{B}}_1}\sum_{\vec{\mathbf{B}}_2}\sum_{\vec{\mathbf{B}}_3}\cdots\sum_{\vec{\mathbf{B}}_k}\int_0^t dt_0\int_{t_0}^t dt_1\cdots\int_{t_{i-1}}^t dt_i\int_t^T dt_{i+1}\int_{t_{i+1}}^T dt_{i+2}\cdots\int_{t_{k-1}}^T dt_k$$
$$\times D(\vec{\mathbf{B}}_i)S(\vec{\mathbf{B}}_0, t_0)T(\vec{\mathbf{B}}_0, t_0 \to t_1)C(t_1; \vec{\mathbf{B}}_0 \to \vec{\mathbf{B}}_1)\cdots C(t_i, \vec{\mathbf{B}}_{i-1} \to \vec{\mathbf{B}}_i)$$
$$\times [1 - F_s(\vec{\mathbf{B}}_i, t_i, t)]T^*(\vec{\mathbf{B}}_i, t_i \to t_{i+1}|t)\cdots$$
$$\times T(\vec{\mathbf{B}}_{k-1}, t_{k-1} \to t_k)C(t_k; \vec{\mathbf{B}}_{k-1} \to \vec{\mathbf{B}}_k)[1 - F_s(\vec{\mathbf{B}}_k, t_k, T)],$$

and when the summation over k is explicitly entered,

$$\langle\xi(T)\rangle = \sum_{i=0}^{\infty}\sum_{\vec{\mathbf{B}}_0}\sum_{\vec{\mathbf{B}}_1}\sum_{\vec{\mathbf{B}}_2}\sum_{\vec{\mathbf{B}}_3}\cdots\sum_{\vec{\mathbf{B}}_i}\int_0^t dt_0\int_{t_0}^t dt_1\cdots\int_{t_{i-1}}^t dt_i\, D(\vec{\mathbf{B}}_i)S(\vec{\mathbf{B}}_0, t_0)$$
$$\times T(\vec{\mathbf{B}}_0, t_0 \to t_1)C(t_1; \vec{\mathbf{B}}_0 \to \vec{\mathbf{B}}_1)\cdots C(t_i, \vec{\mathbf{B}}_{i-1} \to \vec{\mathbf{B}}_i)[1 - F_s(\vec{\mathbf{B}}_i, t_i, t)]$$
$$\times \left\{T^*(\vec{\mathbf{B}}_i, t_i \to T\,|t) + \sum_{\vec{\mathbf{B}}_{i+1}}\int_t^T T^*(\vec{\mathbf{B}}_i, t_i \to t_{i+1}|t)C(t_{i+1}, \vec{\mathbf{B}}_i \to \vec{\mathbf{B}}_{i+1})\right.$$
$$\times [1 - F_s(\vec{\mathbf{B}}_{i+1}, t_{i+1}, T)]\,dt_{i+1}$$
$$+\cdots+\sum_{\vec{\mathbf{B}}_{i+1}}\cdots\sum_{\vec{\mathbf{B}}_k}\int_t^T dt_{i+1}\int_{t_{i+1}}^T dt_{i+2}\cdots T^*(\vec{\mathbf{B}}_i, t_i \to t_{i+1}|t)C(t_{i+1}, \vec{\mathbf{B}}_i \to \vec{\mathbf{B}}_{i+1})\cdots$$
$$\left.\times\cdots\int_{t_{k-1}}^T dt_k\, T(\vec{\mathbf{B}}_{k-1}, t_{k-1} \to t_k)C(t_k; \vec{\mathbf{B}}_{k-1} \to \vec{\mathbf{B}}_k)[1 - F_s(\vec{\mathbf{B}}_k, t_k, T)] + \cdots\right\}.$$

The infinite sum in the curly brackets again has an interpretation similar to that of the event type estimator. The first term is the probability that a history reaching time t in state $\vec{\mathbf{B}}_i$ will proceed in this state up to the time limit T with no further events. The next term describes the probability that following crossing the t time surface, a single additional event will occur before reaching the time limit, etc. Thus, it is the probability of any possible continuation of the history following the crossing of t, and hence this probability is unity. The expected value then takes the form

$$\langle\xi(T)\rangle = \sum_{i=0}^{\infty}\sum_{\vec{\mathbf{B}}_0}\sum_{\vec{\mathbf{B}}_1}\sum_{\vec{\mathbf{B}}_2}\sum_{\vec{\mathbf{B}}_3}\cdots\sum_{\vec{\mathbf{B}}_i}\int_0^t dt_0\int_{t_0}^t dt_1\cdots\int_{t_{i-1}}^t dt_i\, S(\vec{\mathbf{B}}_0, t_0)$$
$$\times T(\vec{\mathbf{B}}_0, t_0 \to t_1)C(t_1; \vec{\mathbf{B}}_0 \to \vec{\mathbf{B}}_1)\cdots C(t_i; \vec{\mathbf{B}}_{i-1} \to \vec{\mathbf{B}}_i)$$
$$\times [1 - F_s(\vec{\mathbf{B}}_i, t_i, t)]D(\vec{\mathbf{B}}_i, t). \tag{6.2.31}$$

Comparing equation (6.2.31) with the Neumann expansion of the event rate (6.2.23) indicates that for each given state $\vec{\mathbf{B}}_i$ and time t_i, the expansion yields the event rate

$\psi(\vec{\mathbf{B}}_i, \vec{i}, t_i)$ so that (6.2.31) can be written as

$$\langle \xi(T) \rangle = \sum_{\vec{\mathbf{B}}_i} \int_0^t \psi(\vec{\mathbf{B}}_i, t_i)[1 - F_s(\vec{\mathbf{B}}_i, t_i, t)] D(\vec{\mathbf{B}}_i, t)\, dt_i.$$

However, the integral $\int_0^t \psi(\vec{\mathbf{B}}_i, t_i)[1 - F_s(\vec{\mathbf{B}}_i, t_i, t)]\, dt_i$ is merely the probability of the system to be in state $\vec{\mathbf{B}}_i$ at time t, $P(\vec{\mathbf{B}}_i, t)$, so that

$$\langle \xi(T) \rangle = \sum_{\vec{\mathbf{B}}_i} P(\vec{\mathbf{B}}_i, t) D(\vec{\mathbf{B}}_i, t) = I(t).$$

This completes the proof that the surface crossing estimator (6.2.29) is an unbiased estimator of (6.2.30).

6.3 Variance Reduction Considerations

The preceding sections described in some details the system transport process, the random walk and unbiased estimators adopted for the random walk. The efficiency of a given calculation remains a problem because if the probability of the system to enter a desired state is very small (a rare event). The number of histories required to obtain a reasonable PRSD may be so large as to render the calculation impractical.

Variance Reduction Methods (VRM) are often based on the distortion of the 'analog' random walk process so as to enhance the probability of the appearance of a desired rare event. Analysis of VRMs requires analysis of the second moment of the estimators with distorted random walks (see, for example, Dubi, A. (1985). General statistical model in geometical splitting I & II. *Transport Theory and Statistical Physics*, **14**(2), pp. 167–221). Indeed, as can be seen above, the handling of the first moment is non trivial; the second moment of weighted estimators under distorted random walks poses a serious analysis problem, and is, as yet, at a rather embryonic stage in system engineering.

Forced event estimators Consider again the averaging process of equation (6.2.20).

$$\langle \xi \rangle = \sum_{k=0}^{\infty} \sum_{\vec{\mathbf{B}}_0} \sum_{\vec{\mathbf{B}}_1} \sum_{\vec{\mathbf{B}}_2} \sum_{\vec{\mathbf{B}}_3} \cdots \sum_{\vec{\mathbf{B}}_k} \int_{t_0} \int_{t_1} \cdots \int_{t_k} \xi(c_k) S(\vec{\mathbf{B}}_0, t_0) T(\vec{\mathbf{B}}_0, t_0 \to t_1) C(t_1; \vec{\mathbf{B}}_0 \to \vec{\mathbf{B}}_1) \cdots$$
$$\times\, T(\vec{\mathbf{B}}_{k-1}, t_{k-1} \to t_k) C(t_k; \vec{\mathbf{B}}_{k-1} \to \vec{\mathbf{B}}_k)[1 - F_s(B_k, t_k, T)]\, dt_0\, dt_1 \cdots dt_k. \tag{6.3.1}$$

Expression (6.3.1) suggests that upon entering a state $\vec{\mathbf{B}}_i$ at time t_i, the value of t_{i+1} is sampled from the PDF $T(\vec{\mathbf{B}}_i; t_i \to t_{i+1})$. If $t_{i+1} > T$ the history is terminated, otherwise sampling of the next state is done. Obviously, if T is very small, relative to the mean time to next event, there will be very few events. A small number of events is one contributor to rare events. For example, if a system is constructed of 10 components in parallel, then a system failure requires at least 10 events. If, on the other hand, during the calculation time T the average number of events is, say, 0.77, then the probability of a system failure could be extremely small, and the system failure becomes a rare event. This problem can be overcome by quite simple means. One can enforce events on the system to any required degree.

Replace the kernel, $T(\vec{\mathbf{B}}_i; t_i \to t_{i+1})$ in expression (6.3.1) by

$$F_s(\vec{\mathbf{B}}_i, t_i, T)\frac{T(\vec{\mathbf{B}}_i, t_i \to t_{i+1})}{F_s(\vec{\mathbf{B}}_i, t_i, T)},$$

where $F_s(\vec{\mathbf{B}}_i, t_i, T) = \int_{t_i}^{T} T(\vec{\mathbf{B}}_i, t_i \to x)\, dx$, i.e. the CDF of the free flight kernel or the probability that the next event will occur prior to T. This does not change the averaging process, which now takes the form

$$\begin{aligned}\langle\xi\rangle = &\sum_{k=0}^{\infty}\sum_{\vec{\mathbf{B}}_0}\sum_{\vec{\mathbf{B}}_1}\sum_{\vec{\mathbf{B}}_2}\sum_{\vec{\mathbf{B}}_3}\cdots\sum_{\vec{\mathbf{B}}_k}\int_{t_0}\int_{t_1}\cdots\int_{t_k}\xi(c_k)S(\vec{\mathbf{B}}_0, t_0)F_s(\vec{\mathbf{B}}_0, t_0, T)\frac{T(\vec{\mathbf{B}}_0, t_0 \to t_1)}{F_s(\vec{\mathbf{B}}_0, t_0, T)}\\
&\times C(t_1; \vec{\mathbf{B}}_0 \to \vec{\mathbf{B}}_1)F_s(\vec{\mathbf{B}}_1, t_1, T)\frac{T(\vec{\mathbf{B}}_1, t_1 \to t_2)}{F_s(\vec{\mathbf{B}}_1, t_1, T)}C(t_2; \vec{\mathbf{B}}_1 \to \vec{\mathbf{B}}_2)\cdots\\
&\times F_s(\vec{\mathbf{B}}_{k-1}, t_{k-1}, T)\frac{T(\vec{\mathbf{B}}_{k-1}, t_{k-1} \to t_k)}{F_s(\vec{\mathbf{B}}_{k-1}, t_{k-1}, T)}C(t_k; \vec{\mathbf{B}}_{k-1} \to \vec{\mathbf{B}}_k)\\
&\times [1 - F_s(\vec{\mathbf{B}}_k, t_k, T)]\, dt_0\, dt_1 \cdots dt_k. \end{aligned} \tag{6.3.2}$$

Equation (6.3.2) implies a slightly changed sampling process. First, sample the (discrete) event {Next event will occur prior to time T}. If the opposite occurs, terminate the history, otherwise sample the event point from the conditional PDF:

$$T_N(\vec{\mathbf{B}}_i, t_i \to t_{i+1}) = \frac{T(\vec{\mathbf{B}}_i, t_i \to t_{i+1})}{F_s(\vec{\mathbf{B}}_i, t_i, T)}, \tag{6.3.3}$$

which is normalized in the range $[t_i, T]$. One may note that the number of events is controlled by the values of the 'continuation' probabilities $F_s(\vec{\mathbf{B}}_i, t_i, T)$. Small probabilities will lead to a small number of events, and *vice versa*. These probabilities may be altered by rewriting the average in the form:

$$\begin{aligned}\langle\xi\rangle = &\sum_{k=0}^{\infty}\sum_{\vec{\mathbf{B}}_0}\sum_{\vec{\mathbf{B}}_1}\sum_{\vec{\mathbf{B}}_2}\sum_{\vec{\mathbf{B}}_3}\cdots\sum_{\vec{\mathbf{B}}_k}\int_{t_0}\int_{t_1}\cdots\int_{t_k}\xi(c_k)S(\vec{\mathbf{B}}_0, t_0)\left[\frac{F_s(\vec{\mathbf{B}}_0, t_0, T)}{P_0^*}\right]P_0^*\frac{T(\vec{\mathbf{B}}_0, t_0 \to t_1)}{F_s(\vec{\mathbf{B}}_0, t_0, T)}\\
&\times C(t_1; \vec{\mathbf{B}}_0 \to \vec{\mathbf{B}}_1)\left[\frac{F_s(\vec{\mathbf{B}}_1, t_1, T)}{P_1^*}\right]P_1^*\frac{T(\vec{\mathbf{B}}_1, t_1 \to t_2)}{F_s(\vec{\mathbf{B}}_1, t_1, T)}C(t_2; \vec{\mathbf{B}}_1 \to \vec{\mathbf{B}}_2)\cdots\\
&\times \left[\frac{F_s(\vec{\mathbf{B}}_{k-1}, t_{k-1}, T)}{P_{k-1}^*}\right]P_{k-1}^*\frac{T(\vec{\mathbf{B}}_{k-1}, t_{k-1} \to t_k)}{F_s(\vec{\mathbf{B}}_{k-1}, t_{k-1}, T)}C(t_k; \vec{\mathbf{B}}_{k-1} \to \vec{\mathbf{B}}_k)\\
&\times \left[\frac{Q_s(\vec{\mathbf{B}}_k, t_k, T)}{Q_k^*}\right]Q_k^*\, dt_0\, dt_1 \cdots dt_k, \end{aligned} \tag{6.3.4}$$

where P_i^* is an artificial or external probability which will be used to sample the continuation of the history. Also, $Q_s(\vec{\mathbf{B}}_k, t_k, T)$ is a short notation for $1-F_s(\vec{\mathbf{B}}_k, t_k, T)$, the history termination probability. $Q_k^* = 1 - P_k^*$ is the external defined probability for the termination of the history. The above averaging implies the following sampling process:

1. With probability $1 - P_i^* = Q_i^*$, terminate the history, otherwise (with probability P_i^*).
2. Sample the next event point from the PDF $T_N(\vec{\mathbf{B}}_i, t_i \to t_{i+1})$ of expression (6.3.3).

Observing the process in equation (6.3.4), it is obvious that to maintain an unbiased estimate, the estimator must by multiplied by a factor

$$\left[\frac{F_s(\vec{\mathbf{B}}_{i-1}, t_{i-1}, T)}{P^*_{i-1}}\right]$$

upon entering the ith event. A weight function $W(\vec{\mathbf{B}}_i, t_i)$ is defined such that $W(\vec{\mathbf{B}}_0, t_0) = 1$, and

$$\begin{aligned} W(\vec{\mathbf{B}}_i, t_i) &= W(\vec{\mathbf{B}}_{i-1}, t_{i-1}) \times \left[\frac{F_s(\vec{\mathbf{B}}_{i-1}, t_{i-1}, T)}{P^*_{i-1}}\right] \\ &= \prod_{m=0}^{i-1}\left[\frac{F_s(\vec{\mathbf{B}}_m, t_m, T)}{P^*_m}\right], \quad \text{if } i \neq k, \end{aligned} \tag{6.3.5}$$

if $i = k$, i.e. it is the last event in the history, then the multiplying factor is

$$\left[\frac{Q_s(\vec{\mathbf{B}}_k, t_k, T)}{Q^*_k}\right].$$

The concept of the weight is useful in all biasing (VRM) methods. It provides a visual interpretation of the biasing method, in the sense that initially a unit particle (or system) enters the transport process. At each point where the probability for an event is increased (decreased) the 'weight' of the system/particle is decreased (increased) by the same proportion so as to maintain the product of the weight, and the probability unchanged. The estimators are affected by the weight such that at each point in which a score is made, the score is multiplied by the system weight at that point. The event type estimator will now take the form

$$\xi(c_k) = \sum_{i=0}^{k} W(\vec{\mathbf{B}}_i, t_i) D(\vec{\mathbf{B}}_i, t_i) \chi(\vec{\mathbf{B}}_i). \tag{6.3.6}$$

For the surface crossing estimator a score is made only when the event points t_i and t_{i+1} are known to include the time point of interest, hence the score is multiplied by the weight $W(\vec{\mathbf{B}}_{i+1}, t_{i+1})$, where according to (6.3.5),

$$W(\vec{\mathbf{B}}_{k+1}, t_{k+1}) = W(\vec{\mathbf{B}}_k, t_k) \times \left[\frac{Q_s(\vec{\mathbf{B}}_k, t_k, T)}{Q^*_k}\right].$$

By choosing large values of P^*_i one can increase the number of events at will, while at the same time decreasing the weight of the system. The introduction of biasing requires use of external parameters, such as the P^*_is in this case. This is the case for every biasing method. As in the above case, the external parameters allow some control over aspects of the transport process, the number of events in the present case. Yet they allow only partial and often non-coherent control. Recall that the quantity calculated has a Neumann series expansion (equation (6.2.26)), and the relative size of each term in this expansion can be taken as the importance of the term. Much in the same sense as was discussed in the biasing of one-dimensional integrals, except that now the importance cannot be plotted or calculated easily. Still, any biasing method should preserve the relative importances of the Neumann series terms (for a formal proof of this point, see Dubi, A. and Dudziak, D. J. (1979). Optimal choice of parameters for exponential biasing in Monte Carlo, *Nuclear Science and Engineering*, **70**). If, for example the main contribution to the quantity calculated comes from the 20–30 Neumann series expansion terms, then the above method does not guarantee, under any choice of parameters, that the biased transport will indeed concentrate on these terms. It can only ensure that, if chosen correctly, these terms will get 'better attention' than in the analog (unbiased) transport. The phase space of the system is a complex space. Biasing methods provide tools to influence the navigation in this space, but never to fully control it. To a large extent, biasing methods are similar to moving in a deep forest from point A to point B with a blindfold. Optimization of a biasing method is worth any effort, as otherwise it is up to the experience and 'intuition' of the user to choose, seemingly, appropriate parameters. Incorrect choice of these parameters may bring about all the ill effects discussed in Section 3.3.

Optimization of VRM means that the product of the variance and the calculation time per history are minimized as a function of the external parameters. Instead, one may assume that the second moment is much larger than the square of the mean (or else the mean value is large enough so that no variance reduction methods are needed at all). Thus, it is sufficient to look at the second moment. Also, one may assume that the processing time is either unchanged or is easy to learn by running a number of calculations and observing the resulting computer time. This may save the need to express the calculation time analytically as a function of the biasing parameters.

In the above case, the method allows an increase of the number of events. It is therefore called 'forced events biasing'. Consider the second moment for the above case when the target of the calculation is the detector response at the endpoint of the calculation T. Using the surface crossing estimator, the second moment is obtained as the mean value of the square of the estimator, and this takes the form

$$\langle\xi^2\rangle = \sum_{k=0}^{\infty}\sum_{\vec{\mathbf{B}}_0}\sum_{\vec{\mathbf{B}}_1}\sum_{\vec{\mathbf{B}}_2}\sum_{\vec{\mathbf{B}}_3}\cdots\sum_{\vec{\mathbf{B}}_k}\int_{t_0}\int_{t_1}\cdots\int_{t_k}\left[D(\vec{\mathbf{B}}_k)\left(\prod_{i=0}^{k}\left[\frac{F_s(\vec{\mathbf{B}}_i, t_i, T)}{P_i^*}\right]\right)\left[\frac{Q_s(\vec{\mathbf{B}}_k, t_k, T)}{1-P_k^*}\right]\right]^2$$
$$\times S(\vec{\mathbf{B}}_0, t_0)P_0^*\frac{T(\vec{\mathbf{B}}_0, t_0\to t_1)}{F_s(\vec{\mathbf{B}}_0, t_0, T)}C(t_1;\vec{\mathbf{B}}_0\to\vec{\mathbf{B}}_1)P_1^*\frac{T(\vec{\mathbf{B}}_1, t_1\to t_2)}{F_s(\vec{\mathbf{B}}_1, t_1, T)}C(t_2;\vec{\mathbf{B}}_1\to\vec{\mathbf{B}}_2)\cdots$$
$$\times P_{k-1}^*\frac{T(\vec{\mathbf{B}}_{k-1}, t_{k-1}\to t_k)}{F_s(\vec{\mathbf{B}}_{k-1}, t_{k-1}, T)}C(t_k;\vec{\mathbf{B}}_{k-1}\to\vec{\mathbf{B}}_k)Q_k^* dt_0 dt_1\cdots dt_k. \tag{6.3.7}$$

First assume that a single continuation probability parameter is used, i.e. $P_i^* = p_s\ \forall i$. Substituting p_s in expression (6.3.7) for each P_i^*, the second moment can be written as

$$\langle\xi^2\rangle = \sum_{k=0}^{\infty}\frac{\Gamma_k}{p_s^k(1-p_s)}, \tag{6.3.8}$$

where Γ_k is the remaining expression in (6.3.7). Indeed, Γ_k cannot be calculated analytically, but a number of short MC calculations of the second moment can provide a reasonable fit to expression (6.3.8), which can then be used to obtain an optimum or near optimum value of p_s. It should be noted that equation (6.3.8) indicates that $\langle\xi^2\rangle$ diverges for $p_s = 0$ and $p_s = 1$. This also means that the variance diverges, and an infinite statistical error results. Thus, incorrect choice of the parameters is an obvious possibility. It can be observed that, in this scheme, the probability for m events in a history is $p_s^m(1-p_s)$, hence the average number of events will be

$$\langle m_e \rangle = \sum_{k=0}^{\infty} m p_s^m (1-p_s) = \frac{p_s}{1-p_s}.$$

In a reliability calculation the situation is somewhat different. For an unreliability calculation a failed system state is an absorbing state. In this case, $D(\vec{\mathbf{B}}_k) = 1 - S(\vec{\mathbf{B}}_k)$ in expression (6.3.7). If no system failure is encountered in the history, then the estimator is zero; if a system failure is encountered, the absorption probability is unity, and $F_s(B_k, t_k, T) = 1$, and no bias is applied on that step. Thus, the expression

$$\left[\frac{Q_s(B_k, t_k, T)}{Q_k^*}\right] Q_k^*$$

is unity in (6.3.4), and the term

$$\left[\frac{Q_s(\vec{\mathbf{B}}_k, t_k, T)}{1-P_k^*}\right]^2$$

does not appear in the second moment. In that case, the second moment takes the form

$$\langle \xi^2 \rangle = \sum_{k=0}^{\infty} \frac{\Gamma_k}{p_s^k}$$

with an obvious minimal second moment at $p_s = 1$. This result applies to any calculation in which the probability of first system entry into a state or a subset of states is of interest, such as in risk analysis problems.

If multiple values are allowed for the external parameters, then the second moment takes the form

$$\langle \xi^2 \rangle = \sum_{k=0}^{\infty} \frac{\Gamma_k}{P_0^* P_1^* \cdots P_{k-1}^* (1-P_k^*)} = \sum_{k=0}^{\infty} \frac{\Gamma_k}{P_e(k)}, \tag{6.3.9}$$

where $P_e(k) = P_0^* P_1^* \cdots P_{k-1}^*(1-P_k^*)$ is the probability that a history will have exactly k events, and $P_e(k+1) = P_e(k)(P_k^*/(1-P_k^*))(1-P_{k+1}^*)$. Optimal values of $P_e(k)$ can be found using Lagrange multipliers taking a derivative of (6.3.9) with the constraint $\sum_{k=0}^{\infty} P_e(k) = 1$, yielding

$$\frac{\partial}{\partial P_e(k)} \left[\sum_{k=0}^{\infty} \frac{\Gamma_k}{P_e(k)} + \lambda \left(\sum_{k=0}^{\infty} P_e(k) - 1\right)\right] = 0,$$

which results in $P_e(k) = \sqrt{\Gamma_k/\lambda}$. Substituting back into the normalization condition yields $\lambda = \sum_{k=0}^{\infty} \sqrt{\Gamma_k}$, with the final result for the optimal parameters as

$$P_e^{\mathrm{Opt}} = \frac{\sqrt{\Gamma_k}}{\sum_{k=0}^{\infty} \sqrt{\Gamma_k}}. \tag{6.3.10}$$

The leakage estimator Substituting the event type estimator (6.2.27) into the biased averaging process of equation (6.3.4), for the unavailability at time T (the endpoint of the calculation) results in

$$\langle\xi\rangle = \sum_{k=0}^{\infty}\sum_{\vec{\mathbf{B}}_0}\sum_{\vec{\mathbf{B}}_1}\sum_{\vec{\mathbf{B}}_2}\sum_{\vec{\mathbf{B}}_3}\cdots\sum_{\vec{\mathbf{B}}_k}\int_{t_0}\int_{t_1}\cdots\int_{t_k}\sum_{i=1}^{k}\left\{\left[\prod_{m=1}^{i}\left(\frac{F_s(\vec{\mathbf{B}}_{m-1},t_{m-1},T)}{P^*_{m-1}}\right)\right]\right.$$

$$\left.\times[1-S(\vec{\mathbf{B}}_i)]\times[1-F_s(\vec{\mathbf{B}}_i,t_i,T)]\right\}$$

$$\times\, S(\vec{\mathbf{B}}_0,t_0)P^*_0\frac{T(\vec{\mathbf{B}}_0,t_0\to t_1)}{F_s(\vec{\mathbf{B}}_0,t_{01},T)}C(t_1;\vec{\mathbf{B}}_0\to\vec{\mathbf{B}}_1)P^*_1\frac{T(\vec{\mathbf{B}}_1,t_1\to t_2)}{F_s(\vec{\mathbf{B}}_1,t_1,T)}C(t_2;\vec{\mathbf{B}}_1\to\vec{\mathbf{B}}_2)\cdots$$

$$\times\, P^*_{k-1}\frac{T(\vec{\mathbf{B}}_{k-1},t_{k-1}\to t_k)}{F_s(\vec{\mathbf{B}}_{k-1},t_{k-1},T)}C(t_k;\vec{\mathbf{B}}_{k-1}\to\vec{\mathbf{B}}_k)\left[\frac{Q_s(B_k,t_k,T)}{Q^*_k}\right]Q^*_k dt_0 dt_1\cdots dt_k, \tag{6.3.11}$$

where equation (6.2.25a) is used for the unavailability estimation. Instead of the estimator $[1-S(\vec{\mathbf{B}})],[1-F_s(\vec{\mathbf{B}},t,T)]$ one may insert, any estimator of the type $D(\vec{\mathbf{B}},t)=Y(\vec{\mathbf{B}})[1-F_s(\vec{\mathbf{B}},t,T)]$.

In the above equation, one may choose the set $\{P^*_i\}$ such that for every $i\le K_0<\infty$, $P^*_i=1$, and for $i>K_0$ the analog process is resumed, i.e. $P^*_i=F_s(\vec{\mathbf{B}}_i,t_i,t)$. This means that at least up to the K_0th event, which can be taken as large as desired, the history continues with probability 1. This leads to the following random walk process: at $t_0=0$ a contribution $[1-S(\vec{\mathbf{B}}_0)][1-F_s(\vec{\mathbf{B}}_0,0,T)]$ is tallied. This score presents the contribution to the unavailability at T from histories starting in an unavailable state and maintaining this state (with probability $[1-F_s(\vec{\mathbf{B}}_0,0,T)]$) up to T. The 'weight' of the history will now be $F_s(\vec{\mathbf{B}}_0,0,T)$, and this fraction of the history continues the transport. The next event point is sampled from the normalized distribution $(T(\vec{\mathbf{B}}_0,t_0\to t_1))/(F_s(\vec{\mathbf{B}}_0,t_{01},T))$ and is therefore guaranteed to fall before T. The process so far can be interpreted as if the history is split into two parts according to its analog probabilities. A part of weight $[1-F_s(\vec{\mathbf{B}}_0,0,T)]$ 'leaks' all the way up to T (and contributes to the tally), and a remaining part of weight $W_0=F_s(\vec{\mathbf{B}}_0,0,T)$ continues the transport. Upon reaching the state $(\vec{\mathbf{B}}_1,t_1)$, the same process is repeated. A fraction $[1-F_s(\vec{\mathbf{B}}_1,t_1,T)]$ of the system, which now has a weight W_0, leaks up to time T, contributing a score of $W_0[1-S(\vec{\mathbf{B}}_1)][1-F_s(\vec{\mathbf{B}}_1,t_1,T)]$. The remaining part of weight $W_1=W_0F_s(\vec{\mathbf{B}}_1,t_1,T)$ continues the transport. This process repeats itself such that, in general, upon entry into state $(\vec{\mathbf{B}}_i,t_i)$ the system, of weight W_{i-1}, is split into two parts. A leaking part of weight $W_{i-1}[1-F_s(\vec{\mathbf{B}}_i,t_i,T)]$ leaks up to time T, making a contribution $W_{i-1}[1-F_s(\vec{\mathbf{B}}_i,t_i,T)][1-S(\vec{\mathbf{B}}_i)]$, and a fraction $W_i=W_{i+1}F_s(\vec{\mathbf{B}}_i,t_i,T)$ continues the transport. Note that weight is preserved such that in every event, the sum of the leaking weight and the continuing weight is equal to the weight entering the event. The continuing part proceeds to the next event which, with probability 1, is guaranteed to happen at $t_{i+1}<T$. The estimator obtained in this process is called the 'leakage estimator'. It should be noted that the leakage process can be stopped at any point, i.e. K_0 can be set to any value, and may even change from one history to another. In this

method, K_0 is the external parameter. The advantage of this method is obviously that one can have as many events as are needed to reach a system failed state. Thus, K_0 can be dependent on the events in the history. For example, one may revert to the analog process after the first, or second, etc. failure of the system. The disadvantage is that if K_0 is set too high, the number of events will be large, resulting in a high processing time of systems with very low weights.

The above discussed VRM methods address only the control over the number of events in the system. However, lack of events is only one factor in the problem of rare events, and is not usually the most important one. The second factor is the situation in which there may be enough events in the system, but these events do not lead to the desired (detector) state. For example, consider a system with 10 components in parallel, each having a failure rate of 10^{-4} and repair rates of the order of 0.5. If $m < 10$ components have already failed, the probability that an additional failure will occur in the next event is of the order of $(10 - m) \times 10^{-4}/((10 - m) \times 10^{-4} + 0.5m)$, while that of a repair is $0.5m/((10 - m) \times 10^{-4} + 0.5m)$. For $m = 2$ this means that the probability of a failure is 0.0008 of the probability of repair. The system will have a much higher tendency for repair rather than failure. In fact, the probability for 10 consecutive failures, which will result in a system failure, is 7.289×10^{-31}; this value means that in an analog process, it is totally impractical to see any failure with any reasonable number of histories. If the calculation time range is 300 000 hours there will be enough events (about 30, on average), but the system will flip among partially failed states, and will never reach a system failed state. Obviously, adding events will do little good. Adding events with the above method is useful mainly for unrepairable systems. A VRM that deals with the above problem must alter the collision kernel so as to increase the probability of failures. This subject is briefly considered in Section 7.3.

7
Realization of System Transport

In Chapter 6 the transport process of a system was discussed, with emphasis on the formalism of the process. Here, the implementation of this process for some basic practical cases will be considered.

7.1 Single Component Analysis

The case of a single component is simple. The state vector is one-dimensional with a single element $\vec{\mathbf{B}} = (b_1)$. b_1 takes two possible values, 1 for an operational state, and 0 for the failed state. Denoting by $f(b, t)$ the PDF in state b, the transport kernel is simply the PDF of the component, namely $T(b; t \to t') = f(b, t - t')$. The collision kernel is quite trivial, since only a single transfer can occur in an event, i.e. $C(t; 1 \to 0) = 1$ and $C(t; 0 \to 1) = 1$. Let T_m denote the end point of the calculation (the service time), and let $T_1, T_2, \ldots, T_n, (T_n = T_m)$ be an ascending set of time points at which the availability (or any other time-dependent quantity) is to be calculated. Procedure 7.1.1 describes the flow of a single history using the surface crossing estimator. The process is a detailed account of the analog MC calculation process shown in procedure (6.2.1), and includes partial repair. Conveniently, the PDF in the 'up' state is denoted by $f(t)$, the failure distribution, and in the down state it is $g(t)$, the repair distribution.

Procedure 7.1.1: A single component history

1. Initialize transport quantities: step number $i = 0$, current time $t_i = 0$, previous time $t_p = 0$, birth time $t_b = 0$, set initial age $A_g = 0$ (age is attributed only to the 'up' state). Set initial state $b_i = 1$.
2. Calculate, P_e, the probability that the system will have the next event before T_m. P_e is given by

$$P_e = 1 - \frac{F(T_m - t_b) - A_g}{1 - A_g} \quad \text{for } b_i = 1 \quad \text{and} \quad P_e = 1 - G(T_m - t_i) \quad \text{if } b_i = 0.$$

3. Let ξ be a random number. If $\xi \leq P_e$ the system continues the transport, go to step 4. Otherwise, the system has the next event beyond T_m. Go to step 3.1.
   3.1 History terminated. Tally for each $T_j > t_i$ the estimate $e(T_j) = b_i$ and terminate the history.
4. History continues: sample next event point from the normalized distribution. The next event point is obtained from the solution for t of

$$\frac{F(t-t_b)-A_g}{(1-A_g)P_e} = \xi \quad \text{for } b_i = 1 \qquad \text{and} \qquad \frac{G(t-t_i)}{P_e} = \xi \quad \text{for } b_i = 0.$$

   Denote the next event point as t_n. Proceed to tally the surface crossing estimator for time points in (t_i, t_n).
5. Tally for each $t_i < T_j \leq t_n$ the estimate $e(T_j) = b_i$.
6. Transfer the system to the next state, update the previous time, $t_p = t_i$ increase $i \to i+1$, update the event time, $t_i = t_n$, update the component state; if $b_{i-1} = 0$ update to $b_i = 1$ the next event is a repair, move to step 6.1; otherwise update $b_i = 0$ the event is a failure, proceed to step 6.2.
   6.1 *Failure event*: register the age at the time of failure, $A_g = F(t_i - t_b)$. This is the age of the component just at the point of failure.
   6.2 *Repair event*: update age of component. If repair is 'as good as new' set age $A_g = 0.0$, for the 'as bad as old' case set $A_g = F(t_i - t_b)$, i.e. the age that was recorded upon the last failure. For the general case set $A_g = \alpha F(t_i - t_b)$. Once the age is established, the birth time must be recalculated from the equation $A_g = F(t_i - t_b)$ solved for t_b.
7. Proceed the transport. Go to step 2.

It is to be noted that the above procedure provides a general solution to equation (2.2.2). For any value of α, the age reduction factor, it is a solution of expression (2.2.21), where the event rates are solutions of the two state age inclusive transport equations (2.2.40a,b), with the age transfer function given by $U(t', t, t'^b, t^b) = \delta\{t^b - [t - F^{-1}(\alpha \times F(t' - t'^b))]\}$, i.e. the new birth point results from the previous age multiplied by α. This completes the cycle opened with equation (2.2.2).

To adopt the above procedure for the event type estimator, the following changes must be made. The tally process must be omitted from steps 3.1 and 5. Instead, a tally must be made upon entry into a state (and before the next state is sampled). This is best done by adding a step 1.1 right after step 1, involving

1.1 Tally for each $t_i < T_j$, update the estimate, $e(T_j) = e(T_j) = b_1 \times (F(T_j - t_b) - A_g)/(1 - A_g)$.

Also, the transfer from step 7 should be made into step 1.1.

To introduce a uniform distribution for failure, assume that the uniform distribution is defined by two constants, c_1 and $c_2 > c_1$ such that if $c_1 > 0$ no failure can occur up to c_1. The following adaptations should be included in Procedure 7.1.1 due to the finite range of this distribution. In calculating the event probability P_e, it should be noted that if $t - t_b < c_1$, then $F(t - t_b) = 0$ if $c_1 \leq t - t_b < c_2$ then $F(t - t_b) = (t - t_b - c_1)/(c_2 - c_1)$ and for $(t - t_b) \geq c_2$ $F(t - t_b) = 1$. Also, upon recalculating the birth point at a repair event, it is noted that if the age upon repair is zero (full repair), then $t_b = t_i$; otherwise $t_b = t_i - c_1 - A_g(c_2 - c_1)$.

In Figure 7.1.1, the availability, as calculated with the above procedure, is shown for four cases. In all cases the average repair time is 50 hours and the average time to failure

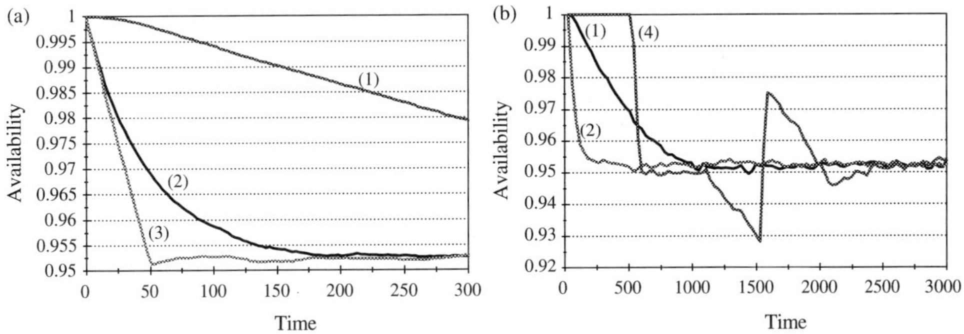

Figure 7.1.1 Availability for three failure distributions with identical MTTF. Short range (a) and steady state (b)

is 1000 hours, so that the steady state availability must be 0.953 for all cases. Full repair is assumed. In line (1) the failure distribution is a Weibull with a scale parameter $\lambda = 7.85404 \times 10^{-7}$ and shape parameter $\beta = 2.0$ [`Program carloww.for`]. Line (2) corresponds to an exponential failure and repair distributions. Line (3) corresponds to an exponential failure distribution and fixed time of repair [`Program carlowf.for`], and line (4) to a uniform failure distribution with $c_1 = 500$ h and $c_2 = 1500$ h, so that no failure can occur before 500 hours of operation, and fixed repair time [`Program carlouf.for`]. Results are obtained with 500 000 histories for Figure 7.1.1(a) in the range of (0–300 h) and 100 000 for Figure 7.1.1(b) in the range (0–3000 h). It can be seen that, although all cases reach the same steady state availability, considerable differences appear in a time range comparable with the MTTF.

Line (2) is the well known smooth shaped exponential case with

$$A(t) = \frac{\mu}{\mu + \lambda} + \frac{\lambda}{\mu + \lambda} e^{-(\mu+\lambda)t}.$$

The Weibull case (1) approaches a steady state much slower (at about the MTTF); the fixed repair case has a sharp discontinuity of the availability derivative at the fixed repair time, and reaches a steady state at about the same time as the exponential case. The uniform failure case (4) maintains considerable variations, and reaches a steady state only at about 2500 hours. Observing the actual availability of a sample of important identical components, and using the above simple procedure, may provide support in identifying the failure distribution.

Minimal repair and maintenance In Figure 7.1.2 the availability is shown for the Weibull case (1), with three values of the age reduction parameter α, namely $\alpha = 0.0$ corresponding to full repair, $\alpha = 0.5$ corresponding to partial repair, and $\alpha = 1.0$ which corresponds to minimal repair (as bad as old). Considerable differences appear between the availabilities, with the notice that for minimal repair the steady state availability is zero. In many cases, the nature of the repair is one of the logistical or design parameters. One may decide to replace a failed unit by a new one upon each failure, resulting in full repair. This may result in a high cost of repair, less failures and higher overall availability. At the other extreme, minimal repair of the same unit may be used, resulting in low repair costs, more failures and an overall lower availability. The

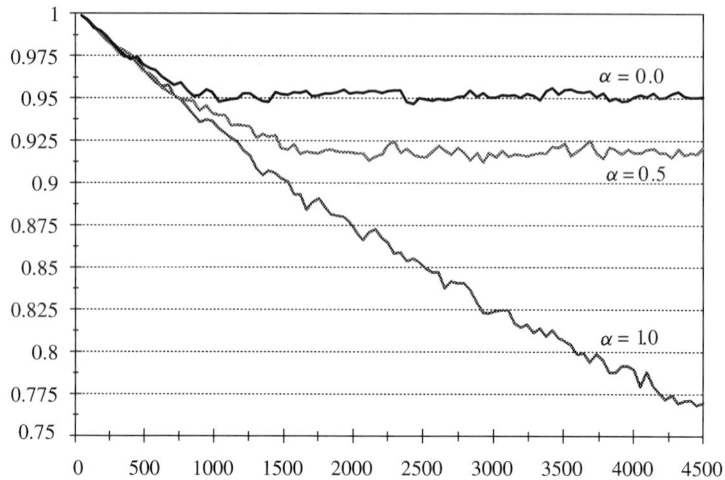

Figure 7.1.2 Availability with minimal, partial and full repair

optimal choice is dependent on the reality of the component, but in any case, the calculations presented in Figure 7.1.2 are necessary for the optimization process. The next example demonstrates this point.

Example 7.1.1 Repair mode optimization

An industrial chemical production line involves a large number of identical pipe sections through which corrosive materials are flowing. The set of the pipes is assumed to be a single component, having a Weibull failure distribution the same as above, namely $\lambda = 7.85404 \times 10^{-7}$ and $\beta = 2.0$. Upon failure of a pipe section, one may either replace only the failed section, which corresponds to a minimal repair, or one may replace all the pipe sections (full repair), or apply, upon failure, a test to all of the pipe sections such that, depending on the current state of each tested section, it will be replaced. Such tests may be conducted with varying threshold levels, so that the fraction of replaced sections can vary continuously from 0.0 to 1. It is further assumed that $1 - \alpha$ is proportional to the fraction of replaced pipe sections. The cost of repair is a function of α in the following form. The cost of minimal repair is $C_r = \$630$ per failure. Applying partial repair has a cost function of the form $C_r = \$[529\,200 \times (1 - \alpha) + 4000]$ per repair. Also, the cost of one hour of downtime due to a single pipe section failure is estimated at $C_d = \$2566$, corresponding to a total yearly production worth \$22.5 M. It is desired to optimize the repair policy with respect to $1 - \alpha$ for a period of one year.

Of special interest is the behavior of the component with minimal repair. It should be noted that, in this case, the average age is a simple geometric progression, starting from 0.0; after the first failure it is 0.5, then 0.75, etc., thus the average age, following the ith failure, is given by

$$\text{Age}(i) = \sum_{k=1}^{i} \left(\frac{1}{2}\right)^k,$$

and will reach 0.9999999 upon the 12th event. Obviously, as the age approaches unity the MTTF decreases, so that eventually failures occur immediately following the repair. It is thus expected that at a certain point the availability will drop to zero (it should be noted that in a computer realization of this process, special measures must be taken when the age approaches unity, as this

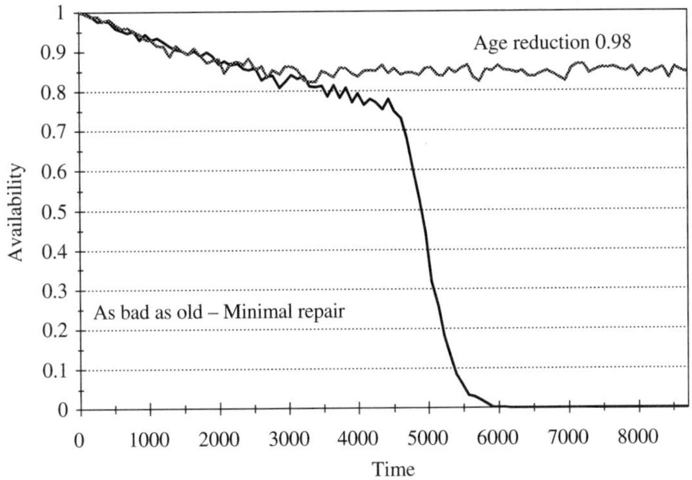

Figure 7.1.3 The collapse of the availability in minimal repair

may result in an overflow in the finite numerical computer calculation). This collapsing phenomena is shown in Figure 7.1.3 for the above case. The case of age reduction of 0.98 is also shown for comparison. It can be seen that for the 0.98 case, a non-zero steady state availability is obtained.

The collapse of the availability resembles realistic situations. A new car behaves in a similar way. For a certain period it is considered 'new'; this is a period that can be seen in Figure 7.1.2. In this period, no significant differences are observed between full repair and minimal repair (up to about the MTTF). Following that period, the availability gradually deteriorates; the car is termed 'used' and its MTBF drops. Failures become more and more frequent. Most people replace the car somewhere along this deterioration curve. If not replaced, the car will reach the 'end of life' point, and a rather sudden drop of availability is experienced. In this period it spends more time in repairs than on the road, and becomes unusable.

For the current example, Procedure 7.1.1 is used with values of α ranging from zero to unity. In each calculation, the average unavailability is evaluated (by simply summing

$$\bar{U} = \sum_{i=1}^{n} \frac{u_{i-1} + u_i}{2},$$

where u_i is the unavailability at the ith time points) and multiplied by 8760 hours to obtain the average downtime. The number of failures is obtained by defining and adding an estimator, which scores unity upon each failure. The results of this analysis are summarized in Table 7.1.1.

The cost curves are shown in Figure 7.1.4. It is interesting to note that with minimal repair, production loss is extreme (over \$28 M). However, a small improvement, amounting to a test which results in a repair of about (2–10)% of the pipe sections through inspection, improves the production considerably. Note also that repair cost at full repair is lower than that with 0.9 repair, because although the cost of a single repair is higher, the reduction in the number of failures offsets the increased cost.

The total cost described above can be expressed in terms of the event rates as

$$C_T = C_f\langle n_f\rangle + C_d T_d = C_f \int_0^T \psi_0(t)\,dt + C_d\left[1 - \int_0^T\int_0^t dt'\psi_1(t')[1 - F(t - t')]\,dt\right], \tag{7.1.1}$$

Table 7.1.1 Optimization table of the repair type

Fraction repaired $1-\alpha$	$\langle A \rangle$ average availability	$\langle n_f \rangle$ average number of failures	Cost of downtime $k	Cost of repairs $k	Total Cost $k
0.0 (minimal)	0.2434	88.2	28539.1	55.5	28594.6
0.02	0.8580	49.4	5356.2	526.4	**5882.7**
0.05	0.8718	44.7	4835.7	1187.0	6022.7
0.1	0.8844	40.6	4360.4	2152.0	6512.4
0.3	0.9071	32.3	3504.2	5138.2	8642.5
0.5	0.9210	27.5	2979.8	7293.7	10273.6
0.7	0.9319	23.8	2568.7	8838.9	11407.7
0.9	0.9425	20.1	2168.9	9578.5	11747.4
1.0 (Full)	0.9530	16.2	1772.8	8592.9	10365.7

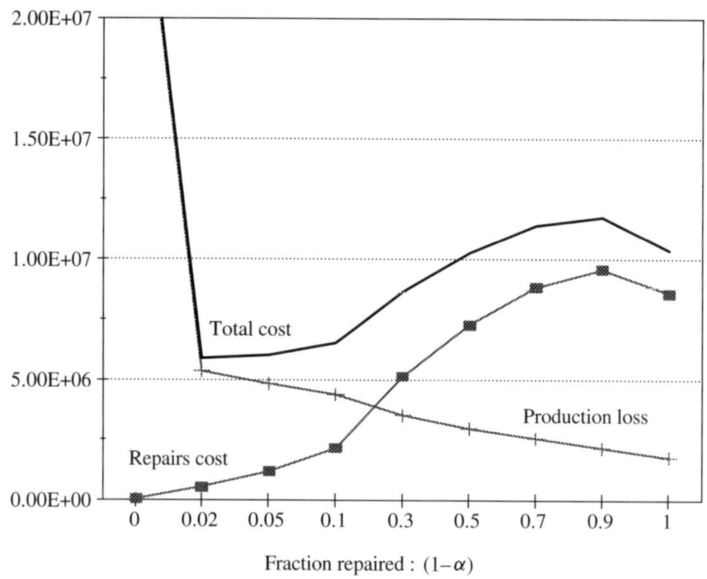

Figure 7.1.4 The optimization cost curves

where the event rates $\psi_1(t)$ and $\psi_0(t)$ are the solutions of equations (2.2.40a,b). In this sense, the Monte Carlo method is solving the transport equations of the system.

It should be noted that in practical applications, the solution of the transport equation and the optimization done above is merely a first step in the process. The industrial reality includes both data uncertainties and varying operational conditions. Thus, a follow-up procedure is essential. However, the above solution makes this follow-up feasible and coherent, because detailed time-dependent quantities are predicted: the number of expected failures as a function of time, the availability and accumulating costs, etc. If the future reality falls well within the above prediction, then some confidence is acquired in the prediction process. If discrepancies arise then the source of such mismatches may be investigated. The source could be in the data, or in the model used (a single component with a single PDF describing a set of hundreds of pipe segments and the assumption of only two possible states), or it may be in the reality; for example, the inspection process is not carried out according to specifications. The calculation is then repeated with the corrected data or model, and the process is reiterated. A number of such cycles guarantees a reliability culture that combines detailed transport calculations with shop floor practices that result in optimal operation. In general, data collection is a unidirectional process. Data is collected by one body in the organization, with no reference to the question of how this data will ever be used. This results in situations in which, for example, only the number of failures per time period is collected, leaving out important elements such as the time of failure, the time of the component's entry into operation and the nature of the repair. Without these items it is impossible to perform point or interval estimates of the PDFs. The above process also results in a bi-directional interaction of analysis, operations and data collection. The follow-up process indicates the quantities essential for prediction verification. The model used affects the parameters that are to be observed and collected. The ability to solve realistic system transport equations makes data collection worthwhile, and serves as an indicator of the important data elements.

Event type estimation and biasing Figure 7.1.5 shows the results for a calculation with exponential distributions with $\lambda = 0.001$ and $\mu = 0.05$ for 20 hours with 10 000 histories.

The PRSD of the event type estimator [`Program carloe.for`] seems to be about a factor of 2–4 smaller than that of the surface estimator. The calculation time of the event type estimator is about three times longer, resulting in a benefit factor of about six in favor of the event type estimator. Still, it can be seen that this difference decreases as the time increases, and at $T = 20$ hours the difference is only about a factor of 1.6. This trend continues as the calculation is extended to higher time points. For 1000 hours, for example (already in a steady state), the surface estimator yields a PRSD of 0.148 (with 10 000 histories), and the event type estimator is a PRSD of 0.648; with the same relationship between the calculation times, the FOM of the surface estimator is 68 times smaller (better) than that of the event type estimator.

This situation can be viewed in the following way. In any calculation the detector response may be divided into two parts: first, the probability of the system reaching the detector; then the contribution a history makes upon reaching the detector. This can be denoted in the form of the product $I = P_d\langle\vartheta\rangle$, where P_d is the probability of reaching the detector and $\langle\vartheta\rangle$ is the average conditional contribution made in the detector. The second moment is given by $S^2 = P_d\langle\vartheta^2\rangle$. In the case of the surface estimator, the detector zone is defined by the condition {Passing through time point T_j in state 1}.

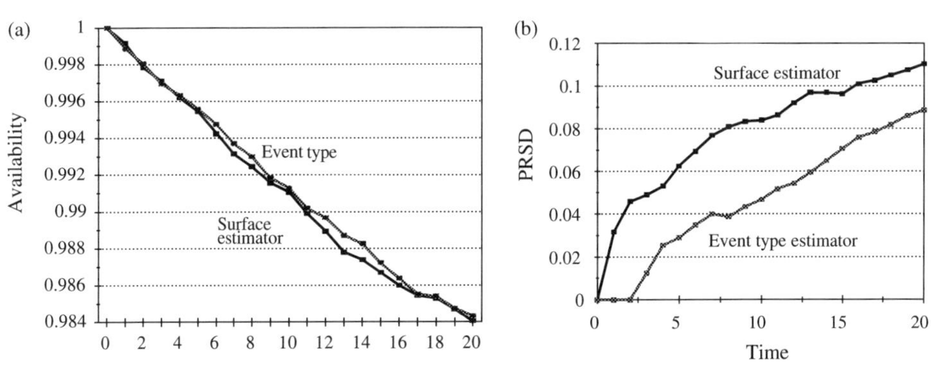

Figure 7.1.5 Availability and PRSD with the surface and event type estimators

The probability for that is exactly the availability, i.e. $P_d = A(T_j)$. The contribution made in the detector is strictly unity, i.e. there is no uncertainty in the score made at the detector. The uncertainty is fully contained in reaching the detector zone, thus the second moment of ϑ is also unity, resulting in the well-known binomial relation $S^2 = P_d$, so that the variance is $V = S^2 - I^2 = P_d(1 - P_d)$. The alleged advantage of variance reduction methods and special estimators is that they increase the probability of reaching the detector, thereby reducing the uncertainty involved. Thus, P_d is increased and the uncertainty in it decreases. Note that the second moment is proportional to P_d. However, such an increase must be complemented by a decrease in the average value of ϑ in order to maintain an unbiased calculation. For the event type estimator, the probability of reaching the detector is unity: it is the exact opposite of the surface crossing estimator, as there is no uncertainty in reaching the detector. ϑ now becomes a random variable, and bears the full impact of the uncertainty. It is the balance between these two uncertainties that determines the efficiency of the process. Indeed, in the above example there is no compelling reason to assume that any estimator is better than any other. With the relations shown above, and noting the virtue of simplicity of the surface crossing estimator, it may be the preferred option.

It is worthwhile applying biasing only when the unavailability is very small. Consider the unavailability at 0.5 hours. For the above case, its value is 4.937E−05, thus in an unbiased calculation about 20 000 histories are required to obtain a detector event. The PRSD is readily known, since the surface estimator is a binomial estimator, and the PRSD for 100 000 histories is 45%. The calculation time is 12 seconds, resulting in a FOM of 540.

Forced event biasing with a single fixed event probability [`Program carlob.for`] was used for that calculation. The second moment, with p_s in the range 0.01–0.98, is shown (solid line) in Figure 7.1.6, and is compared to equation (6.3.8) (cross markers). The constant $\sum_{k=0}^{\infty} \Gamma_k$ was obtained from a single calculation to be 2.431E−09. It can be seen that a minimal second moment is obtained for a relatively wide range of probabilities, which is typical of many 'good' variance reduction schemes. In fact, a

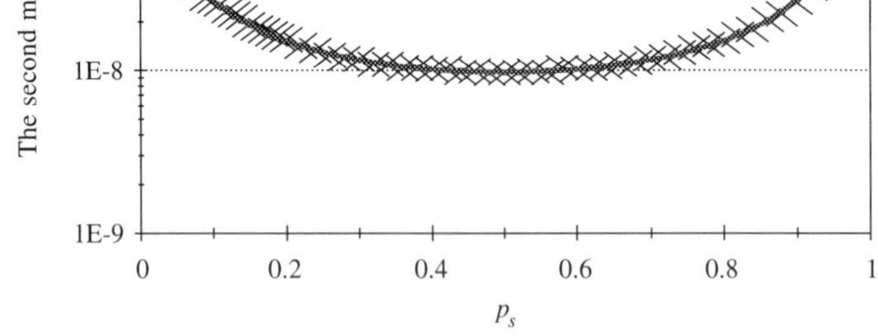

Figure 7.1.6 The second moment as a function of the fixed next event probability

good variance reduction scheme may be characterized so that it yields a high benefit over a wide range of parameters. Still, a choice of p_s above 0.98 (which will ensure a large number of induced events) will result in a very high variance.

The optimum is at $p_s = 0.5$, resulting in an estimate of 4.927×10^{-5} with a PRSD of 1.7% with 10 000 histories. The calculation time is 2.2 seconds, and the resulting FOM is thus 6.35, yielding a benefit of 85 over the analog calculation. The forced event biasing is effective in this case, since a single event results in a failure. This is not the case when systems with redundancies are considered, as will be demonstrated in the next section.

7.2 Systems with Exponential Distributions

Consider a general system with n components. Let the failure and repair rates of the ith components be denoted by λ_i and μ_i, respectively. Given a system state $\vec{\mathbf{B}} = (b_1, \ldots, b_n)$, the rate of change of the ith component is β_i, where

$$\beta_i = \begin{cases} \lambda_i & b_i = 1 \\ \mu_i & b_i = 0. \end{cases}$$

The hazard of the system is given by

$$\alpha(\vec{\mathbf{B}}) = \sum_{i=1}^{n} \beta_i. \tag{7.2.1}$$

The Free flight kernel (6.2.10) takes a particularly simple form of an exponential PDF:

$$T(\vec{\mathbf{B}}; t \to t') = \alpha e^{-\alpha(t'-t)}, \tag{7.2.2}$$

and the collision kernel (6.2.12) can be expressed in terms of the probability $P(i)$ that the ith component will change its state in an event, namely

$$P(i) = \frac{\beta_i}{\alpha}. \tag{7.2.3}$$

The above transport kernels implies the following procedure for a single history.

Procedure 7.2.1: A general system history

1. Initialize transport quantities: event number $i = 0$, current time $t_i = 0$, previous event time $t_p = 0$. Set initial state $\vec{\mathbf{B}}_i$ with $b_{j,i} = 1$, (i=event number, j=component number).
2. Update the system hazard $\alpha(\vec{\mathbf{B}}) = \sum_{i=1}^{n} \beta_i$.
3. Calculate P_e, the probability that the system will have the next event before T_m. P_e is given by

$$P_e = 1 - e^{\alpha(T_m - t_i)}.$$

4. Let ξ be a random number. If $\xi \leq P_e$ the system continues the transport, go to step 5. Otherwise, the system has the next event beyond T_m. History terminated. Go to step 4.1.
   4.1 *History terminated.* Tally surface estimator for each $T_j > t_i$ ($\{T_j\}, j = 1, \ldots, k$ is a set of time points at which the estimated quantity is to be calculated) the estimate $e(T_j) = S(\vec{\mathbf{B}}_i)$ and terminate the history.

5. *History continues*: sample next event point from the normalized distribution. The next event point is obtained from the solution for t of $1 - e^{-\alpha(\vec{\mathbf{B}})(t-t_i)} = P_e \times \xi$. Denote the next event point as t_n. Proceed to tally the surface crossing estimator for time points in (t_i, t_n).
   5.1 Tally for each $t_i < T_j \leq t_n$ the estimate $e(T_j) = S(\vec{\mathbf{B}}_i)$. Note that the term $S(\vec{\mathbf{B}}_i)$ is related to any system function or system response of interest. Thus, there may be several such functions.
6. Transfer the system to the next state, update the previous time, $t_p = t_i$ increase $i \rightarrow i + 1$, update the event time, $t_i = t_n$.
7. Sample the next state of the system, sample the event kernel. Sample m, with probability $P(m) = \beta_m/\alpha$, where β_m is the hazard of the mth component in its current state, and α is the sum of hazards over all components. m is the number of the trigger component. Change the state indicator of the mth component to its new state. Proceed to step 7.1 for additional logic.
   7.1 *Additional logic and contingency rules*: if as a result of the change in the state of the trigger component additional changes in the system are occurring, execute them at this point. At the end of steps 7 and 7.1 a new state $\vec{\mathbf{B}}_i$ of the system is established.
8. Proceed to step 2.

The above procedure is first applied to the system shown in Figure 4.1.6. The system is shown again in Figure 7.2.1, with the MTTFs given above each component. The MTTRs are 48 h for components 1–5, 16 h for 6–8, 32 h for 9–10, 48 h for 11–16, 32 h for 17–18, 16 h for 19–21, 48 h for 22–27, 16 h for 28–30 and 48 h for 31–32. The time dependent unavailability in the range 0–200 hours is shown in Figure 7.2.2. [`Program relab1.for`]

The calculation was performed with 200 000 case histories. The PRSD is in the range of 5–10%. Each history contains, on average, 38.9 events (failure and repairs), and the

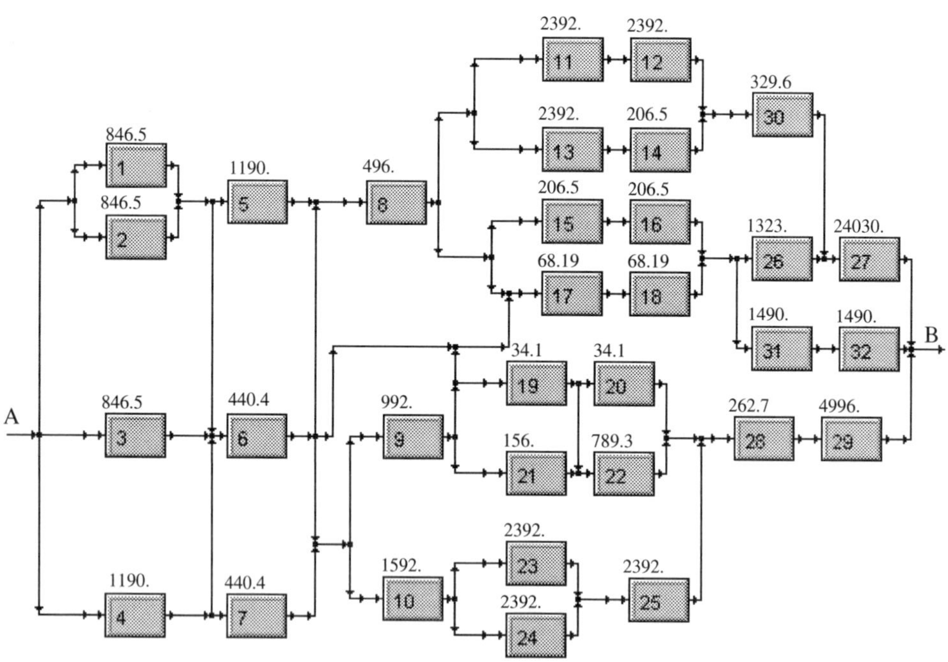

Figure 7.2.1 The system structure

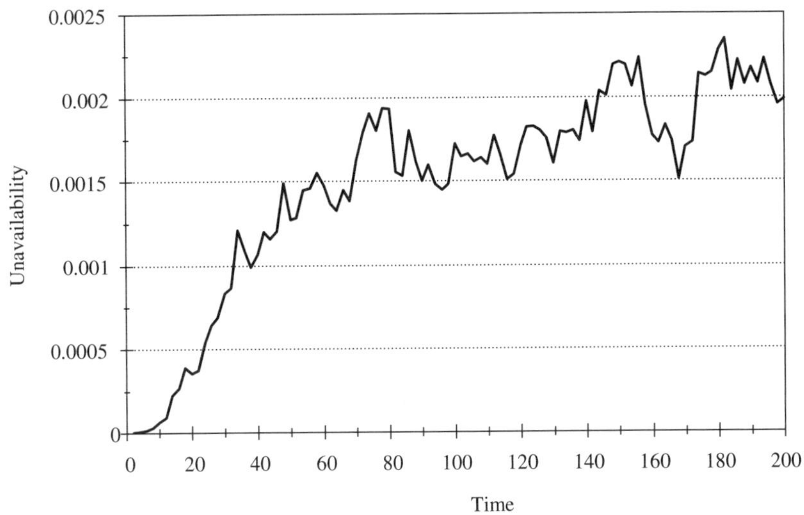

Figure 7.2.2 The unavailability of the system of Figure 7.2.1

average processing time of a single history is 0.00628 seconds. There are two additional quantities worth observing which may support the design improvement of the system. They are presented in Figure 7.2.3. The top of the figure shows the average number of failures of each component in the system. This quantity is related only to the failure and repair distributions of each component, and not to its position in the topology of the system. It can be seen that components 17–21 have the highest number of failures. A large number of failures indicates a high load on the crash maintenance effort, since in this model it is assumed that each component is repaired upon failure, i.e. the failure is detected, and all the resources needed for repair are available.

In the bottom of the figure the 'unavailability sensitivity' is shown. This sensitivity, also denoted as U sensitivity, is best defined in an operational form as follows:

1. Each time the *system* fails, denote the down time of the system as τ_i.
2. Identify the components 'responsible' for this system failure, as follows. Scan all the components, for each component j; if the component is not failed, skip it. If the component is failed, check the state of the system if this component would have been operational (change its state to 1 and check $S(\vec{\mathbf{B}})$ with this change). If a repair of the component results in a repair of the system, then and only then, attribute the down-time τ_i to that component. The sensitivity is the ratio of the contributions to each component and the sum of the down times. It yields the fraction of down time for which each component is 'responsible'. If, instead of the down time, unity is scored at each failure, a reliability sensitivity results.

It can be seen above that there is little correlation between the number of failures and the actual performance. Components 19 and 20 have a large number of failures with a very small effect on the system's unavailability. Components 8 and 28 have few failures with a critical impact on the performance.

The average unavailability is $U_0 = 0.00157$. Increasing the failure rates of components 6 and 7 (with U-sensitivities of about 0.03) by a factor of two increases the

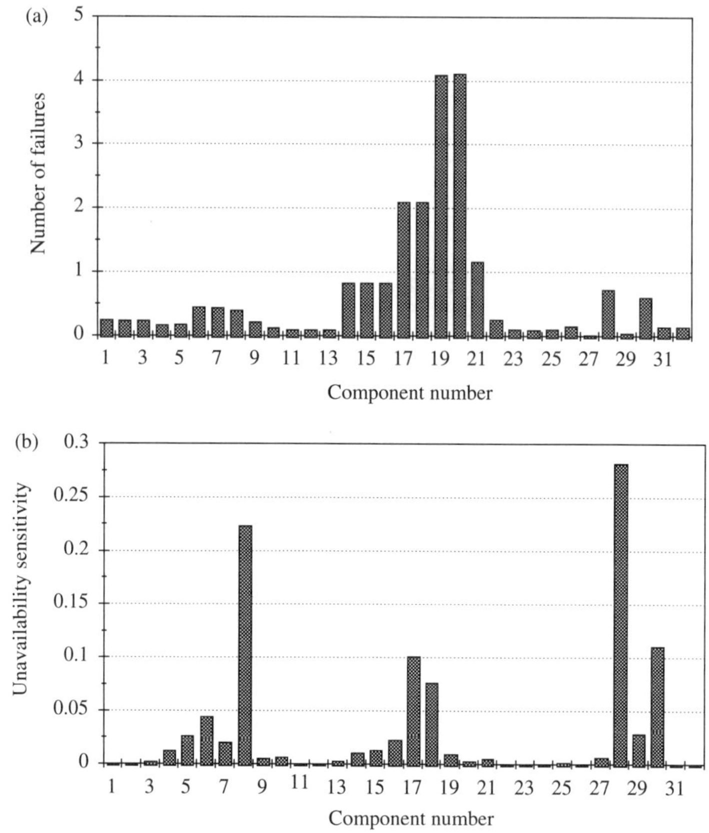

Figure 7.2.3 Average number of failures and unavailability sensitivity

unavailability to 0.00186, thus the ratio of increase is $\Delta U/U_0 = 0.184$. A similar increase in the failure rates of components 8 and 28 (with U-sensitivity of about 0.25) yields an average unavailability of 0.0044, with $\Delta U/U_0 = 1.8$, hence the relative increase in the unavailability is proportional to the sensitivity. The sensitivity is an indicator of the performance 'bottlenecks' of the system. Other elements of the design may be investigated, for example, removing the line connecting component 6 to components 9 and 10 could save money in the design, but it will result in an increase by a factor of two in the unavailability.

Example 7.2.1: Design options of a conveyer

A conveyer belt is designed to carry heavy loads on a continuous basis. Two design options are considered. In the first option, it is a simple belt made of 22 identical elements in series. The failure rate of each element is 3.33×10^{-4} and the repair rate is 0.4. The cost of one hour of down time of the system is $c_d = \$687$; the cost of each repair is $c_f = \$840$. The second option involves improved units with a reduced failure rate of 2.1×10^{-4}, designed such that a system failure occurs only if two adjacent elements have failed. Thus, in this design single failed units are not repaired, and repair is applied only when a system failure occurs. Consequently, the repair time is slightly higher and the repair rate is 0.125. The second design is \$18 000 more expensive than the first. Both designs operate for one year, at which time the whole system is replaced. Which design is more economic?

For the first design, the system function is simply the product of the components state indicators, i.e. $S(\vec{\mathbf{B}}) = \prod_{i=1}^{22} b_i$.The steady state availability can in fact easily be calculated analytically. A binomial estimator is used to estimate the average number of system failures (this estimator accumulates unity upon each failure of the system). For the second design, use of additional logic as suggested in step 7.1 of the general procedure is made. Each component has a third state, 'failed not being repaired', with a value (−2). In this state the hazard of the component is zero. Upon failure of a component, the state of the system is checked. If the system is operational, the component is set into state −2. If the system is failed, then all the components that are in state −2 are set to state (0) so that repair is resumed. Indeed, the fact that all the failed components are set to state 0 means that there are enough repair teams to treat all of the failed components simultaneously. Also, the system function reflects the fact that a system failure occurs only when two adjacent components are failed; this is achieved with the function

$$S(\vec{\mathbf{B}}) = \prod_{i=1}^{21}(a_i + a_{i+1} - a_i a_{i+1}) \quad \text{where } a_i = \begin{cases} 1 & \text{if } b_i = 1 \\ 0 & \text{if } b_i < 0. \end{cases}$$

The results of the availability as a function of time are shown in Figure 7.2.4. The average availability for the first design is 0.9818, and for the second option it is 0.99827. This means an average downtime of 159.4 hours at a cost of \$109 529 for the first design, and a downtime of 15.15 hours at a cost of \$10 411 for the second design. The average number of failures in the first design is 2.876 failures per component, or a total of 63.27 failures with a repair cost of \$53 148. The number of system failures is also 63.27, since each component failure is also a system failure. For the second design the average number of failures per component is only 2.584, because a component may remain failed for a long period and is not repaired while the system is operational, so its 'lifetime' is shorter and fewer failures are expected. The total number of component failures is 56.848, with a repair cost of \$47 752. In this design, the number of system failures is not the same as the number of component failures, and only 22.8 system failures occur. The total average cost of the first design is \$162 677, while that of the second is \$58 163. The second design

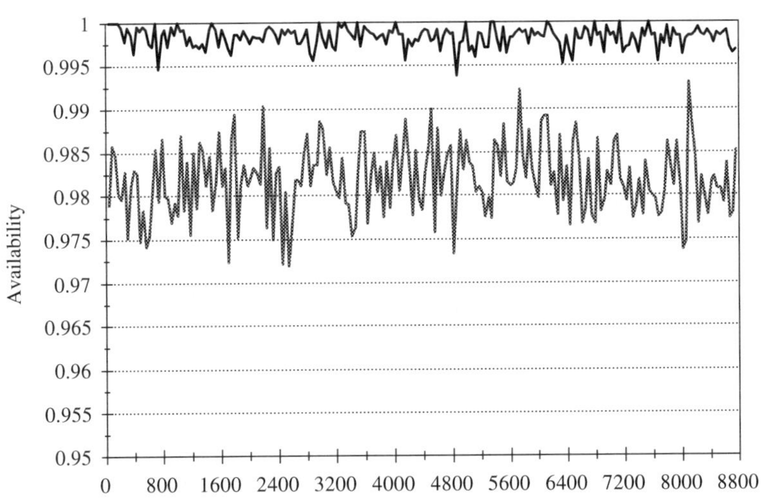

Figure 7.2.4 Availability as a function of time for two conveyer design options

is superior to the first, and will bring a saving of \$86 514 after including the initial higher expense. The reduced number of system failures is a fringe benefit of the second design.

Example 7.2.2: Stand-by relations, load considerations, induced failures and throughput

Step 7.1 in Procedure 7.2.1 is the key to the versatility of the modeling enabled by MC calculations. A multitude of realistic scenarios can be modeled upon the occurrence of a trigger event. Some of these are demonstrated in this example.

The system involves two parallel branches, each containing a capacitor and an electronic valve, as shown in Figure 7.2.5.

The failure rate of each capacitor is $\lambda_c = 6.25 \times 10^{-4}$, that of the valves is $\lambda_v = 1.45 \times 10^{-4}$ and that of the lines (1 & 2) is $\lambda_L = 8.22 \times 10^{-5}$. The repair rates are 0.0625 for capacitors and valves and 0.0161 for the lines. The availability system function for the subsystem of capacitors and valves is

$$S_1(\vec{\mathbf{B}}) = b(1)b(3) + b(2)b(4) - b(1)b(2)b(3)b(4).$$

However, in this system (as in many other production systems), one is mainly interested in the production rate rather than the availability. The production rate is often not proportional to the availability. Consider the section of the two line components. When both lines are operational, the production rate is 1800 units/hour. When only one branch is operational the production reduces to 1200 units/hour. Under these conditions, the availability gives a misleading result, and the parallel drawing of the two line components is not really a network. The system function to be used in this case is

$$S_2(\vec{\mathbf{B}}) = S_1(\vec{\mathbf{B}}) \times \{1800 \times [b(5)b(6)] + 1200^*[b(5) - b(5)b(6)] + 1200^*[b(6) - b(5)b(6)]\}.$$

A different way to write the same system function is

$$S_2(\vec{\mathbf{B}}) = \begin{cases} 1800 & \text{if } b(5)b(6) > 0 \text{ and } S_1(\vec{\mathbf{B}}) > 0 \\ 1200 & \text{if } b(5)b(6) = 0 \text{ and } [b(5) + b(6)] > 0 \text{ and } S_1(\vec{\mathbf{B}}) > 0 \\ 0 & \text{Otherwise.} \end{cases}$$

[`Program relab5.for`].

Induced failures The conditions of the system are such that if any of the capacitors fails, an immediate failure may occur to any of the valves; an induced failure. The

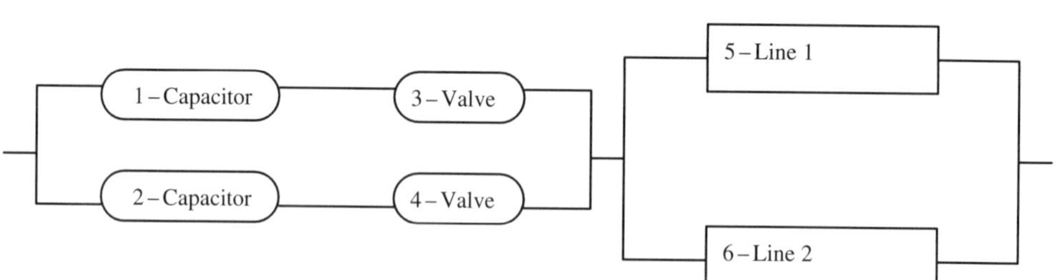

Figure 7.2.5 A parallel standby system

induced failure probability is a non-linear function of time and is given by $P_{id}(t) = 0.005 + 3.23 \times 10^{-7} \times t^{1.5}$. This probability starts at 0.5% and reaches 0.269% at the end of the year. A general maintenance is applied at the end of each year which reduces this probability to its original value. Since standby logic is also used in this example, it is assumed that a passive state is denoted by -1. In a passive state the induced failure probability is a factor of 10 smaller than when the component is active. The introduction of this logical element is trivial, since at step 7.1 of the calculation algorithm, the states and events of all of the components are known. The following procedure captures this logic.

Procedure 7.2.2: Induced failure logic

1. If {components 1 or 2 failed now}? Denoting the previous states of components by $b_p(i)$ and the current state by $b(i)$, the condition is fulfilled if {for either $i = 1$ or $i = 2$, $b_p(i) = 1.0$ and $b(i) = 0.$} Proceed to step 2. Otherwise proceed to step 4. Steps 2 and 3 should be repeated for components 3 and 4.
2. If component $j = 3, 4$ is in state 1 (operational and active), assign the probability $P = P_{id}(t)$. If the component is in state -1 (operational and passive), assign the probability $P = (P_{id}(t))/10$. If the component is failed, assign $P = 0.0$.
3. Sample a random number ξ. If $\xi \leq P$, induce failure on the component, i.e. set its current state to 0. Proceed either to next valve or step 4.
4. Logic terminated.

Load factors As explained above, if any line fails the other line can increase its output to a level of 1200 units. This comes at a price, because the line is not designed to operate under such stress. The failure rate may increase considerably, which again is a function of the time with a reset at the end of every year. The factor by which the failure rate increases is called the *load factor*, and is given by $L(t) = 2. + 0.32 \times t^{0.5}$. This logic is easily implemented in step 7.1. It should be noted that in this case it is not necessary to look for the event that occurred to components 5 and 6. It is enough to enquire into their states at every event. The following procedure captures this logic.

Procedure 7.2.3: Load factor logic

1. If both components are operational, i.e. $b(5) = 1.0$ and $b(6) = 1.0$. Reset the failure rates of both components to their nominal value. Go to step 3. Otherwise proceed to step 2.
2. Multiply the nominal failure rates by $L(t)$ and assign the increased failure rates to the components.
3. Logic terminated.

Figure 7.2.6(a) displays the average production as a function of time for the first three cases.

The average annual production rate for the three above cases is 1773.3, 1768.5 and 1756.2, respectively. [Program `relab6.for`]

Stand-by relations Stand by relations are often used when one unit suffices for a particular operation and a second unit is set as a back up to go into operation if the first one fails. The benefit of such a design is that, since passive units do not operate most of the time, fewer failures will occur. In a symmetric back-up structure the unit

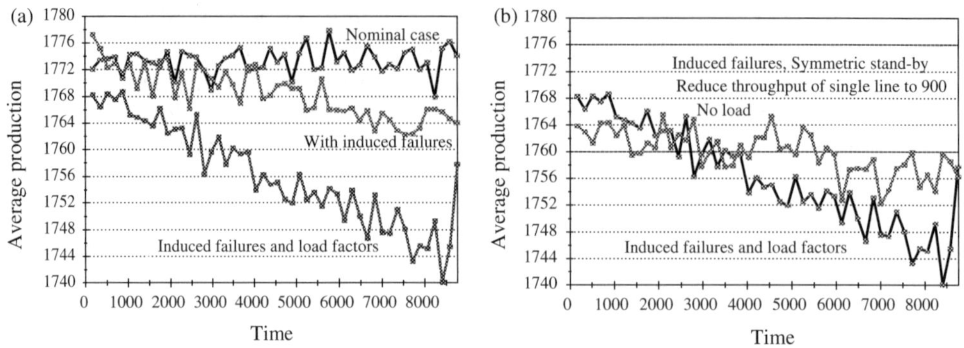

Figure 7.2.6(a,b) Average production as a function of time

that goes into operation remains in operation until it fails. A symmetric standby, in most cases, has about half the failures of the active design. If aging occurs, it is distributed evenly between the components, and less aging is expected. In the present case, aging is of no importance since the failure distributions are exponential. An additional advantage is obtained since the induced failure probability to a passive component is smaller. In the present example, a standby relation can be arranged for the two branches of the capacitors and valves.

The standby relation is realized by setting the components of the lower branch to state -1 at the beginning (at $t = 0$). It is also assumed that passive components have a zero failure rate (cold standby). The logic is outlined in the following procedure:

Procedure 7.2.4: Symmetric standby: (to be inserted at step 7.1 in Procedure 7.2.1)

1. *Failure of upper branch*: if the upper branch now failed (i.e. if $b_p(1) = 1.0$ and $b_p(3) = 1.0$, and $b(1) \times b(3) = 0$) then go to step 1.1.
   1.1 If $b(2) \times b(4) = 1$ (lower branch not failed) activate lower branch, set $b(2) = 1.0$ and $b(4) = 1.0$.
2. *Repair of upper branch*: if the upper branch is now repaired (i.e. if $b_p(1) \times b_p(3) = 0$ and $b(1) \times b(3) = 1$) then go to step 2.1.
   2.1 If $\{b(2) = 1 \cap b(4) = 1\}$ (lower branch active) set upper branch passive, set $b(1) = -1$ and $b(3) = -1$. Otherwise, set upper branch active, set $b(1) = 1$ and $b(3) = 1$.
3. *Failure of lower branch*: if the lower branch now failed (i.e. if $b_p(2) = 1.0$ and $b_p(4) = 1.0$ and $b(2) \times b(4) = 0$) then go to step 3.1.
   3.1 If $b(1) \times b(3) = 1$ (upper branch not failed), activate upper branch, set $b(1) = 1.0$ and $b(3) = 1.0$.
4. *Repair of lower branch*: if the lower branch is now repaired (i.e. if $b_p(2) \times b_p(4) = 0$ and $b(2) \times b(4) = 1$) then go to step 4.1.
   4.1. If $\{b(1) = 1 \cap b(3) = 1\}$ (upper branch active), set lower branch passive, set $b(2) = -1$ and $b(4) = -1$. Otherwise, set upper branch active, set $b(2) = 1$ and $b(4) = 1$.

[`Program relab7.for`]

This case is used together with a reduction of the throughput of a single line to 900 units/hour, which reduces the load factor to unity. The comparison of this case with the last case is shown in Figure 7.2.6(b). The average production is 1759.9, higher than the case with the load (and higher single line production). It is noted that, since the load factor is small at the beginning of the year, in that region the case with the load yields a higher production. The situation is reversed after about 3000 hours of operation. This may suggest the possibility of a mixed logic, i.e. use increased production up to 3000 hours and then reduce to 900 units with a single line.

Example 7.2.3: Two data relay computers with standby, delay and spare parts options

This example follows the case discussed in the introduction to this book. Two data computers, (A) and (B), are set in parallel. Machine (A), being more advanced and reliable (and consequently considerably more expensive), is used as the main system. (B) serves as a back-up in a Master/Slave logic. Initially, (B) is passive and not connected to the system. Upon failure of (A), (B) is activated following a time delay. (A) is shipped for repairs to an external location. Upon the return of A, it is installed in the system and (B) is set back to a passive state. The activation of (A) upon its installation does not involve any delay. If (B) fails, it is also sent to a remote repair, and returns to the site after a given recycling time. The data related to the problem is as follows: the failure time distributions of (A) and (B) are Weibull distributions with scale parameters $\lambda_A = 5.67 \times 10^{-5}$ and $\lambda_B = 1.498 \times 10^{-4}$, respectively, and a shape parameter $\beta = 1.4$ is identical for both. The delay time is uniformly distributed between 1.5 hours and 2.5 hours. The turn-around times of the two machines are uniformly distributed between 168 and 240 hours. The replacement time of a repaired machine in its position is exponentially distributed with a MTTR = 10 hours.

The task in this case is to first evaluate the performance level of the system, and then suggest and study options for its improvement. To model this problem, five components are defined. The components are machine A, machine B, a delay clock, a component modeling a spare of A and a component modeling a spare for B. The states and distributions attached to the components are detailed in Table 7.2.1.

A state vector of the system is a combination of the above states. The initial state is $\vec{\mathbf{B}}_0 = (1, -1, -1, -2, -2)$. Since non-exponential distributions are involved in this process, it is necessary to maintain the birth point of each component in each state and update it appropriately. It is

Table 7.2.1 States and distributions

Component name	States number and meaning		PDF in state
1 – Machine A	1	Operational and active	Weibull (5.67×10^{-5}, 1.4)
	0	Failed under replacement	EXP (10)
	−2	Failed waiting for spare	Event at infinity – (No event)
2 – Machine B	1	Operational and active	Weibull (1.498×10^{-4}, 1.4)
	0	Failed under replacement	EXP (10)
	−1	Operational and passive	Event at infinity – (No event)
	−2	Failed waiting for spare	Event at infinity – (No event)
3 – Delay clock	1 – Active		Uniform (1.5, 2.5)
	−1 – Passive		Event at infinity – (No event)
4 – Spare for A	1 – Spare available		Event at infinity – (No event)
	0 – Spare in recycling		Uniform (168, 240)
	−2 – Spare not available and not in cycle		Event at infinity – (No event)
5 – Spare for B	1 – Spare available		Event at infinity – (No event)
	0 – Spare in recycling		Uniform (168, 240)
	−2 – Spare not available and not in cycle		Event at infinity – (No event)

Table 7.2.2 Logical rules of the system

Trigger event	Operations following the trigger event
Failure of A	1. Set state of A to (−2) – Waiting for spare. 2. Set component 3 active (activate the delay clock). Set the birth point to current time. 3. Set component 4 in state 0 – Start recycling. Set the birth point to current time.
Repair of A	1. Set state of A to 1. Set birth point to current time. 2. Set the delay clock passive (state −1). 3. If B is active set it passive (−1).
Failure of B	1. Set state of B to (−2) – Waiting for spare. 2. Set component 5 in state 0 – Start recycling. Set the birth point to current time.
Repair of B	1. If A is in state 1 set B in state −1. Otherwise, set state of B to 1. Set birth point to current time. 2. Set the delay clock passive (state −1).
Event in delay clock	1. This is end of delay – Set delay clock passive, state −1. 2. If B is in state −1 activate it into state 1. Set the birth point to current time.
Event in 4 Arrival of A spare	1. Set A into state 0, set component 4 into state −2.
Event in 5 Arrival of B spare	1. Set B into state 0, set component 5 into state −2.

assumed that, upon repair, the component is as good as new. The updating of the birth times will be part of the logical rules of the system.

The sampling of the next event point is done for this model by sampling the next event point, for each component, according to its current state (and birth point), and then taking the minimal event time. This means that, along with the next event time point, one also obtains the number of the trigger component. The logical rules relevant for this are described in Table 7.2.2.

The availability as function of time, for 1000 hours, is shown in Figure 7.2.7. [`Program relay.for`] The average unavailability, of about 1.5%, is unacceptable: it is required to have an unavailability below 0.5%. To achieve this goal, four alternative design options are considered:

Option 1: add a spare Machine A on location. To realize this design, a sixth component needs to be added to accommodate for the fact that there may be two machines of type A in recycling. This component has the same states and distributions as component 4. The initial state is $\vec{\mathbf{B}}_0 = (1, -1, -1, 1, -2, -2)$. The logic upon failure of A needs to be modified as shown in Table 7.2.3.

Option 2: add a spare Machine B on location. To realize this design a sixth component needs to be added to accommodate for the fact that there may be two machines of type B in recycling. This component has the same states and distributions as component 5. The initial state is $\vec{\mathbf{B}}_0 = (1, -1, -1, -2, 1, -2)$. The logic upon failure of B needs to be modified as shown in Table 7.2.3 with reference to components 2, 5 and 6, respectively.

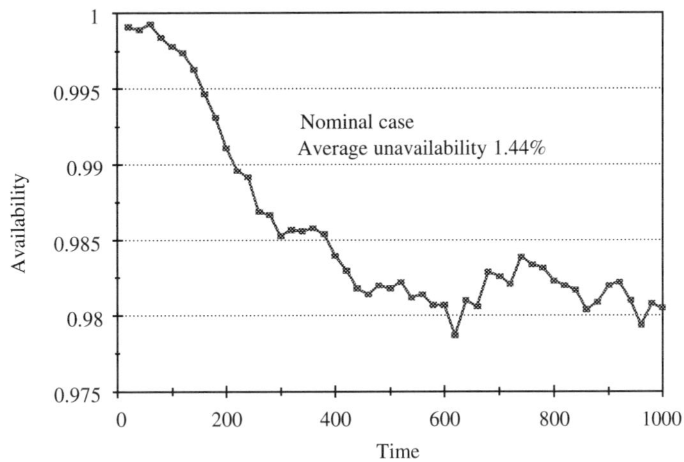

Figure 7.2.7 Availability as a function of time, nominal system

Table 7.2.3 Logic of spare added to A

Failure of A	1. Set component 3 active (activate the delay clock). Set the birth point to current time. 2. If b(4) = 1, set b(1) to 0. Set b(4) to 0. Set both birth points to the current time. Go to step 5. Otherwise go to step 3. 3. If b(6) = 1, set b(1) to 0. Set b(6) to 0. Set both birth points to the current time. Go to step 5. Otherwise go to step 4. 4. No spare available – Set b(1) = −2. If b(4) = −2 set b(4) = 0, set its birth point to current time. Otherwise set b(6) to zero set birth point of component 6 to current time. 5. Exit.

This option is not attractive, although it is much cheaper than option 1, because machine B is a backup. It is expected to operate about 15% of the time (since the steady state availability of A is about 85%), and it would seem unreasonable to purchase a spare for a backup that may be viewed in itself as a spare.

Option 3: replace machine B by a machine of type A. This model is realized by simply using the failure distribution of A for B. This option is expected to yield a benefit, since machine B will be more reliable.

Option 4: eliminate the delay and set B in an active state. This option is expected to yield a benefit, since the delay time attached to each failure of A will be eliminated. Its realization is achieved by minimal changes in the nominal model. All references to component 3 in the logic are omitted and, upon repair, component 2 goes into state 1. The initial state is $\vec{\mathbf{B}}_0 = (1, 1, -1, -2, -2)$. The availability of all five cases as a function of time is shown in Figure 7.2.8, with the average unavailability of each case highlighted in Table 7.2.4.

The results indicate that the most economic design that meets the system requirements is option 2, i.e. adding a spare to the machine B. Incidentally, this was the option that was not suggested when a debate took place on possible systems improvements.

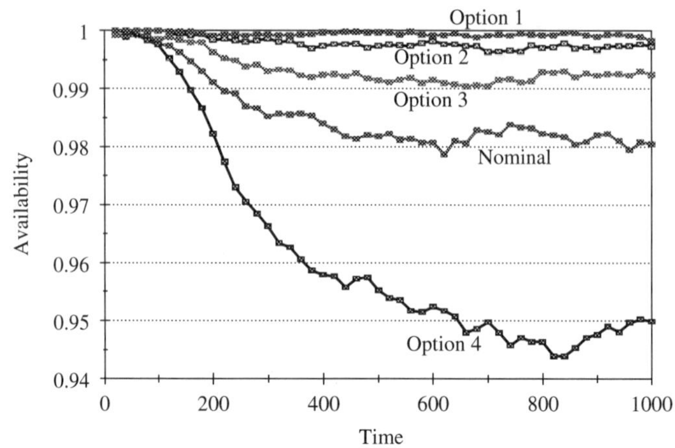

Figure 7.2.8 Availability of five design options

Table 7.2.4 Average unavailabilities

Case	Average unavailability (%)
Nominal	1.44
Option 1	0.056
Option 2	0.21
Option 3	0.65
Option 4	3.7

7.3 Transition Biasing Considerations

Consider the unreliability of two identical active components in parallel. Unreliability calculations follow exactly the same history generation procedures as described in Section 7.2. The estimator takes a value of unity for every time point larger than the point of system failure, and zero otherwise. Upon failure of the system, the history may be terminated. Consider the case in which the failure rate is $\lambda = 0.001$, and the repair rate is $\mu = 2.5$ for both components. The unreliability at 100 hours is 7.958×10^{-5}, as can be calculated from the expression for the reliability (see Van Alven, W. H., Editor, Reliability Engineering, ARINC Research Corporation, Prentice-Hall, 1964):

$$R(t) = \frac{s_1 e^{s_2 t} - s_2 e^{s_1 t}}{s_1 - s_2} \quad \text{with} \quad s_1 = -\frac{1}{2}\left[(3\lambda + \mu) - \sqrt{\lambda^2 + 6\lambda\mu + \mu^2}\right] \quad \text{and}$$

$$s_2 = -\frac{1}{2}\left[(3\lambda + \mu) - \sqrt{\lambda^2 + 6\lambda\mu + \mu^2}\right]. \tag{7.3.1}$$

This equation is easily derived from the Markov equations of the system using the Laplace transform.

In a Monte Carlo calculation, the unreliability represents the probability for a system failure up to a time of 100 hours. Thus, about 12 565 histories are required to see a single system failure. Clearly, at least two events are required to see a system failure. The average number of events encountered in 100 hours is about 0.4. In 81.8% of the histories there will be no events at all, and only about 18% of the histories will have one or more events. In general, increasing the number of events is done with 'forced event biasing'. In this case it will not be used because a lack of events is not essential. The main difficulty has to do with the nature of the events. At least two events are needed, and these must be consecutive failure events. Assuming that at $T = 0$, the system starts in the state $(1, 1)$, it is obvious that in the first event one of the components will fail. In the second event, with probability $P_f = \lambda/(\lambda + \mu)$ the second component will fail, resulting in a desired system failure, and with a probability $P_r = \mu/(\lambda + \mu)$ the failed component will be repaired, bringing the system back into the original state. In most cases,

$$\frac{\mu}{\lambda + \mu} \gg \frac{\lambda}{\lambda + \mu}$$

(in the present case, the ratio is 2500). It is thus the preference of repair over failure that makes the system failure a rare event. To enhance the appearance of this rare event, it seems reasonable to increase the failure probability. This can be done by increasing the failure rates, using a transformed failure rate $\lambda^* = \beta\lambda$, where β is an external parameter called the *transition biasing factor*. The probabilities for a failure and repair become

$$P_f^* = \frac{\lambda^*}{\lambda^* + \mu} \quad \text{and} \quad P_r^* = \frac{\mu}{\lambda^* + \mu},$$

respectively. This transformation will be called *transition biasing*, since bias is applied to the state transition probabilities. Indeed, if β is increased indefinitely, the probability for a failure increases to unity, guaranteeing a failure at every event and a biased estimate.

The case of two components is convenient since, for this simple case, an approximate theory supporting the optimization of the method can be constructed. Let $P_h(k)$ denote the probability that a particular history will have exactly k events, and such that no failure is encountered in the first $(k - 1)$ events and the kth event is a system failure. The unreliability at time t, denoted by $\bar{R}(t)$, can then be expressed as

$$\bar{R}(t) = \sum_{k=1}^{\infty} P_h(k). \tag{7.3.2}$$

Next, it will be assumed that the repair time is negligible. This is justified, since in those cases where transition biasing is required, $\mu \gg \lambda$. When the system is in state $(1, 0)$ or $(0, 1)$, the average time for the next event is $1/(\lambda + \mu)$. This time is negligible compared to the time to the next event from the state $(1, 1)$, which is $1/2\lambda$. With this assumption a 'collision' now involves two simultaneous transfers: first, the transfer $(1, 1) \rightarrow (0, 1)$ or $(1, 0)$; and then an immediate transfer back to state $(1, 1)$. The transport of the system is now viewed as a set of such events. The average time between two consecutive events is $1/2\lambda$, and if a failed system state occurs the system is not repaired (this is true for all reliability calculations). The probability $P_h(k)$ can be written as $P_h(k) = G_h(k) \times L_h(k)$, where $G_h(k)$ is the probability to have exactly k such collisions up to time t, and $L_h(k)$ is the probability that in the first $(k-1)$ collisions the system will transfer back into state $(1,1)$, while in the kth collision the transfer will be into state $(0,0)$. $G_h(k)$ is given by the Poisson probability

$$G_h(k) = \frac{(2\lambda t)^k e^{-2\lambda t}}{k!}.$$

$L_h(k)$ is given by

$$L_h(k) = \left[\frac{\mu}{\lambda+\mu}\right]^{k-1}\left[\frac{\lambda}{\lambda+\mu}\right].$$

Thus, the unreliability can be expressed as

$$\bar{R}(t) = \sum_{k=1}^{\infty} P_h(k) = \sum_{k=1}^{\infty} \frac{(2\lambda t)^k e^{-2\lambda t}}{k!}\left[\frac{\mu}{\lambda+\mu}\right]^{(k-1)}\left[\frac{\lambda}{\lambda+\mu}\right]. \tag{7.3.2a}$$

When transition biasing is applied, the probabilities

$$P_f^* = \frac{\lambda^*}{\lambda^*+\mu} \quad \text{and} \quad P_r^* = \frac{\mu}{\lambda^*+\mu}$$

are used instead of P_f and P_r. To keep the calculation unbiased, a weight $W(k)$ must be attached to the estimator. To see how this weight evolves, one may write the biased process with the weight in the form

$$\bar{R}^*(t) = \sum_{k=1}^{\infty} P_h^*(k) = \sum_{k=1}^{\infty} \frac{(2\lambda t)^k e^{-2\lambda t}}{k!}\left[\frac{\mu}{\lambda^*+\mu}\right]^{(k-1)}\left[\frac{\lambda^*}{\lambda^*+\mu}\right] \times W(k).$$

It can be seen that, to keep the calculation unbiased (i.e. $\bar{R}^*(t) = \bar{R}(t)$), the weight must take the form

$$W(k) = \left(\frac{P_r}{P_r^*}\right)^{(k-1)}\left(\frac{P_f}{P_f^*}\right). \tag{7.3.3}$$

Starting with a unit weight at $t = 0$, denote the weight following the ith event by $W(i)$. Then, if the event sampled, while in states $(1,0)$ or $(0,1)$, is a repair of the failed component, the weight is updated according to $W(i) = W(i-1)(P_r/P_r^*)$. If the sampling results in a failure of the operational component, the weight is updated by $W(i) = W(i-1)(P_f/P_f^*)$. It is easy to see that this procedure leads to the weight given in expression (7.3.3). This procedure can easily be generalized to a system with n components with a set of failure rates $\{\lambda_i\}$ and repair rates $\{\mu_i\}$. In transition biasing the failure rates are replaced by a set $\{\lambda_i^*\}$. Accordingly, when sampling the trigger component (see expression (7.2.3)), a transformed probability $P^*(i) = \beta_i^*/\alpha^*$ is used, where

$$\beta_i^* = \begin{cases} \lambda_i^* & b_i = 1 \\ \mu_i & b_i = 0 \end{cases} \quad \text{and} \quad \alpha^*(\vec{\mathbf{B}}) = \sum_{i=1}^{n} \beta_i^*,$$

instead of the probability $P(i) = \beta_i/\alpha$. The weight of the system is then adjusted by

$$W(i) = W(i-1)\left(\frac{P(j)}{P^*(j)}\right),$$

where j is the number of the sampled trigger component. Note that this will apply together with any other biasing (such as a forced event). In the above, one can see a realization of a general principle of biasing. When biased distributions or probabilities are used, the weight is updated by the ratio of the analog and biased probabilities of the sampled event to maintain an unbiased calculation. For the above case, the first moment of the calculation with transition biasing can be written as the first moment of $W(k)$, namely:

$$\langle W\rangle = \bar{R}(t) = \sum_{k=1}^{\infty} P_h^*(k)W(k) = \sum_{k=1}^{\infty} \frac{(2\lambda t)^k \mathrm{e}^{-2\lambda t}}{k!} P_r^{*(k-1)} P_f^* \times W(k). \tag{7.3.3a}$$

The second moment is given by

$$\langle W^2\rangle = \sum_{k=1}^{\infty} P_h^*(k)W^2(k) = \sum_{k=1}^{\infty} \frac{(2\lambda t)^k \mathrm{e}^{-2\lambda t}}{k!} P_r^{*(k-1)} P_f^* \times W^2(k),$$

which, upon substitution of P_r^*, P_f^* and $W(k)$, yields

$$\langle W^2\rangle = \sum_{k=1}^{\infty} P_h^*(k)W^2(k) = \sum_{k=1}^{\infty} \frac{(2\lambda t)^k \mathrm{e}^{-2\lambda t}}{k!} \left[\frac{\mu}{\lambda+\mu}\right]^{(k-1)} \left[\frac{\lambda}{\lambda+\mu}\right] \left[\frac{\beta\lambda+\mu}{\lambda+\mu}\right]^k \times \left(\frac{1}{\beta}\right). \tag{7.3.4}$$

For $\beta = 1$ the analog case is obtained and $\langle W^2\rangle = \langle W\rangle$, since the unreliability estimator is binomial. As β increases, the term $(1/\beta)$ will cause a reduction of the second moment. However, if β further increases, the second moment (and thus the variance) will increase, and will eventually diverge as $\beta \to \infty$. Although it is nice to see that the second moment contains a 'self-protection' mechanism which tells us when the second moment diverges, one may still wonder as to the mechanism that causes the variance to grow and diverge. This point will be discussed shortly. To check the validity of expressions (7.3.4) and (7.3.3), the PRSD as function of β, namely

$$\mathrm{PRSD} = \frac{100}{\langle W\rangle} \times \sqrt{\frac{\langle W^2\rangle - \langle W\rangle}{N}},$$

was calculated for $N = 300\,000$ using expression (7.3.4), the result is the predicted PRSD. The same PRSD was calculated directly in a Monte Carlo calculation for an unreliability at 100 hours for the above example [`Program relabt.for`]. The results

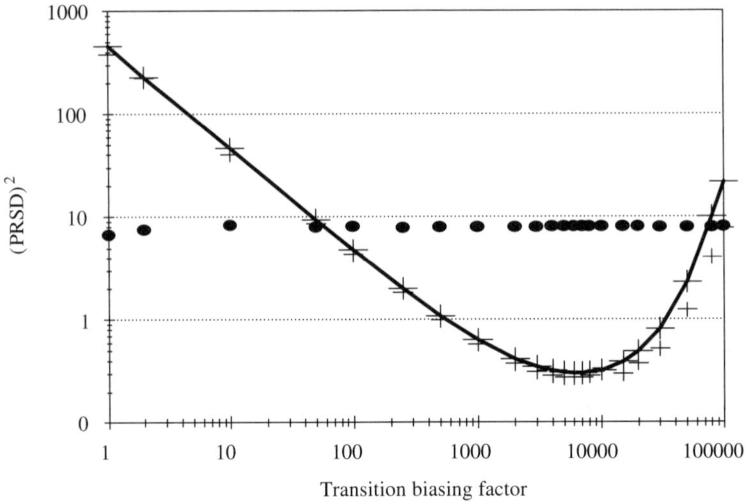

Figure 7.3.1 Comparison of predicted and calculated variance

of both calculations, for the square of the PRSD, are shown in Figure 7.3.1, with the solid line being the predicted values and the plus markers representing the MC calculated results. The circular marker displays the average values for the unreliability obtained in each calculation (multiplied by 10^5).

The comparison indicates that the approximate theory does provide a good estimate both for the optimal range of parameters and the benefit expected. In this example, a benefit of about 1500 is obtained. In general, the analysis of the second moment is a complex subject. Note that the first moment which is the target of the calculation is a solution of a non-trivial multi-dimensional integral equation. The second moment is the solution of a non-linear equation for which very little theory exists. Thus, *a priori*, developing any theory for the second moment would seem to be a discouraging task, yet such theories can be developed based on reasonable approximations, since their main purpose would not be the calculation of the absolute second moment, but a determination of an optimal range for biasing parameters, or even just a 'safe range'. The ability to obtain a good estimate of the expected benefit is a fringe benefit of such theories.

A clue as to the reason for the divergence of the second moment can be found by observing the approximation in 7.3.2(a) and noting that, in fact, it is a Neumann series expansion. It is also important to note that in MC calculations, a history having k events, with a score upon the last event, contributes exclusively to estimation of the kth term of the Neumann expansion of the target quantity. The relative value of each term in the expansion may be viewed as its true importance, much in the same way as was described in Section 3.3 in relation to integration by MC. As β increases, lower terms in the expansion are better estimated. This can be easily seen by looking at the *biased* unreliability,

$$\langle W^*\rangle = \sum_{k=1}^{\infty} \frac{(2\lambda t)^k e^{-2\lambda t}}{k!} P_r^{*(k-1)} P_f^* = \sum_{k=1}^{\infty} \frac{(2\lambda t)^k e^{-2\lambda t}}{k!} \left[\frac{\mu}{\beta\lambda+\mu}\right]^{(k-1)} \frac{\beta\lambda}{\beta\lambda+\mu},$$

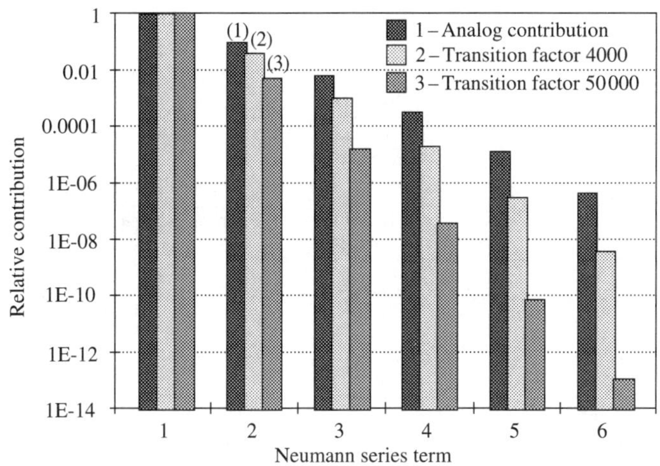

Figure 7.3.2 Attributed importance and true importance

obtained from (7.3.3) by removing the weight factor. Each term in this sum represents the attributed importance to this term in the expansion. As $\beta \to \infty$ all terms except the first are reduced to zero. This is also intuitively plausible since, in this case, the probability of a failure approaches unity and, eventually, there will be only a single event in each history. Recall that events $(1,1) \to (0,1) \to (0,0)$ are viewed in this approximation as a single event, thus no information will be obtained for higher terms in the expansion, leading to an infinite variance. The true importance, along with the attributed importance, for $\beta = 4000$ (a near optimum value) and $\beta = 50\,000$ are shown in Figure 7.3.2 on a logarithmic scale. It can be shown that a variance reduction method that yields a non-zero estimate for each history, and for which the attributed importance is identical to the true importance, is a zero variance method.

In the above example, a single value was used for the transition bias factor. The next example is used to expand this concept to a more general situation.

Example 7.3.1: A fire control system – risk assessment

The system shown in Figure 7.3.3 is a possible design of a fire control system in a critical chemical process. A failure of this system may lead to a severe accident. It is thus designed with a considerable amount of redundancy. The target of the calculation is the unreliability of the system for a period of 2000 hours of operation. The failure rates are given in units of failure per 10^5 hours on the top left of each upper component. The value applies to all the components in a column. The repair rates appear on the right side, given in units of repairs per hour.

An analog calculation with 50 000 histories yields no results at all. The system never fails. The average number of events is 27.1, thus no forced events are needed. A rough estimate of the unreliability can be obtained by noting that the shortest cut set in the system is (13,14) or (13,16), etc. The probability of a failure of component 13 in 2000 hours is 0.0217. The probability of a failure of components 14 or 16 during the 1.56 hours of repair of component 13 is about 0.00022, thus the probability that this minimal cut set will occur is 7.4×10^{-7}. Thus, roughly one failure is expected in a million hours. To obtain any significant result with an analog calculation, 10 million histories are required. This calculation takes 10 500 seconds, and results in an estimate of the

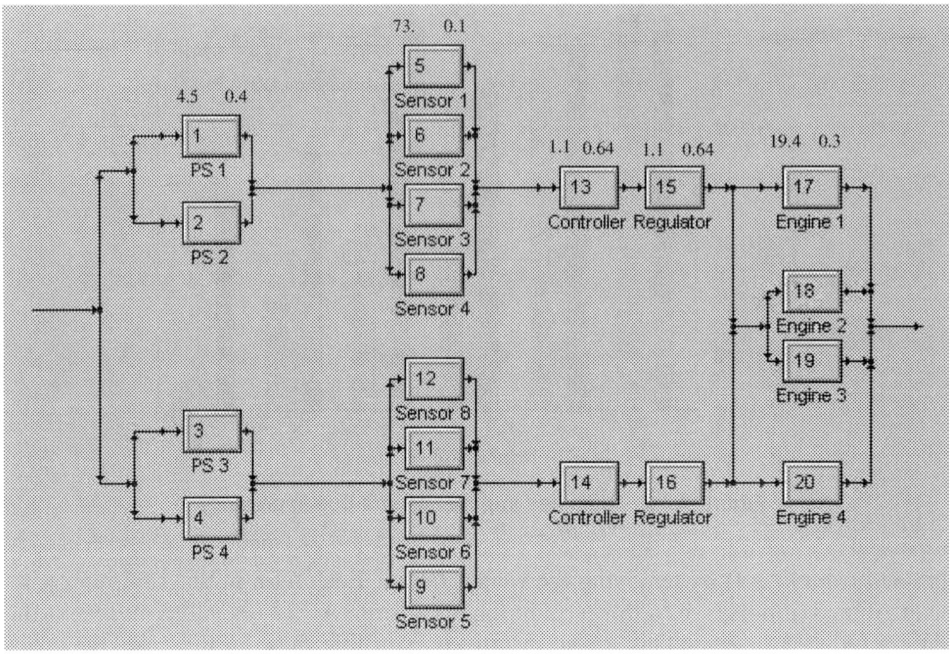

Figure 7.3.3 A fire control system

unreliability at 2000 hours of 3×10^{-6}, with a PRSD of 18.25%. A large FOM of 3 497 150 and 27.1 events per history. This is a border case in which a 5% PRSD calculation will require 38 hours. Transition biasing seems in order for this case.

Pathology of a numerically biased calculation In the former case it was shown that the optimal values of the transition biasing factor are in the range 2000–6000. It is, however, always recommended to start with low values. For the present example, taking $\beta = 50$ and using 10 000 histories yields an estimate of 3.95×10^{-12} with a PRSD of 99.77%. This result is erroneous and worthless. Having no prior information, it may lead to a serious error of six orders of magnitude. It results from grossly inappropriate sampling of the phase space of the system. Furthermore, reducing β to 20 and running 10 000 case histories yields a result of 4.11×10^{-9} (three orders of magnitude off) and a PRSD of 98.9%. Observing the average number of events, one notices that in the case of $\beta = 50$ there are 182.2 events in 2000 hours, and for $\beta = 20$ there are 63.85. Recall that upon a system failure the history is terminated. Also, from the analog calculation it can be seen that most of the contribution to the unreliability comes from the first 27 terms of the Neumann expansion, thus from systems that have up to 27 events. The biased cases are estimating the range around the 182th and 64th terms. The attributed importance is in gross mismatch with the true importance. This is the source of the erroneous results. This mismatch comes about because, in this biasing framework the failure rate of each component is multiplied by the same factor. Clearly, in the above diagram each component contributes a different amount to the unreliability. For example, components 5, 6, 7 and 8 will contribute hardly any significant quantity to the unreliability. Increasing the failure rate of such components causes them to have a large number of failures (thereby increasing the calculation time as an additional negative sideeffect), which contributes nothing to system failures yet results in a shift of the calculation towards evaluating higher expansion terms. This can be avoided if different transition biasing factors are assigned to components, depending on their contribution to the unreliability. To that end, the topological importance of components is defined, using the following algorithm.

Topologic importance procedure

1. Start with a state vector in which all components are 'up', i.e. $b_i = 1 \; \forall \; i$.
2. Choose at random a component number, say j, and set $b_j = 0$. Repeat this process until a system failed state is reached.
3. Upon reaching a system failed state, add unity to a counter for the last failed component (as this component is definitely responsible for the system failure). One can obtain additional information by following the 'sensitivity procedure' and adding unity to the counter of each component that, upon its '*adhoc*' repair, the system is repaired. Repeat steps 2 and 3 N times.
4. Let $k(j)$ denote the score accumulated for the j component, and let $k_{\max}$ denote the maximum score. Define the topological importance as $\gamma_j = k(j)/k_{\max}$.

The topological importance obtained for the system of Figure 7.3.3, with 1000 trials, is shown in Figure 7.3.4. One can see four distinct groups. Components 1–4 with an importance of 0.2, components 5–12 with an importance around 0.03, components 13–16 with a unit importance and components 17–20 with an importance of about 0.1. The statistical fluctuation in these numbers is of minute importance.

The transition bias factor is now expressed in the form:

$$\beta_j = (1 + \gamma_j \beta), \tag{7.3.5}$$

and β_j multiplies the failure rate of component j. With this type of biasing, a set of calculations was executed, with 100 000 histories, for β in the range 2–1500. As can be seen, the use of topological importance in modulating the transition biasing factors amends the pathology observed with the single parameter biasing.

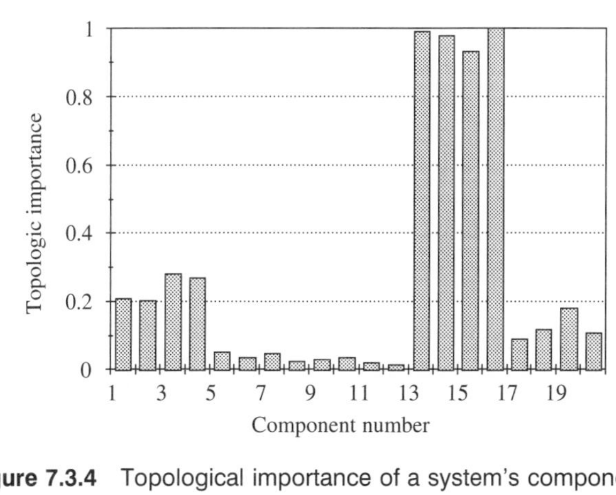

Figure 7.3.4 Topological importance of a system's components

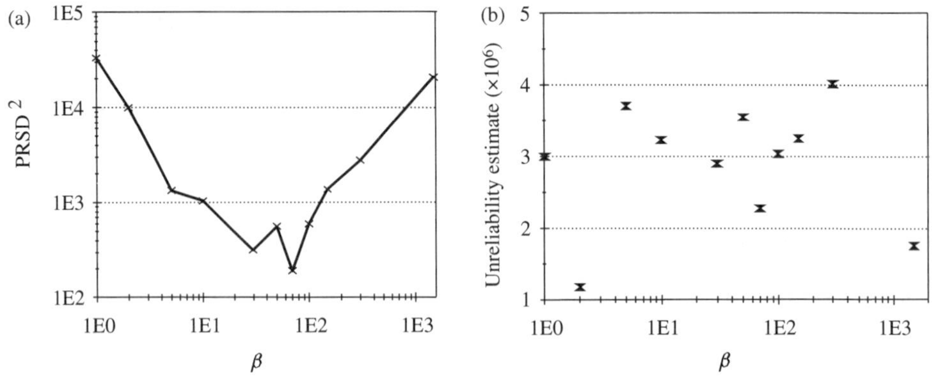

Figure 7.3.5(a,b) Square of PRSD and estimate as a function of transition biasing parameter

The results for the square of the PRSD and the unreliability estimate obtained are shown in Figures 7.3.5(a,b).

In the process of this calculation, special notice should be taken for cases in which the first moment is significantly small (as in the cases of $\beta = 2$, 80 and 1500). In such cases, a possible inadequate sampling occurred and the PRSD may be grossly underestimated. For the above case, the value $\beta = 50$ seems to be near optimal. For this case, the result with 100 000 histories is 3.55×10^{-6}. The PRSD is 23.69% and the calculation time is 132 seconds. This yields a FOM of 74 000. To obtain a more reliable result, it is advisable to go below 10% in the statistical error. The next calculation was done with 1.5 million histories. By $1/\sqrt{N}$ convergence, this should reduce the PRSD to 6.11%. The result obtained is 3.35×10^{-6} with a PRSD of 5.79% and a calculation time of 1980 seconds. The benefit over the analog calculation is 47. The average number of events is 30.1, close to the number in the analog case, so that no serious distortion occurs to the Neumann terms evaluation.

Biasing in the area of rare events and risk assessment currently leaves more questions than answers. Yet the above methods indicate a route through which efficient and reliable biasing methods may be developed.

Index